EUROPEAN INVERTEBRATE SURVEY

ATLAS
OF THE
NON-MARINE MOLLUSCA
OF THE
BRITISH ISLES

edited by

M.P. KERNEY

for the

Conchological Society of Great Britain and Ireland and the
Biological Records Centre, Institute of Terrestrial Ecology, Monks
Wood Experimental Station, Abbots Ripton, Huntingdon.

1976.

The Biological Records Centre at Monks Wood Experimental Station is part of the Institute of Terrestrial Ecology, which is a component body of the Natural Environment Research Council.

Printed in England
by Graphic Art (Cambridge) Ltd.

Published by the Institute of Terrestrial Ecology
68 Hills Road, Cambridge, England.

ISBN 0 904282 02 3

Introduction

The land and freshwater Mollusca are not among the more neglected animal groups. Their scientific study has a long history in Britain, and the restricted distribution of certain species has attracted comment at least since the 17th century. In one respect publication of this atlas in the present year is specially appropriate: the Conchological Society of Great Britain & Ireland was founded exactly a hundred years ago, and it was in the same year (12 October 1876) that its first secretary, W.D. Roebuck, began systematically to accumulate authenticated records of land and freshwater molluscs for what was shortly to become the Society's official *Census of Distribution*. The units of recording adopted by Roebuck were the well-known botanical vice-counties of H.C. Watson (for Great Britain) and C.C. Babington and R.L. Praeger (for Ireland). The *Census* passed through no fewer than seven published editions beginning in 1885. The earlier of these show distributions in tabular form only, but the two last (Roebuck and Boycott, 1921; Ellis, 1951) also include complete sets of vice-comital maps. Earlier versions of a number of these had already appeared in Taylor's *Monograph* (1894-1921) from 1902 onwards. Increments were published annually in the Non-marine Recorder's Report in the *Journal of Conchology,* an arrangement which continues to the present day.

For Ireland, a set of vice-comital maps using the "typomap" system devised by Praeger (1906) was published by Stelfox as long ago as 1911, in what still remains the standard work on the distribution of non-marine molluscs in that country.

In 1961, following the example of the botanists, the Conchological Society decided to adopt the method of grid mapping. Systematic surveys were organized, and by 1976 about 115,000 new 10-kilometre square records had been filed. In addition, all earlier usable records contained in the thirty-odd M.S. volumes of the Society's vice-comital Census have been incorporated, and - for the rarer species - publications and museum collections have been searched. For Ireland, where the recent work remains less complete than in Britain, an attempt has been made to extract exhaustively all the accessible data (Kerney, 1973).

The result may be judged from the pages which follow. The first map makes it clear that already a very good coverage of the British Isles has been achieved. Thinly-worked areas of course remain, but nevertheless the broad differences shown on this map do in fact significantly reflect true differences in regional species diversity ascribable mainly to climatic, topographical or geological factors. The individual distribution patterns may therefore be broadly accepted as "real", apart from a few cases involving recent segregates, special difficulties of identification, the minuteness or elusiveness of particular species, or other similar problems of which field workers will be aware.

The scheme is the most detailed undertaken for any invertebrate group of comparable size. Several molluscs previously unsuspected in the British Isles have been found (e.g., *Vitrea subrimata, Boettgerilla pallens, Sphaerium solidum).* Also, we now have much better information on questions of status, and of changes in distribution with time. Some species which were believed to be rare have turned out to be not uncommon, whilst some are obviously declining, and a few appear to be at serious risk. The value of such data as a guide to taking appropriate measures for protecting endangered species is evident. There have also been a number of less tangible results of the scheme. Collections made in the course of the surveys have focussed attention on problems of taxonomy and of geographical variation, for example in the *Arionidae* and *Limacidae.* In some cases an examination of interim distribution maps has revealed hitherto unsuspected relationships, suggesting further lines of enquiry. Research into field ecology has undoubtedly been stimulated.

It is hoped that publication of this first edition of the atlas will promote further interest in distribution studies, and will contribute towards the eventual compilation of unified maps for the Mollusca of Europe using the methods recommended by the committee of the European Invertebrate Survey (EIS); a few trial maps have already been published (Kerney, 1976a).

Special notes

The nomenclature and taxonomic arrangement adopted here follows the recent check lists by Waldén (1976) and Kerney 1976b), and is also in conformity with that to be used in a forthcoming field-guide to the terrestrial species of north-west Europe (Cameron and Kerney, in preparation). Specific synonyms used in the standard text book *British Snails* (Ellis, 1926; 1969) and in the last non-marine Census list (Ellis, 1951) are given in the index, plus a few others which may sometimes be encountered in 20th century British literature, notably in the classic papers of Boycott (1934, 1936) (The many generic name synonyms will, it is hoped, mostly be obvious and have been disregarded).

A very few taxa are not mapped separately owing to lack of sufficiently detailed information: these are *Arion rufus* (included with *Arion ater,* aggregate), *Limax flavus,* segregate (included with *Limax flavus,* aggregate) and *Euconulus alderi* (included with *Euconulus fulvus,* aggregate).

Naturalized introductions (e.g., *Hygromia cinctella*) are shown on the maps, although rare casuals (e.g., *Cochlostoma septemspirale*) and species established only in greenhouses (e.g., *Zonitoides arboreus*) have been omitted. Three possibly extinct species of doubtful status (*Milax nigricans, Bradybaena fruticum, Cernuella neglecta*) have however been included.

On most of the maps records not confirmed living for particular grid squares since 1950 are shown by a separate symbol. In practice, the great majority of such records fall into the period 1876-1914, whilst those in the "1950 onwards" category are nearly all later than 1961. Concentrations of old records in certain areas (especially in Ireland) should not necessarily be taken as evidence of recent extinction without first consulting the post-1950 coverage map.

Fossil occurrences are shown for certain terrestrial snails where there has been a significant contraction of range during the Flandrian (Postglacial) period; these are *Pomatias elegans, Acicula fusca, Succinea oblonga, Oxyloma sarsi, Truncatellina cylindrica, Vertigo pusilla, V. substriata, V. moulinsiana, V. lilljeborgi, V. alpestris, V. geyeri, V. angustior, Leiostyla anglica, Lauria sempronii, Spermodea lamellata, Ena montana, Zonitoides excavatus, Macrogastra rolphii, Monacha cartusiana, Helicodonta obvoluta* and *Helicigona lapicida.*

Finally, it may be noted that as an aid to the interpretation of the distribution patterns a set of transparent overlays will shortly be published by the Biological Records Centre.

Acknowledgements

The atlas is essentially a corporate effort, and I should like to thank all those people who have helped over the past fifteen years. Their names are as follows:

B.T.Abbott, F.A.Adams, the late Dr W.E.Alkins, K.G.Allenby, A.Allison, P.S.Anderson, Dr R.Anderson, Dr Elizabeth Andrews, S.Angus, S.G.Appleyard, J.Armitage, G.A.Arnold, M.A.Arnold, Mrs A.Avens, R.E.Baker, M.W.Baldwin, G.A.S.Barnacle, J.F.M. de Bartolomé, J.Bass, K.E.Beckett, F.G.Berry, the late Rev. H.E.J.Biggs, Dr E.O.Bishop, Dr M.J.Bishop, Mrs D.Blezard, the late E.Blezard, M.R.Block, Dr H.J.M.Bowen, D. Boxley, P.F.Burns, A. Buse, P.Cambridge, Dr R.A.D.Cameron, J.A.Carman, R.Carr, Miss P.D.Carter, I.M.Cassells, the late C.P.Castell, Mrs J.L. Charlish, Dr June Chatfield, G.M. Collis, Dr B.Colville, Mrs M.E.Conway, J. Cooper, Mrs A.S.Corley, Lord Cranbrook, D.J.Cranmer, G.I.Crawford, T.J.Crawford-Sidebotham, the Rev. Canon C.T.Cribb, R.J.Croome, T.E.Crowley, S.P. Dance, Miss S.M.Davies, Dr D.S.Davis, I.K.Dawson, E.Dearing, B.P.Dennison, T.R.E.Devlin, the late C.R.P.Diver, R.Dixon,D.Doogue, H.E.M.Dott, C.D.Drake, R.J.Driscoll, N.Dudley, Dr J.H.Duffus, W.F.Edwards,A.E.Ellis, I.M.Evans, Dr J.G.Evans, S.T.A.Featherstone, I.D.Finney, Mrs M.Fogan, Mrs M.J.Fox, R.Fresco-Corbu, Dr Larch Garrad, T.S.H.Gibson, C.Gillard, Brig. E.A.Glennie, M. Goodchild, Dr C.B.Goodhart, D.A.Gowans, Dr J.J.D.Greenwood, A.T.Gregg, the late Canon L.W.Grensted, Mrs M.Gunning, Dr J.B.Hall, F.J.Harper, G.H.Harper, J.I.Harris, D.R.B.Harrison, Miss B.E.Hart-Jones, Miss E.I.Harvey, D.M.Hawker, C.Hayton, M.P.Headen, Dr D.Heppell, P.B.Heppleston, R.T.Herbert, Miss S.J.Hewitt, R.C.Higgins, P.M.Horsfield, J.Humphreys, Dr Anne Hurst, T.Huxley, E.G.Ing, Dr J.W.Jackson, F.J.Jennings, A.W.Jones, J.E.L.Jones, Dr J.S.Jones, E.Kellock, M.Kennedy, the late D.K.Kevan, I.J.Killeen, Dr M.Ladle, J.C.Lavin, Lady Christina Letanka, Mrs P.List, Dr L.Lloyd-Evans, D.C.Long, the Rev. G.E.H.Long, R.H.Lowe, Dr T.T.Macan, R.Macdonald, Mrs M.McMillan, Dr P.Makings, Dr P.S.Maitland, V.J.Mallett, the late Miss M.Marklove, F.Marshall, Dr F. May, Lady Sophie Meade, Prof. Dr A.D.J.Meeuse, R.G. Meiklejohn, Dr J.O'N.Millott, C.Moore, Dr P.B.Mordan, B.D.Moreton, C.Moriarty, M.J. Morphy, R.W.Morrell, Dr B.S.Morton, Miss S.Murrell, G.N.Myers, Dr B.Nau, Dr P.Newell, A.Norris, W.J.Norton,

R.J.O'Connor, M.O'Grady, M.O'Meara,
A.P.H.Oliver, A.G.H.Osborn, Mrs B.J.Paton, Mrs
J.A.Paton, Dr C.R.C.Paul, W.A.Pearson, Dr
F.H.Perring, C.W.Pettitt, E.G.Philp, D.G.Pickrell,
G.W.Pitchford, Mrs E.Platts, Dr E.Pollard,
D.E.Pomeroy, R.C.Preece, Miss M.Pugh, A.W.
Punter, the late Dr H.E.Quick, Mrs E.B.Rands,
B.Rawlinson, P.Reavell, E.J.Redshaw, A.Rennie,
S.M.Rhind, T.H.Riley, E. Robinson, Mrs
H.C.G.Ross, M.W.Rowe, Miss J.Royston, Dr
A.J.Rundle, Dr N.W.Runham, Mrs M.Saul,
D.R.Saunders, D.R.Seaward, Miss F.M.Seeley,
H.H.Shephard, Professor F.W.Shotton,
T.B.Silcocks, M.Sinclair, Mrs G.Slater, Mrs
B.M.Smith, Dr Shelagh Smith, G.G.Spencer,
M.Spray, the late A.W.Stelfox, A.McG.Stirling,
the late L.W.Stratton, Dr A.T.Sumner,
C.M.Swaine, D.A.J.Taylor, B.Thistleton,
I.J.L.Tillotson, R.Tindal, Mrs S.M.Turk, Dr
C.Turner, Dr B.Verdcourt, G.J.M.Visser, M.Wade,
Dr H.W.Waldén, R.B.Walker, M.R.Wallis, A.
Walton, P.Warris, Dr T.Warwick, A.R.Waterston,
R.Watkin, G.Whitfield, Mrs V.Wilkin,
R.B.Williams, Dr R.B.G.Williams, P.Wilson,
P.T.Wimbleton, F.R.Woodward, D.Worth, Dr J.O.
Young, Professor J.Z.Young, M.R.Young.

A special debt is due to the staff of the
Biological Records Centre, and in particular to
Miss D.W. Scott, on whom fell the main task of
data processing and map production.

M.P.Kerney
May 1976

References

BOYCOTT, A.E., 1934. The habitats of land
Mollusca in Britain. *J. Ecol.* **22**: 1-38

BOYCOTT, A.E.,1936. The habitats of freshwater
Mollusca in Britain. *J. Anim. Ecol.* **5**: 116-186

CAMERON, R.A.D. and KERNEY, M.P. (in preparation). *A Field-guide to the land Mollusca of
North-west Europe.* London: Collins.

ELLIS, A.E., 1926; revised edition 1969. *British
Snails. A guide to the non-marine Gastropoda
of Great Britain and Ireland.* Oxford:
Clarendon Press.

ELLIS, A.E. 1951. Census of the distribution
of British non-marine Mollusca. *J. Conchol.,
Lond.* **23**: 171-244.

KERNEY, M.P., 1973. Mapping non-marine
Mollusca in north-west Ireland, summer 1972.
Ir. Nat. J. **17**: 310-316.

KERNEY, M.P., 1976a. European distribution
maps of *Pomatias elegans* (Müller), *Discus
ruderatus* (Férussac), *Eobania vermiculata*
(Müller) and *Margaritifera margaritifera*
(Linné). *Achiv. Molluskenk.* **106**: 243-249.

KERNEY, M.P., 1976b. A list of the fresh and
brackish-water Mollusca of the British Isles.
J. Conchol., Lond. **29**: 26-28.

PRAEGER, R.L., 1906. A simple method of
representing geographical distribution.
Ir. Nat. **15**: 88-94.

ROEBUCK, W.D. and BOYCOTT, A.E., 1921.
Census of the distribution of British land and
freshwater Mollusca. *J. Conchol., Lond.* **16**:
166-212.

STELFOX, A.W., 1911. A list of the land and
freshwater mollusks of Ireland. *Proc. R. Ir.
Acad.* B **29**: 65-164.

TAYLOR, J.W., 1894-1921. *Monograph of the
land and freshwater Mollusca of the British
Isles.* 4 volumes (unfinished). Leeds: Taylor
Brothers.

WALDÉN, H.W., 1976. A nomenclatural list of the
land Mollusca of the British Isles. *J. Conchol.,
Lond.* **29**: 21-25.

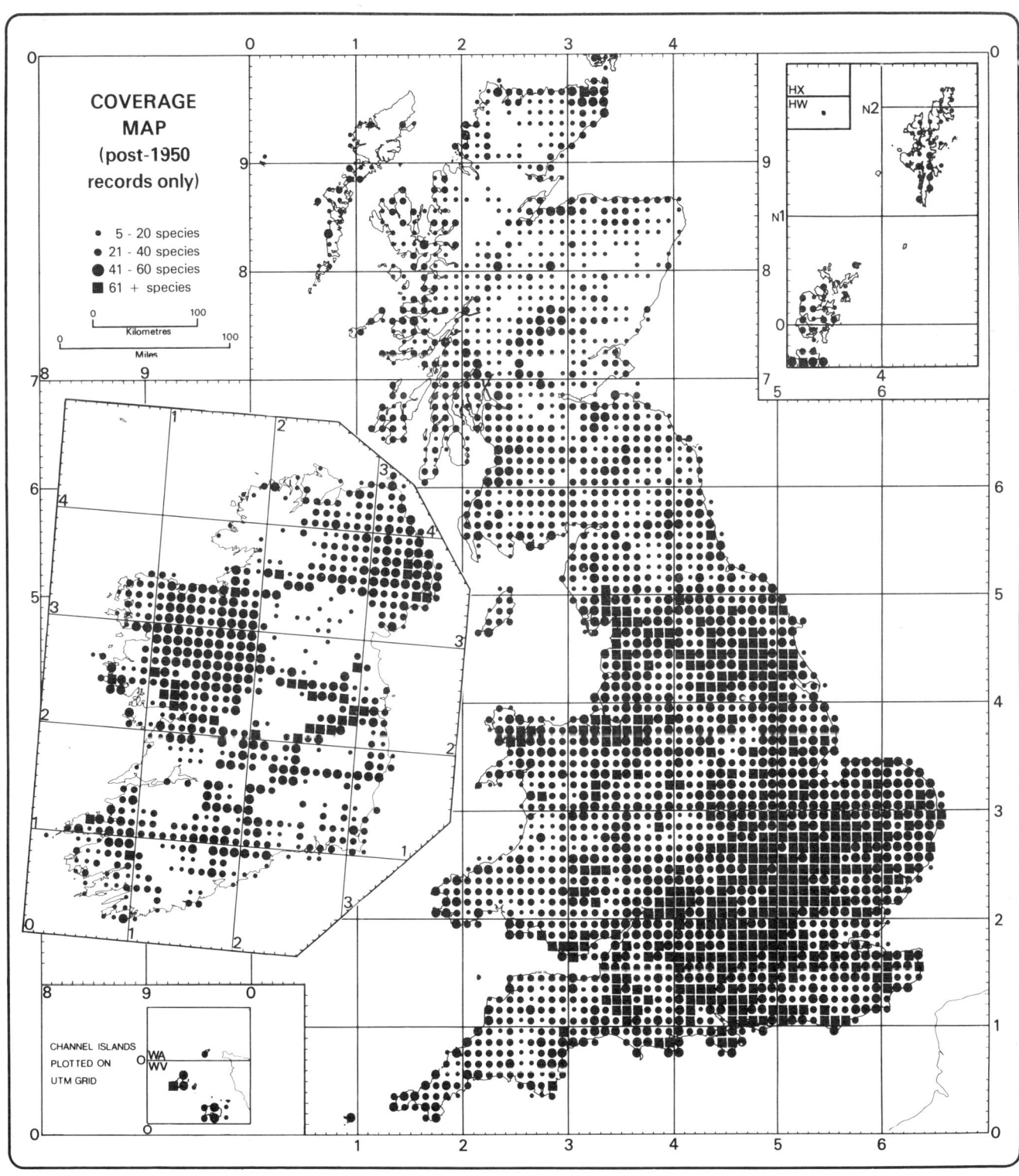

COVERAGE
MAP
(post-1950
records only)

· 5 - 20 species
· 21 - 40 species
● 41 - 60 species
■ 61 + species

Kilometres

Miles

CHANNEL ISLANDS
PLOTTED ON
UTM GRID

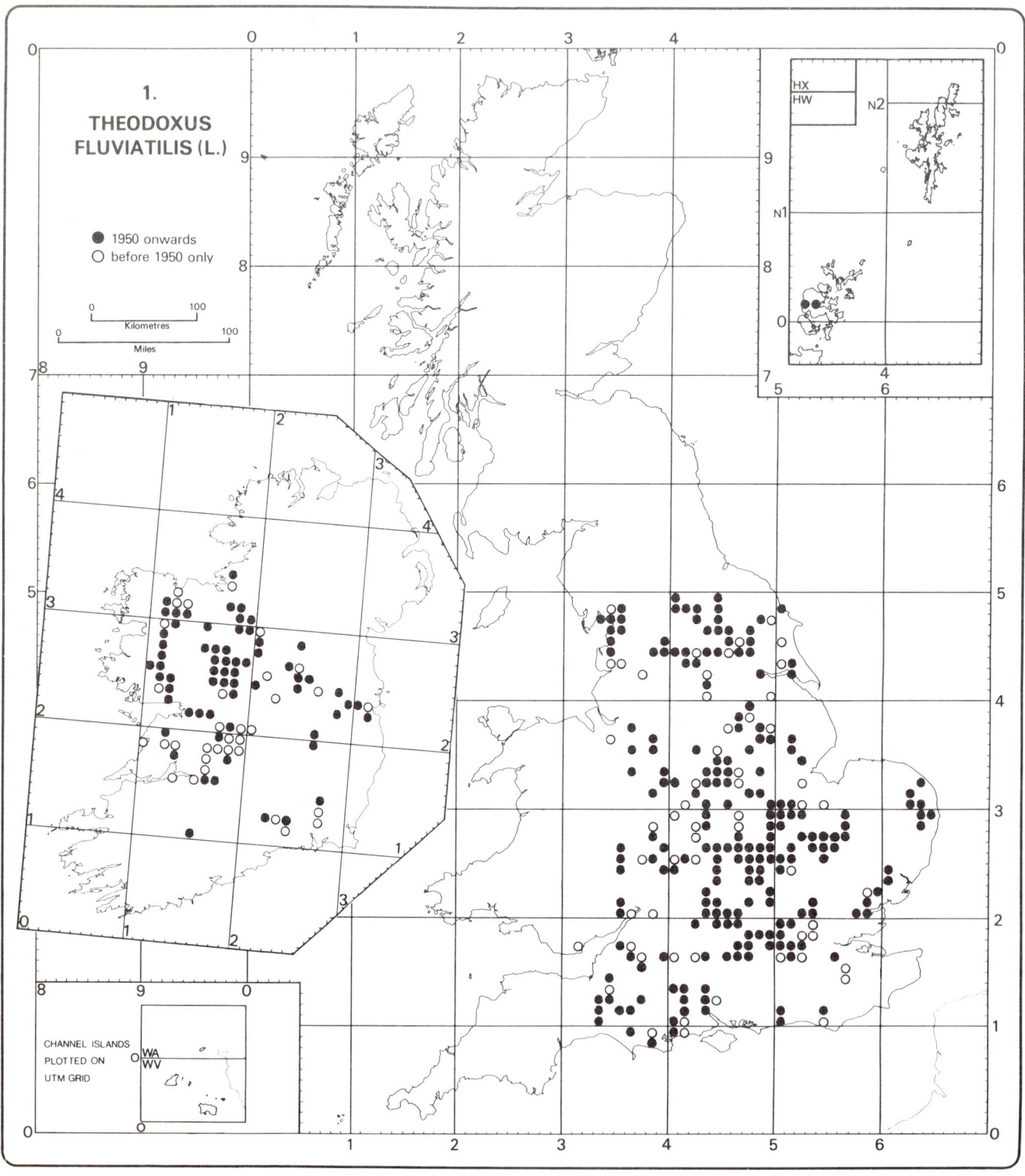

1.
THEODOXUS
FLUVIATILIS (L.)

● 1950 onwards
○ before 1950 only

Kilometres

Miles

CHANNEL ISLANDS
PLOTTED ON
UTM GRID

WA
WV

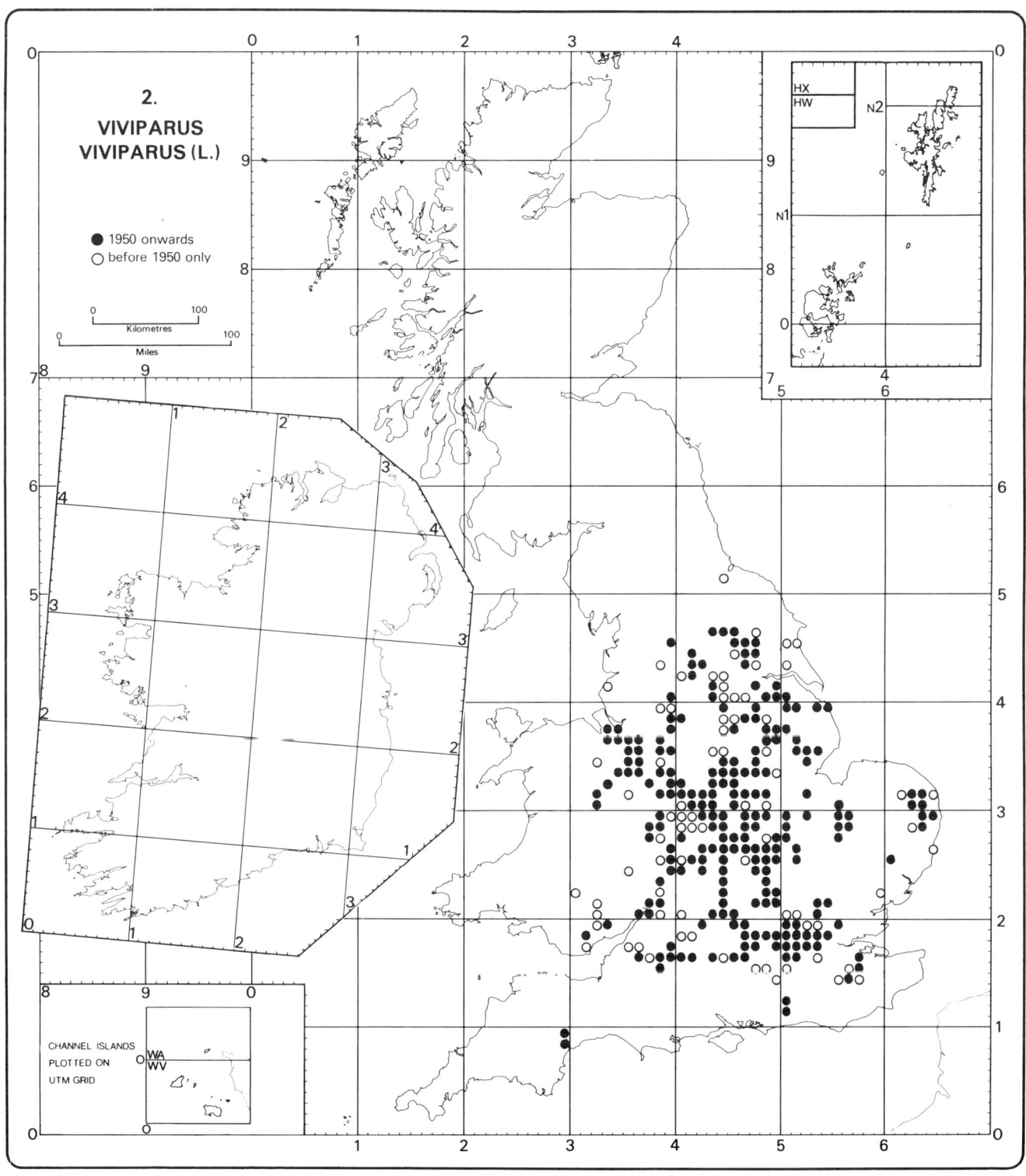

2.
VIVIPARUS
VIVIPARUS (L.)

● 1950 onwards
○ before 1950 only

CHANNEL ISLANDS
PLOTTED ON
UTM GRID
○ WA
WV

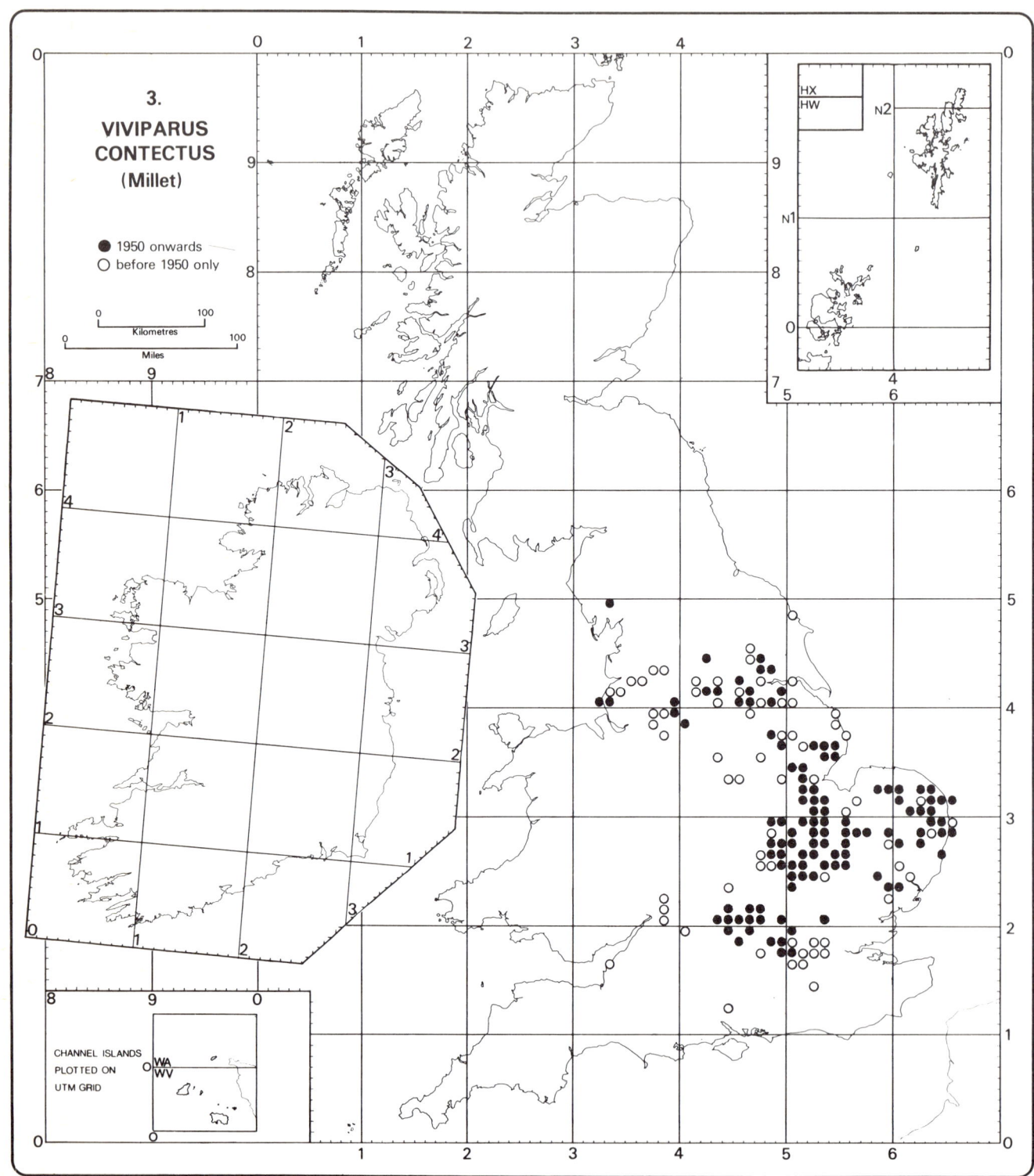

3.
VIVIPARUS
CONTECTUS
(Millet)

● 1950 onwards
○ before 1950 only

0   100
Kilometres
0   100
Miles

CHANNEL ISLANDS
PLOTTED ON
UTM GRID

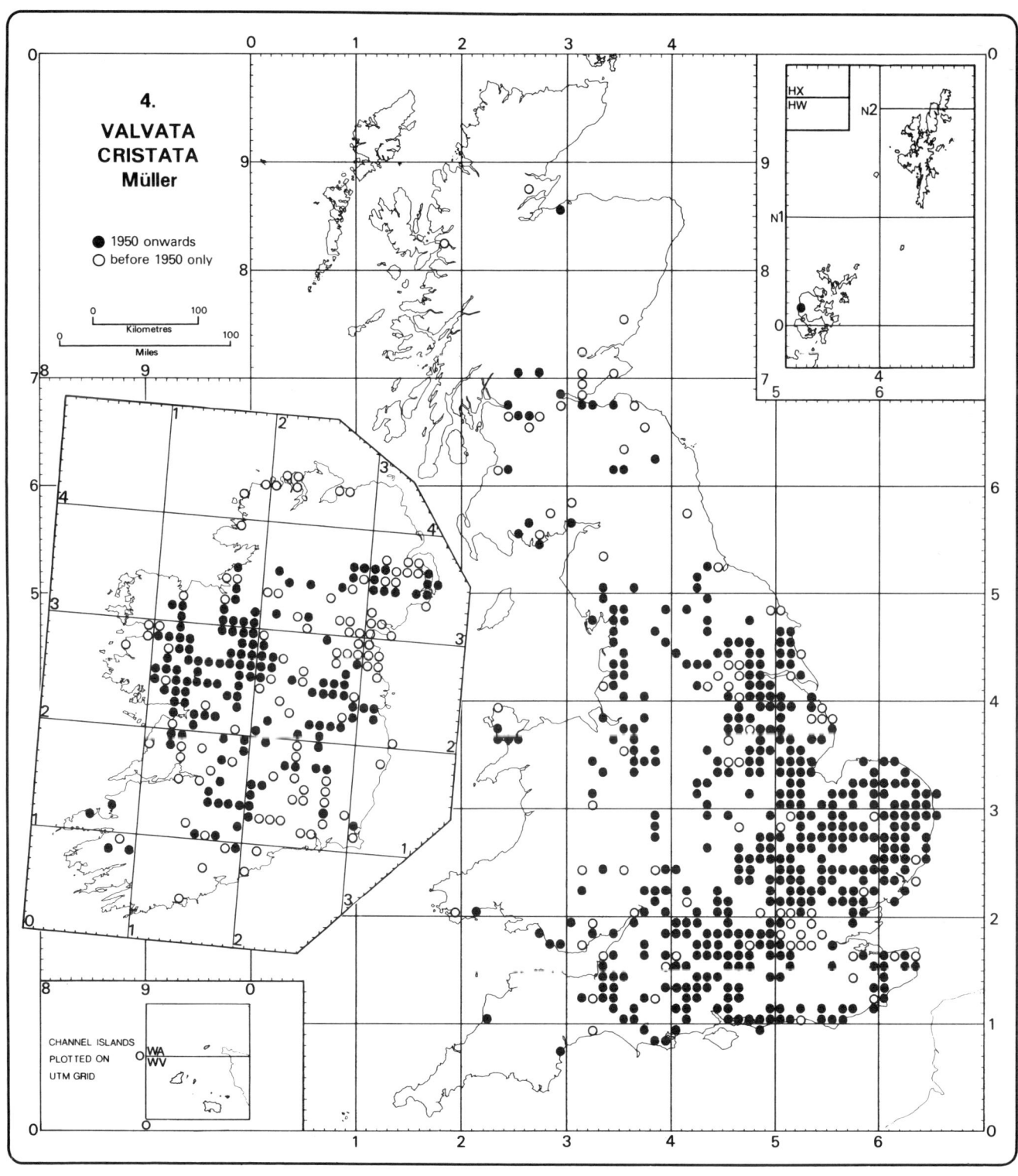

4.
VALVATA
CRISTATA
Müller

● 1950 onwards
○ before 1950 only

Kilometres
Miles

CHANNEL ISLANDS
PLOTTED ON
UTM GRID

WA
WV

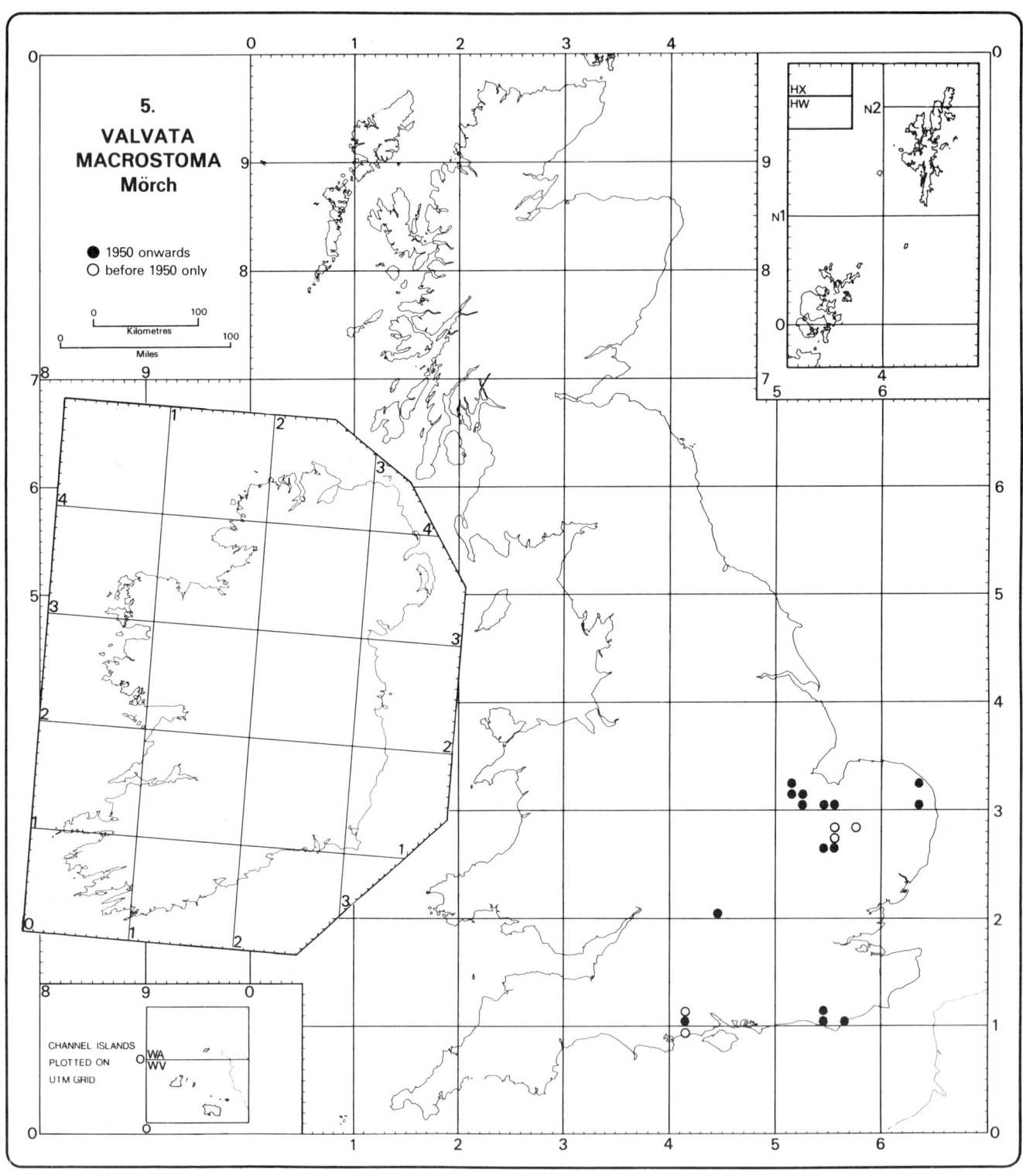

5.

VALVATA
MACROSTOMA
Mörch

● 1950 onwards
○ before 1950 only

0   100
Kilometres
0   100
Miles

CHANNEL ISLANDS
PLOTTED ON
UTM GRID

WA
WV

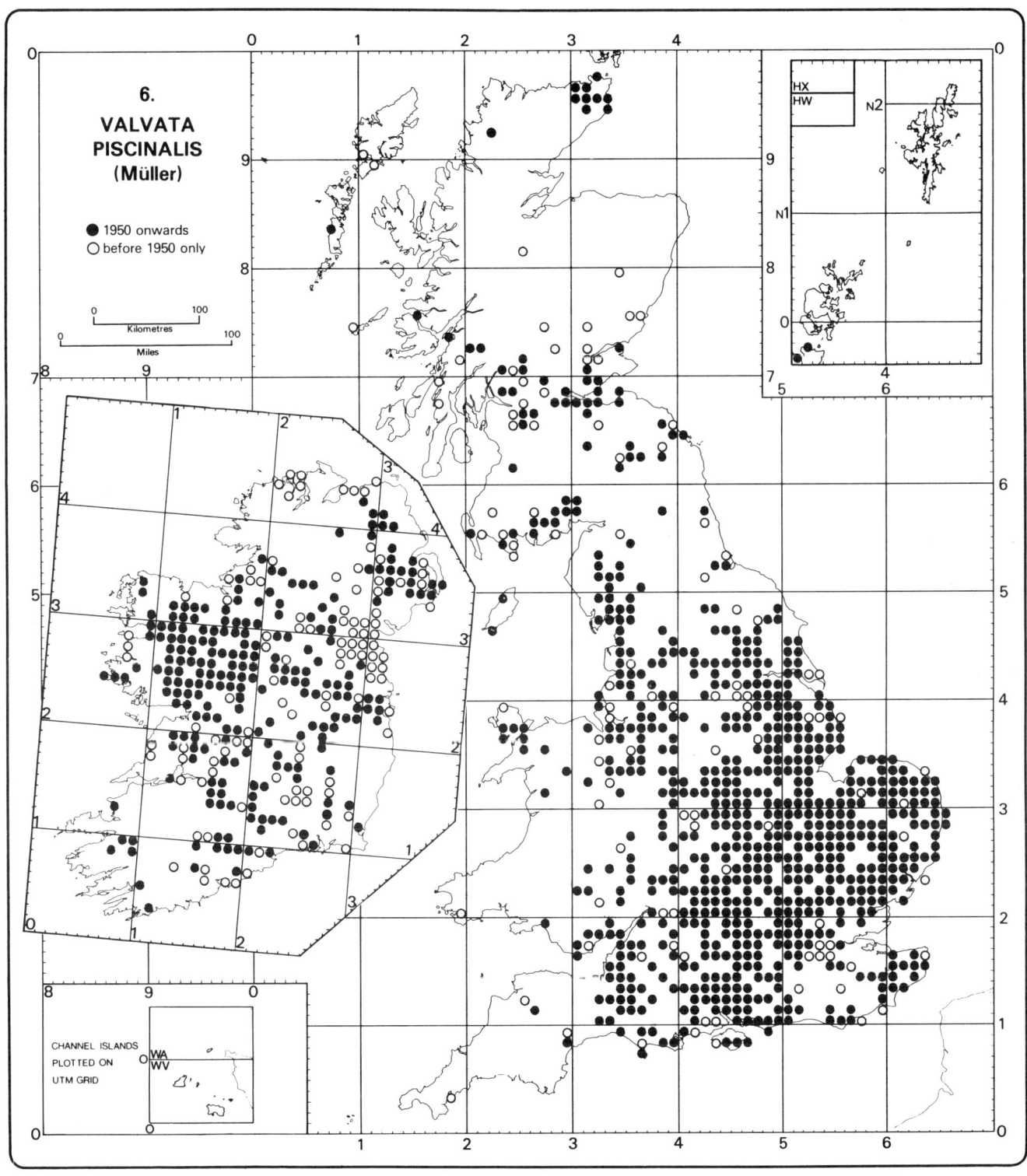

6.

VALVATA PISCINALIS
(Müller)

● 1950 onwards
○ before 1950 only

Kilometres
Miles

HX
HW
N2

N1

CHANNEL ISLANDS
PLOTTED ON
UTM GRID
WA
WV

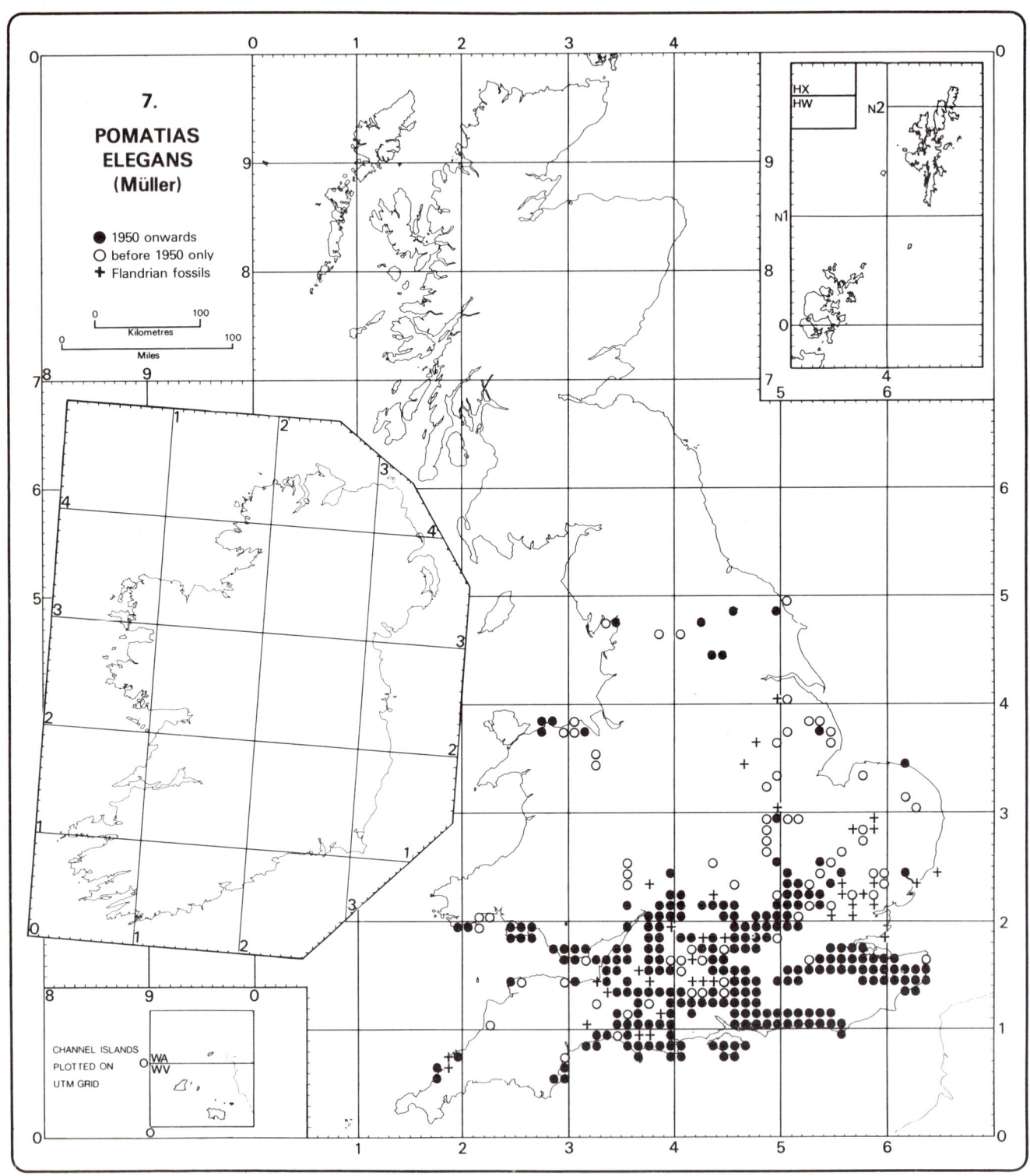

7.

**POMATIAS
ELEGANS**
(Müller)

● 1950 onwards
○ before 1950 only
+ Flandrian fossils

Kilometres
0 — 100

Miles
0 — 100

CHANNEL ISLANDS
PLOTTED ON
UTM GRID

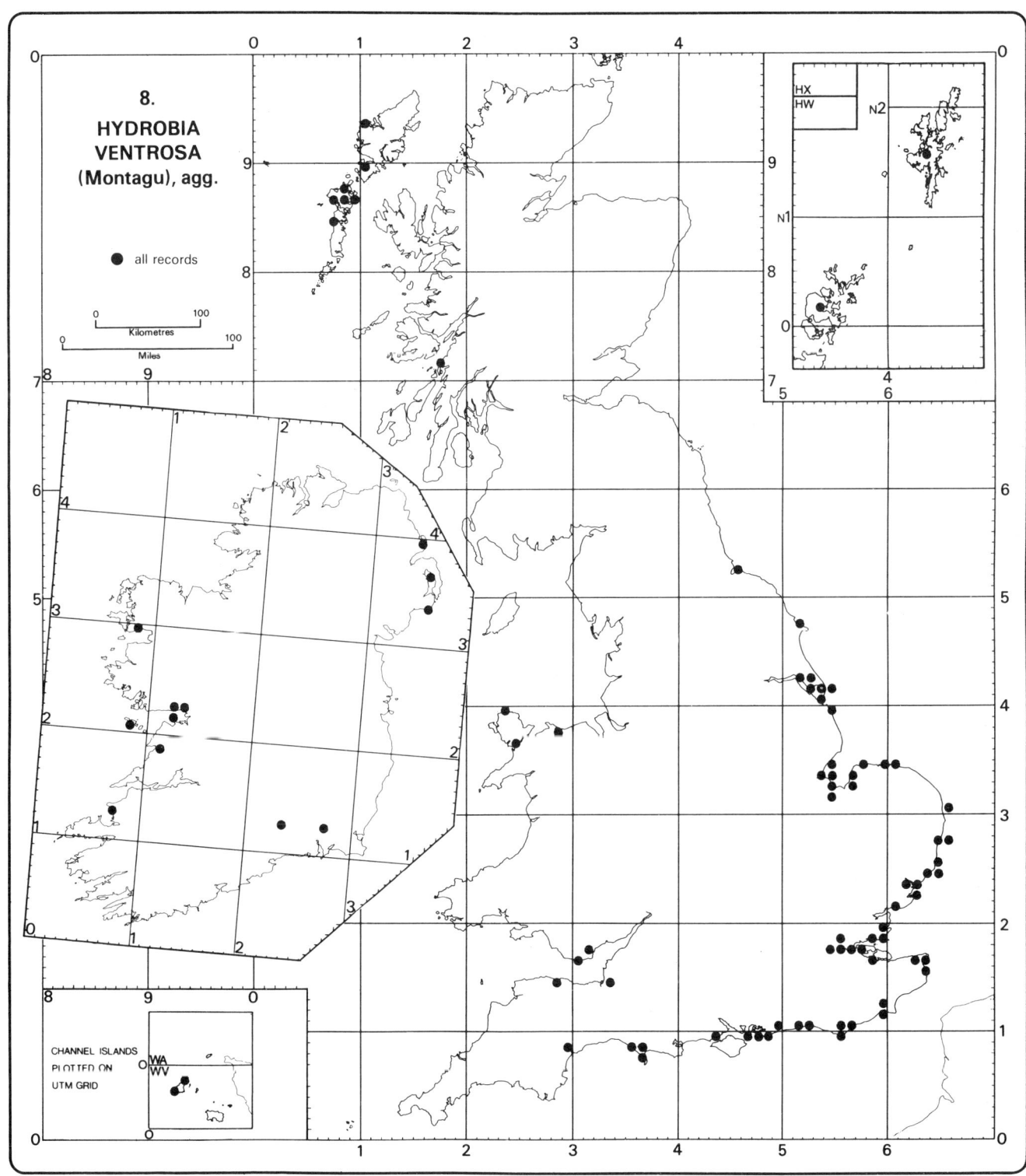

8.

**HYDROBIA
VENTROSA
(Montagu), agg.**

● all records

Kilometres

Miles

CHANNEL ISLANDS
PLOTTED ON
UTM GRID

WA
WV

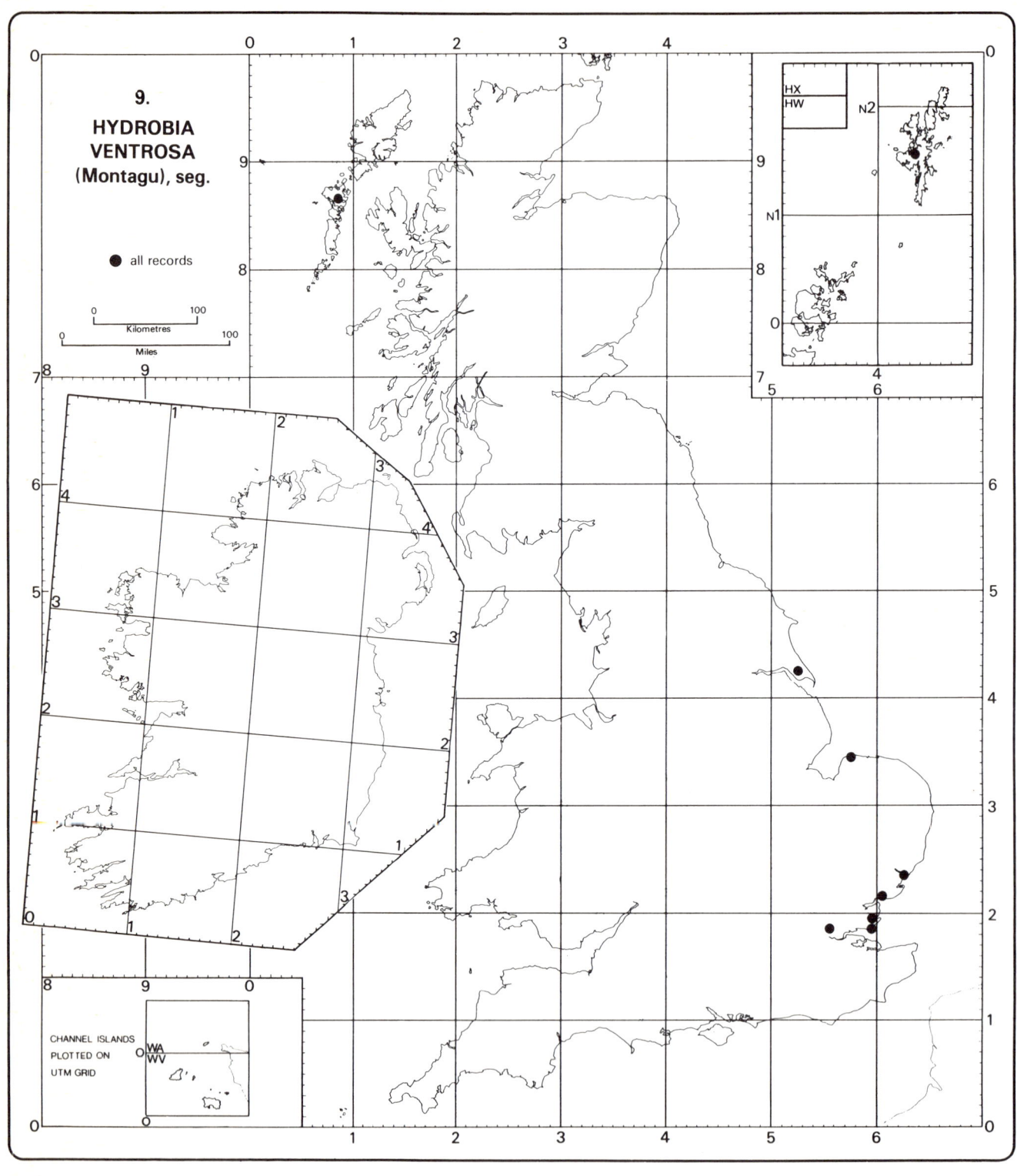

9.
HYDROBIA
VENTROSA
(Montagu), seg.

● all records

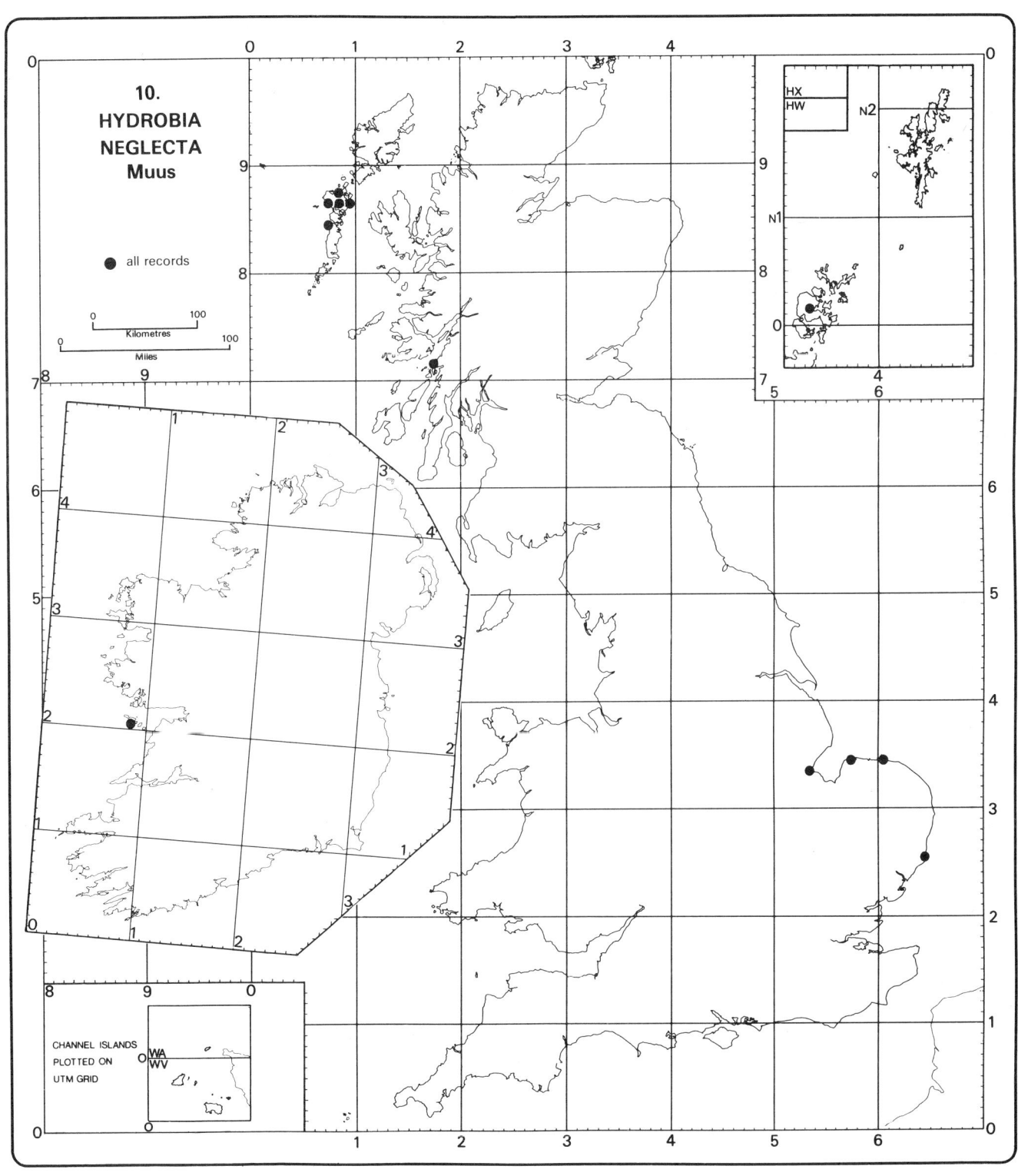

10.
HYDROBIA
NEGLECTA
Muus

● all records

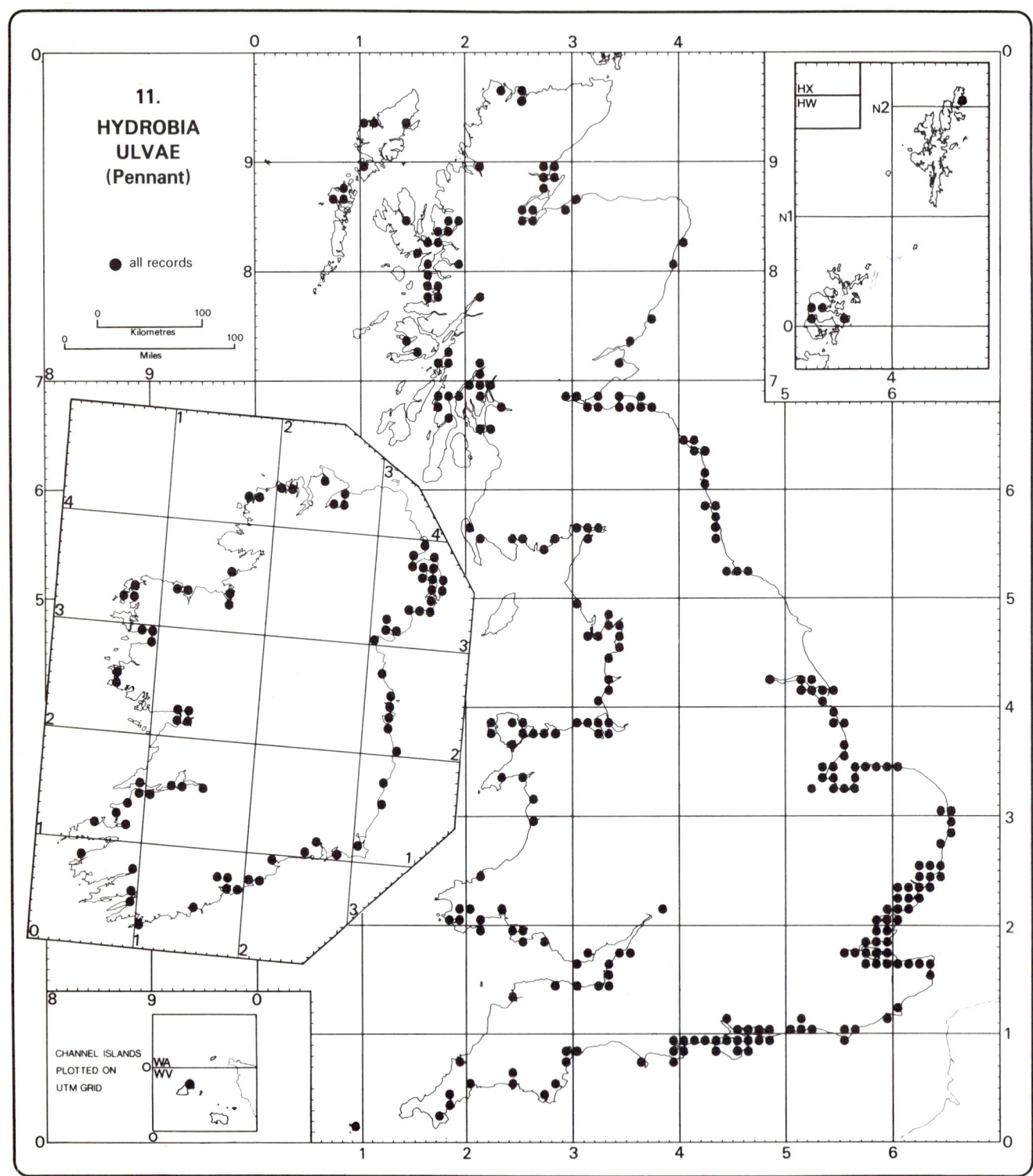

**11.
HYDROBIA
ULVAE
(Pennant)**

● all records

Kilometres

Miles

CHANNEL ISLANDS
PLOTTED ON
UTM GRID

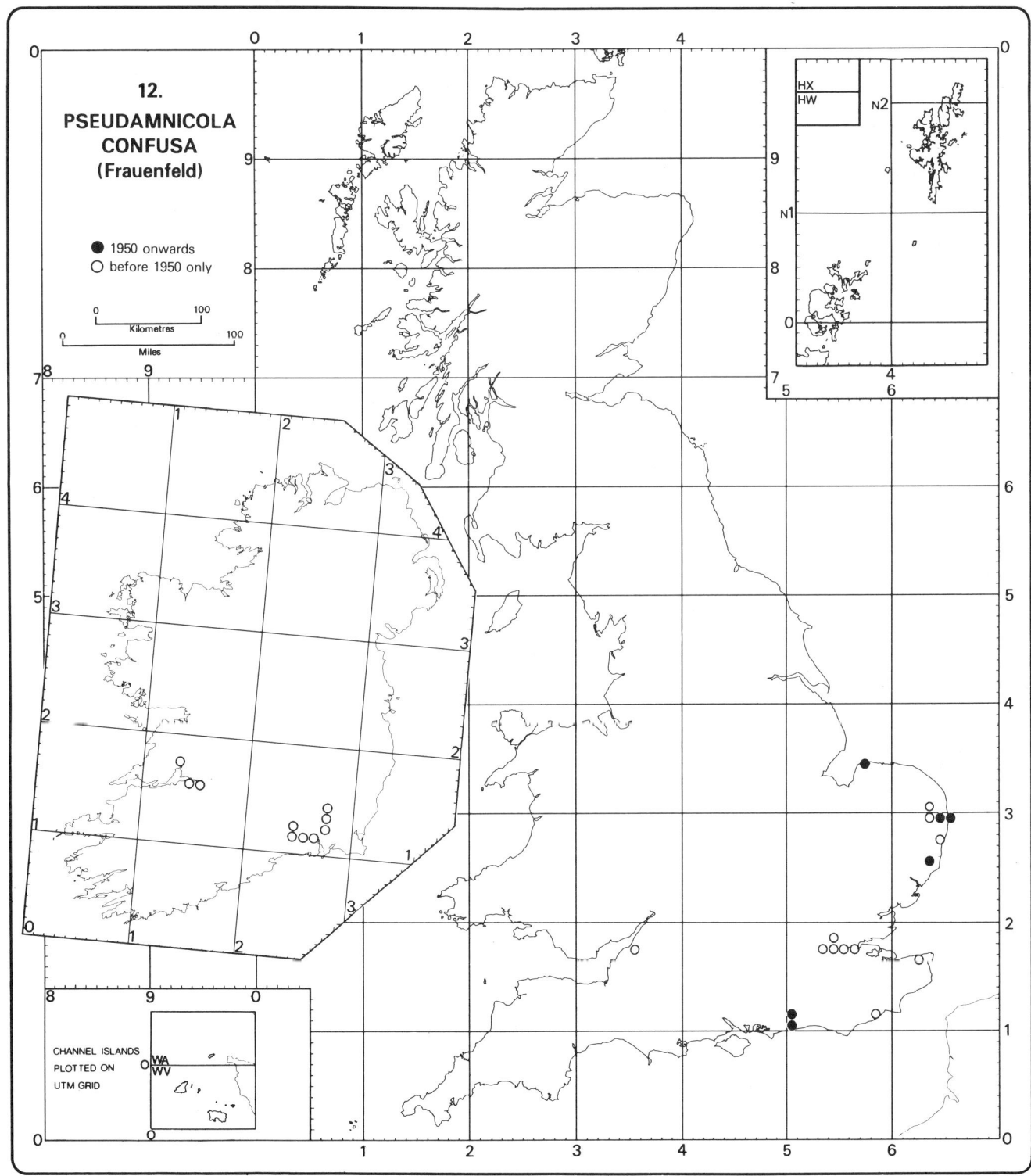

12.
PSEUDAMNICOLA
CONFUSA
(Frauenfeld)

● 1950 onwards
○ before 1950 only

Kilometres
Miles

CHANNEL ISLANDS
PLOTTED ON
UTM GRID

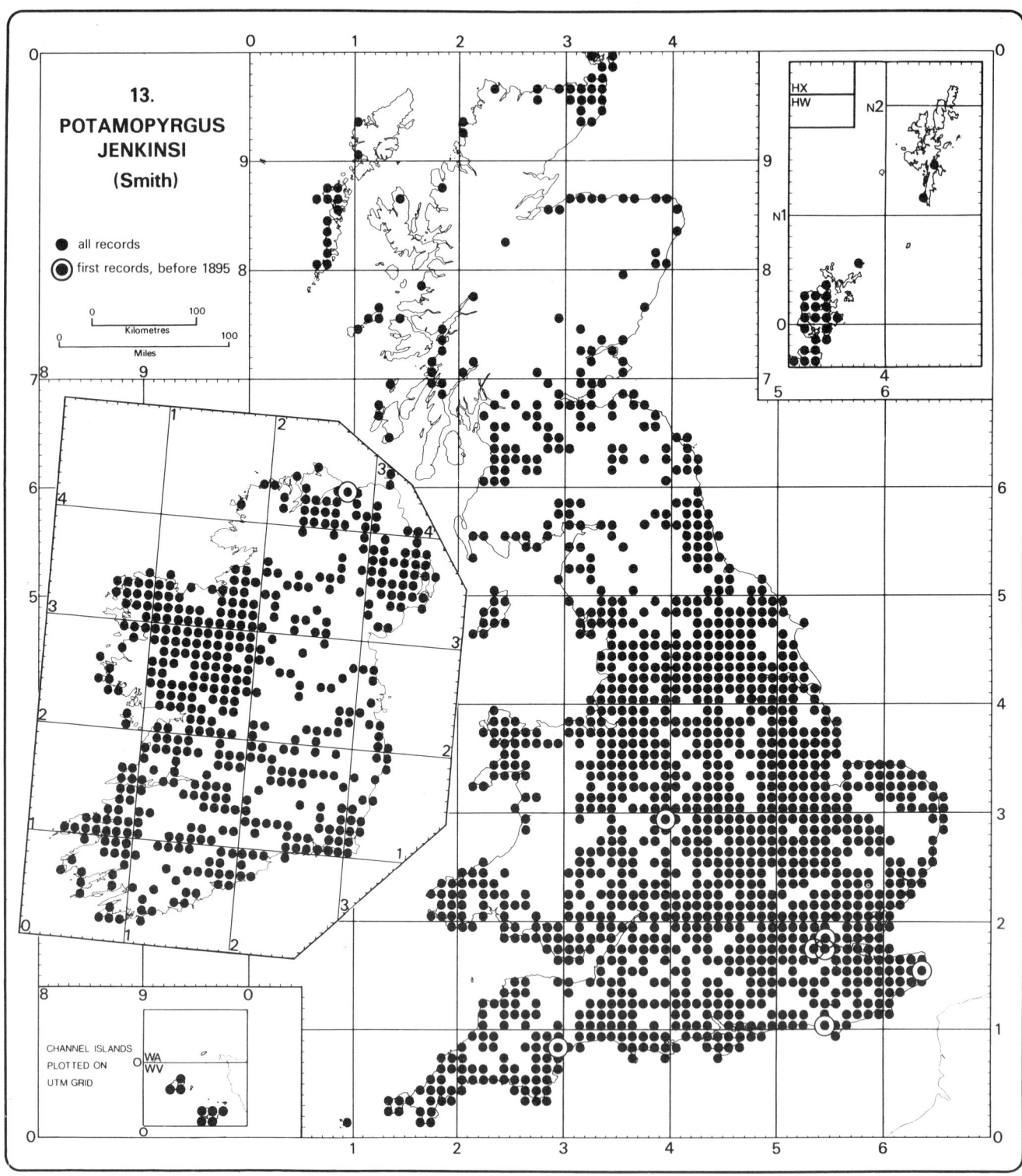

13.
POTAMOPYRGUS
JENKINSI
(Smith)

● all records
◉ first records, before 1895

Kilometres
0   100
Miles
0   100

CHANNEL ISLANDS
PLOTTED ON
UTM GRID

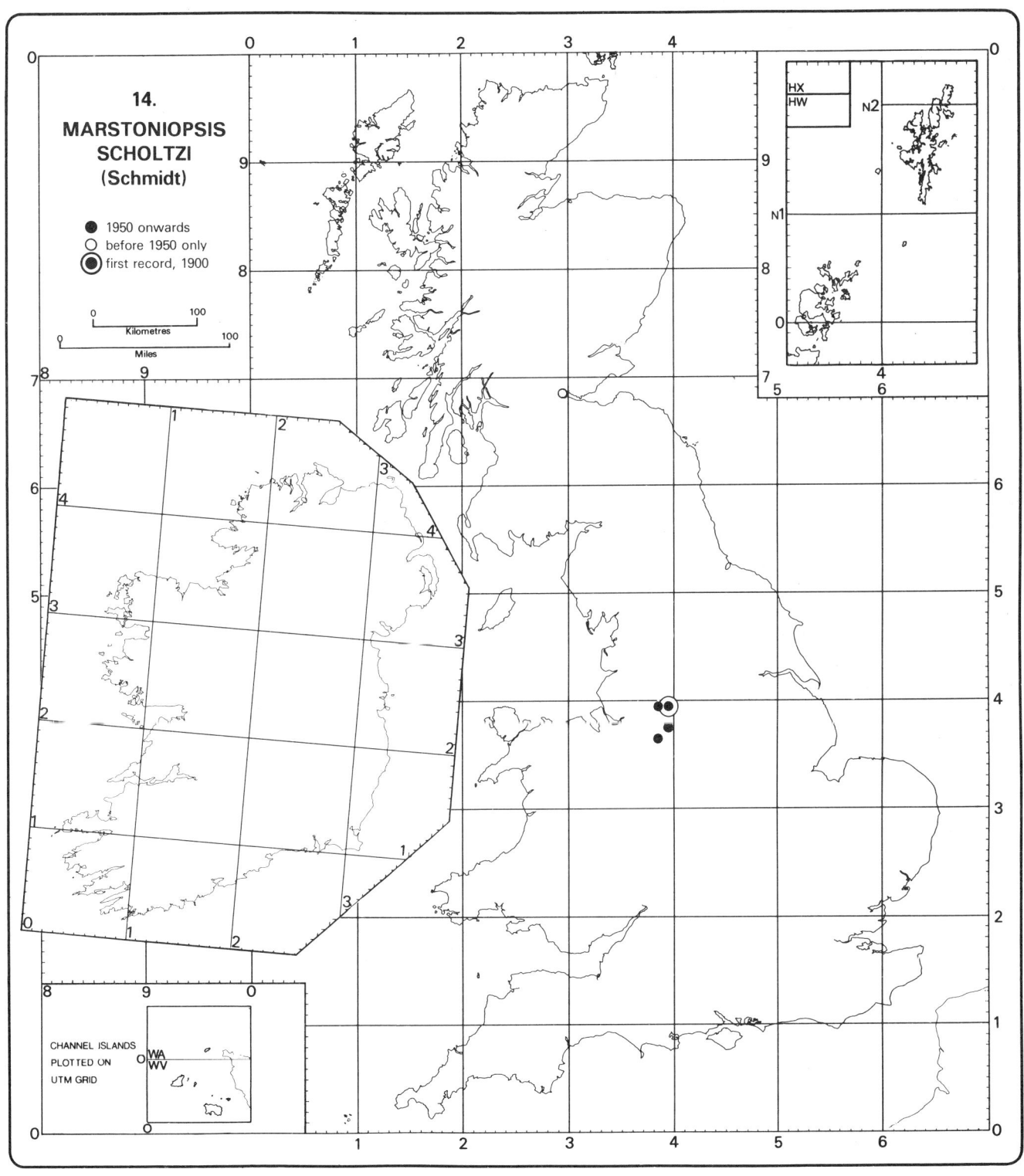

14.

**MARSTONIOPSIS
SCHOLTZI
(Schmidt)**

● 1950 onwards
○ before 1950 only
◉ first record, 1900

Kilometres

Miles

CHANNEL ISLANDS
PLOTTED ON
UTM GRID

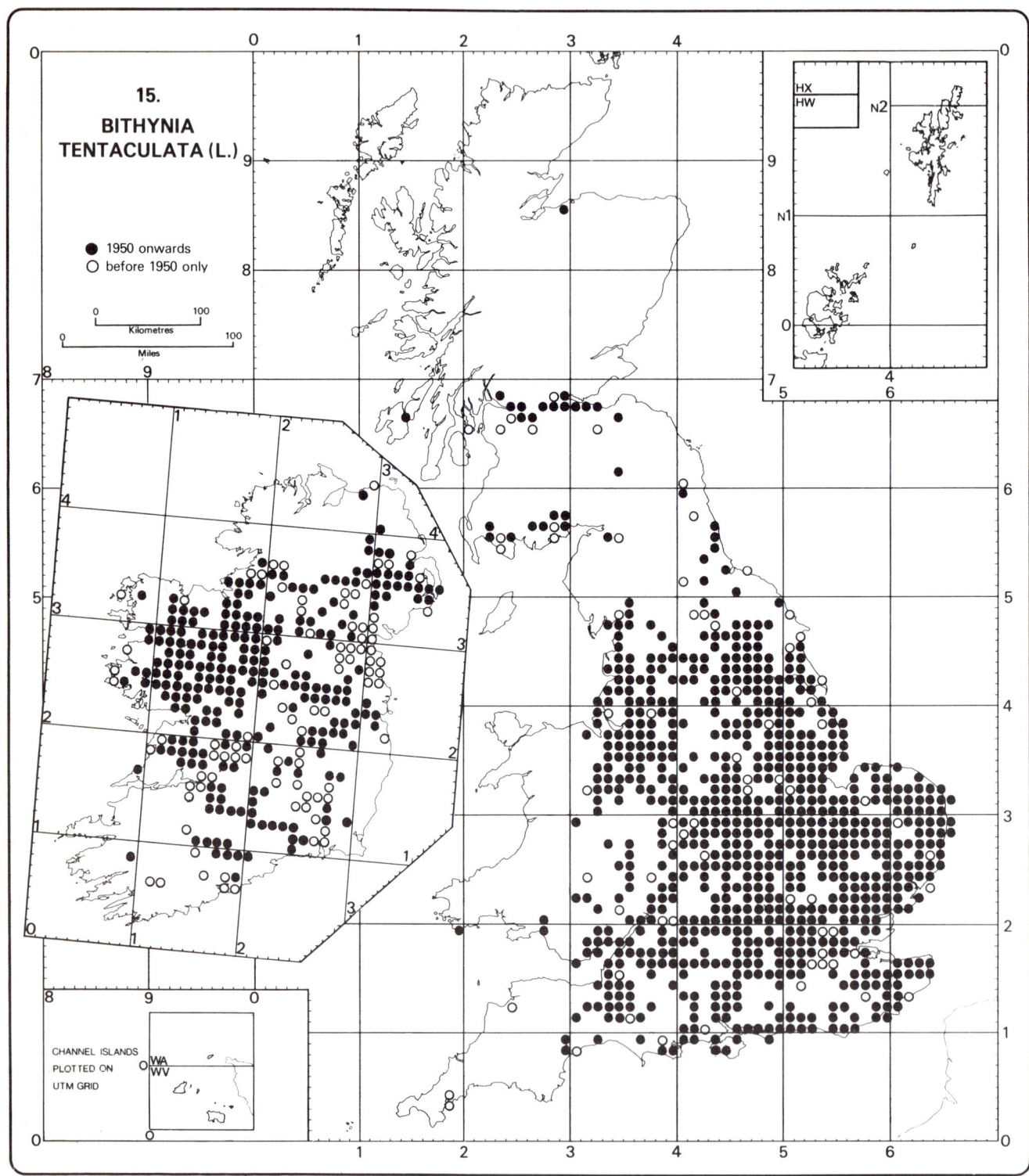

15.

BITHYNIA
TENTACULATA (L.)

● 1950 onwards
○ before 1950 only

0   100
Kilometres
0   100
Miles

CHANNEL ISLANDS
PLOTTED ON
UTM GRID

WA
WV

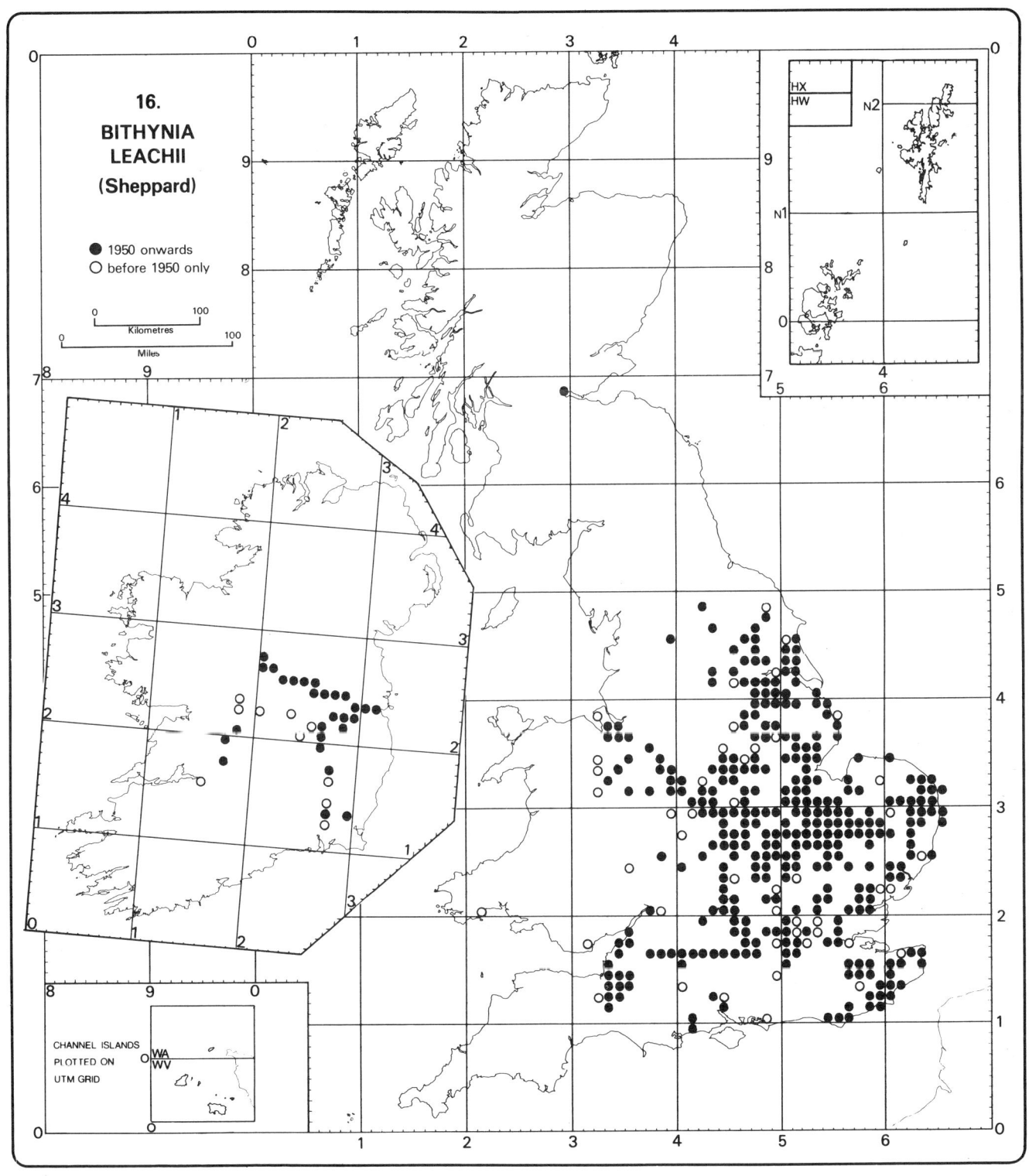

16.
BITHYNIA
LEACHII
(Sheppard)

● 1950 onwards
○ before 1950 only

Kilometres
Miles

CHANNEL ISLANDS
PLOTTED ON
UTM GRID

WA
WV

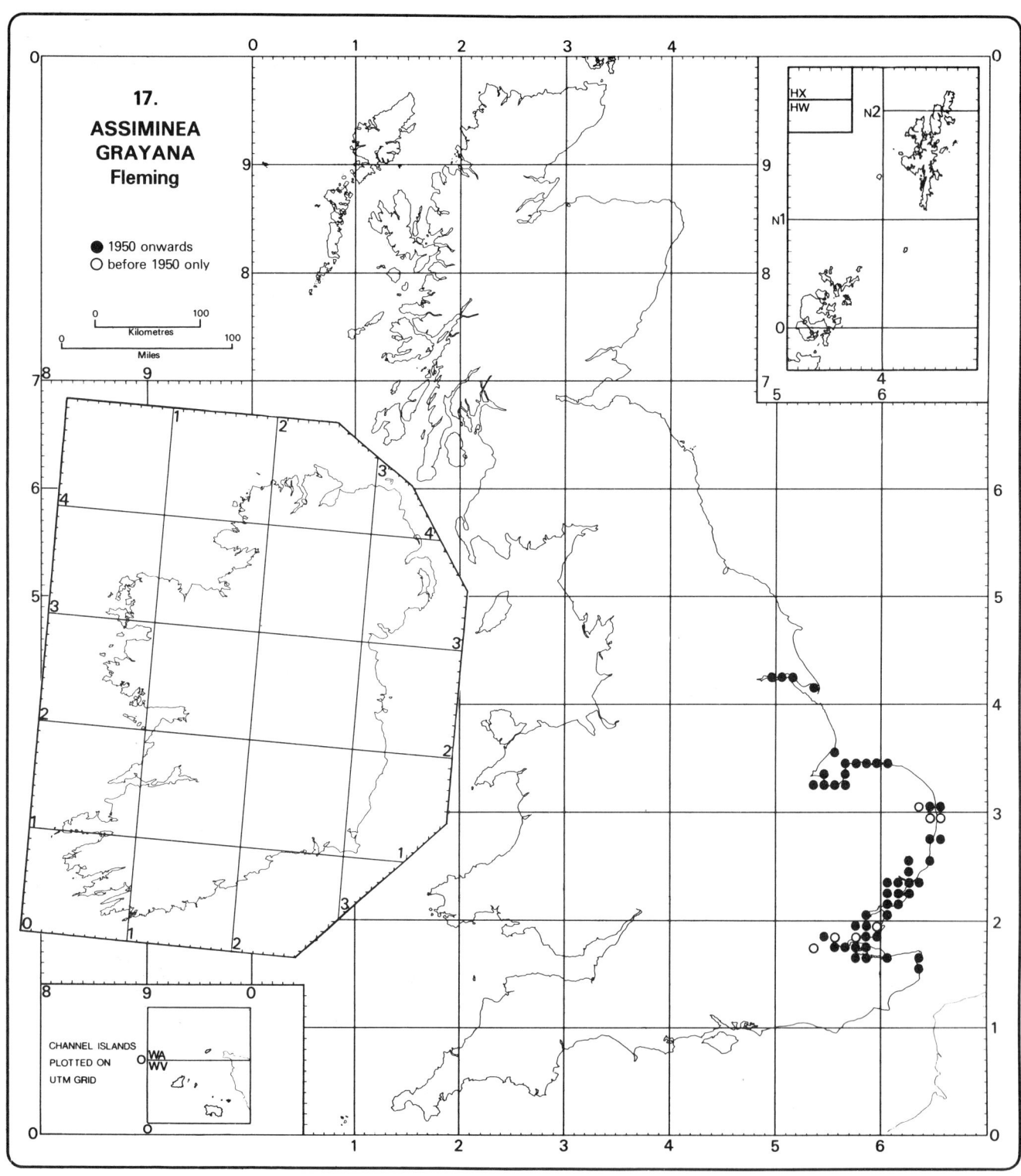

17.
ASSIMINEA
GRAYANA
Fleming

● 1950 onwards
○ before 1950 only

0 ___ 100
Kilometres
0 ___ 100
Miles

CHANNEL ISLANDS
PLOTTED ON
UTM GRID

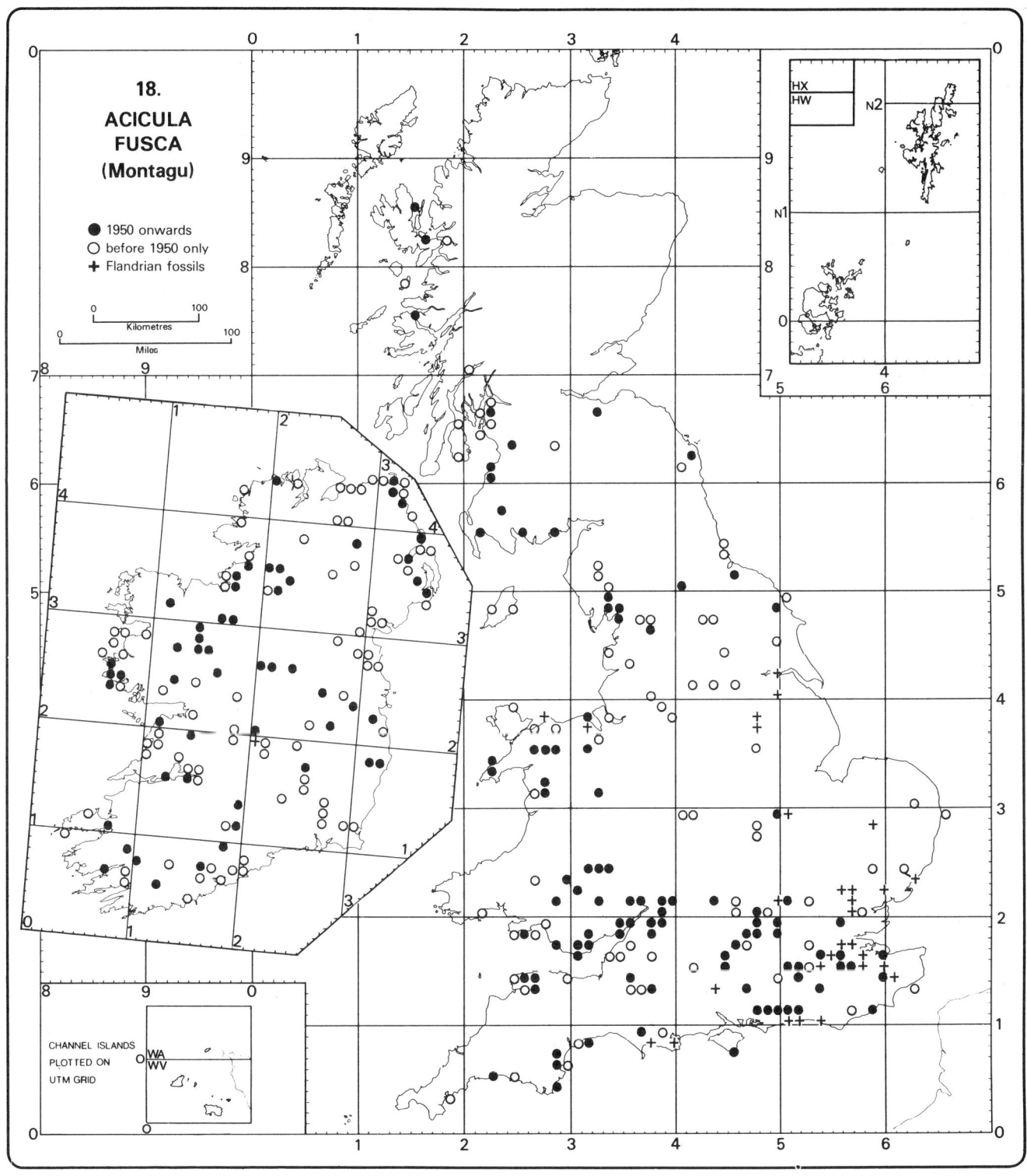

18.
ACICULA
FUSCA
(Montagu)

● 1950 onwards
○ before 1950 only
+ Flandrian fossils

Kilometres

Miles

CHANNEL ISLANDS
PLOTTED ON
UTM GRID

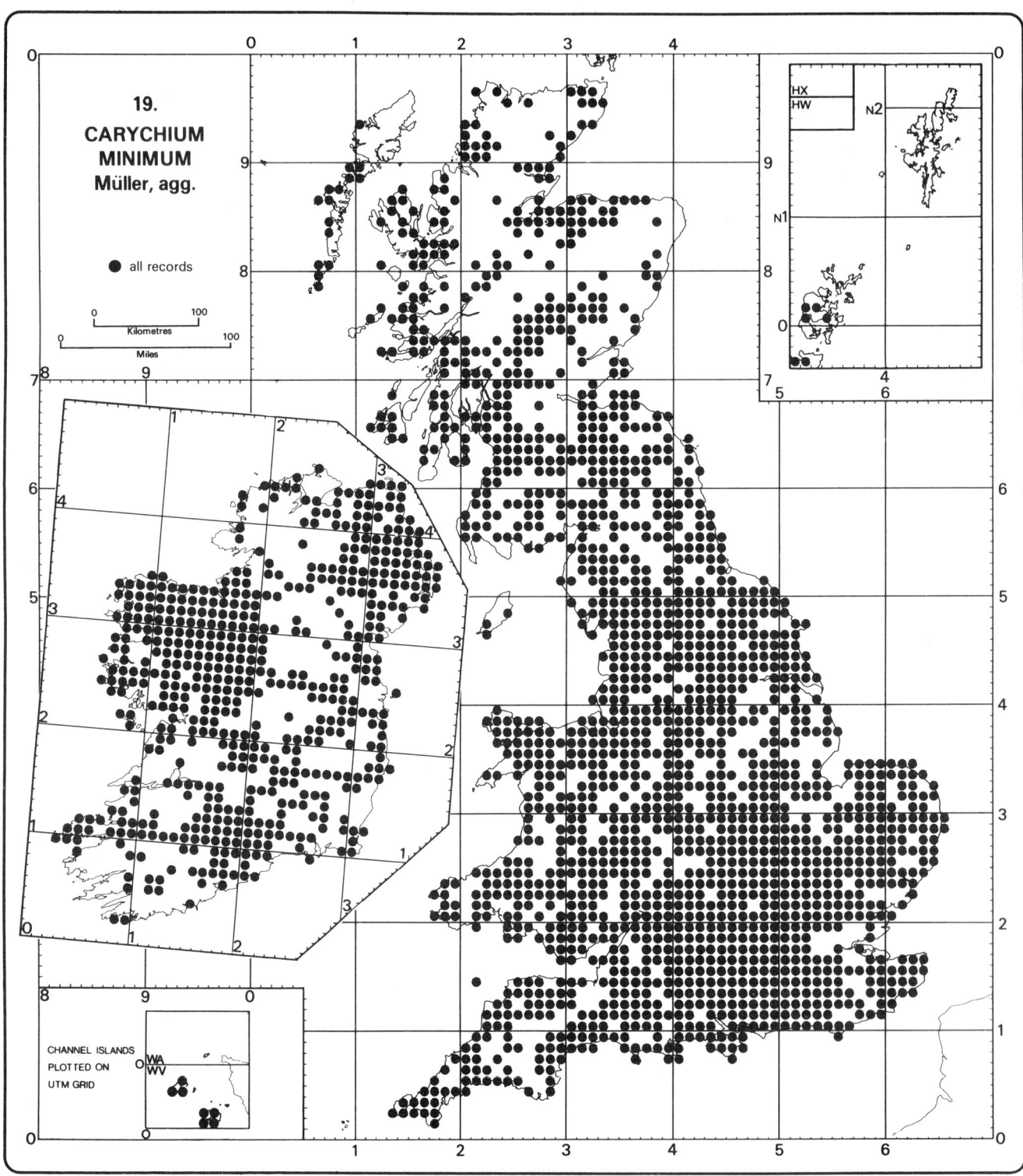

**19.
CARYCHIUM
MINIMUM
Müller, agg.**

● all records

Kilometres
Miles

CHANNEL ISLANDS
PLOTTED ON
UTM GRID

WA
WV

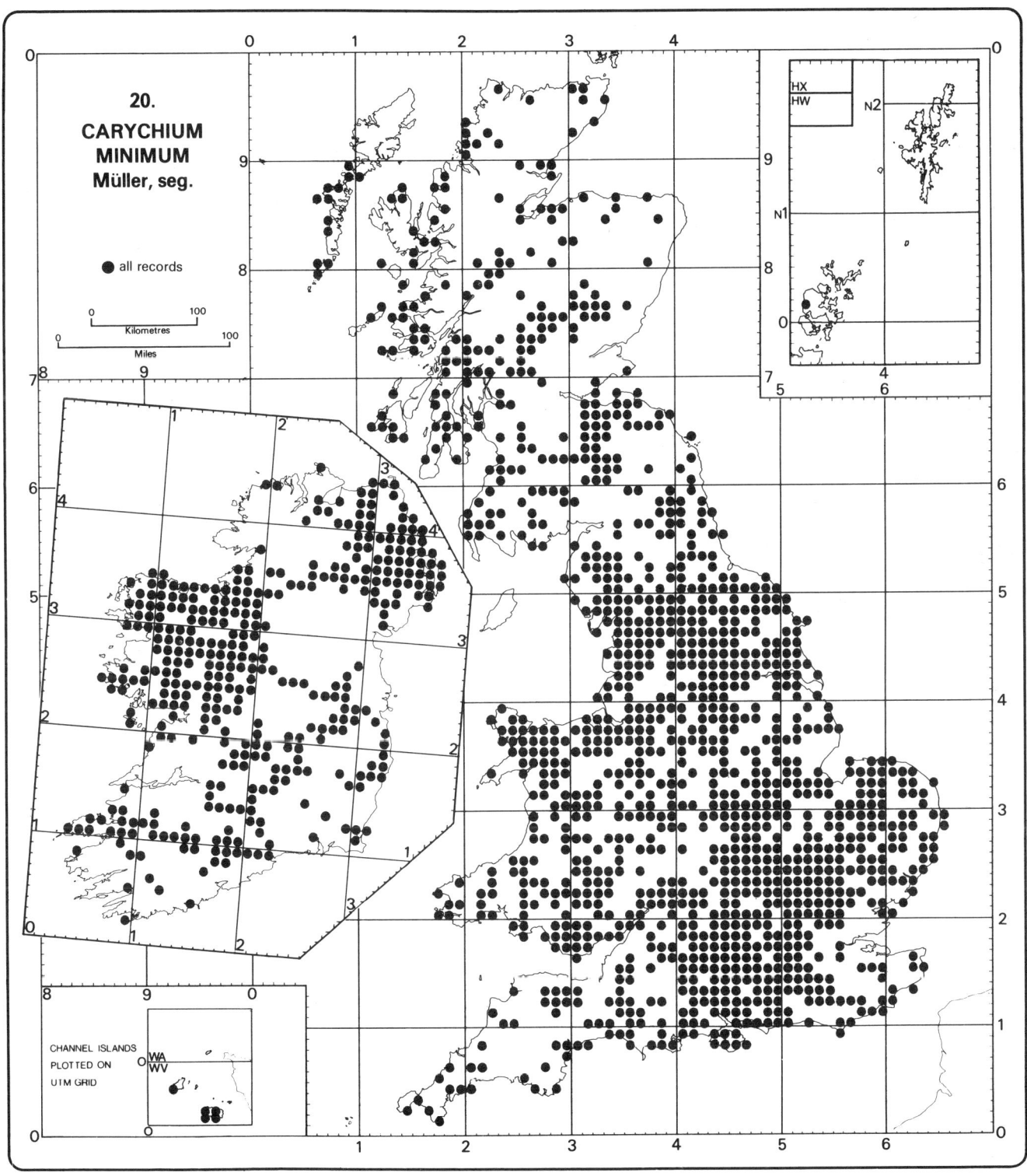

**20.
CARYCHIUM
MINIMUM
Müller, seg.**

● all records

Kilometres

Miles

CHANNEL ISLANDS
PLOTTED ON
UTM GRID

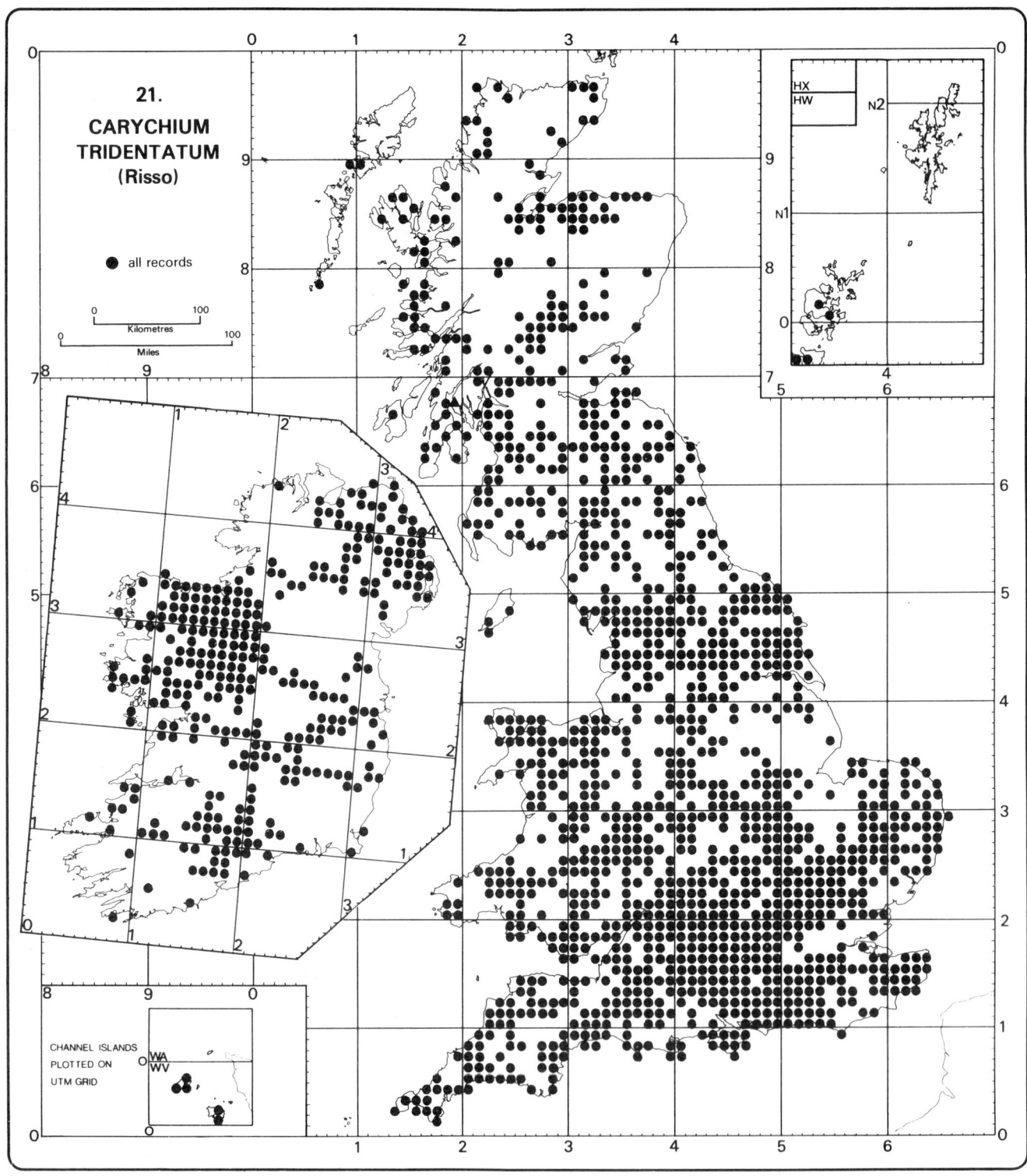

21.
CARYCHIUM
TRIDENTATUM
(Risso)

● all records

0   100
Kilometres
0   100
Miles

HX
HW

N2

N1

CHANNEL ISLANDS
PLOTTED ON
UTM GRID

WA
WV

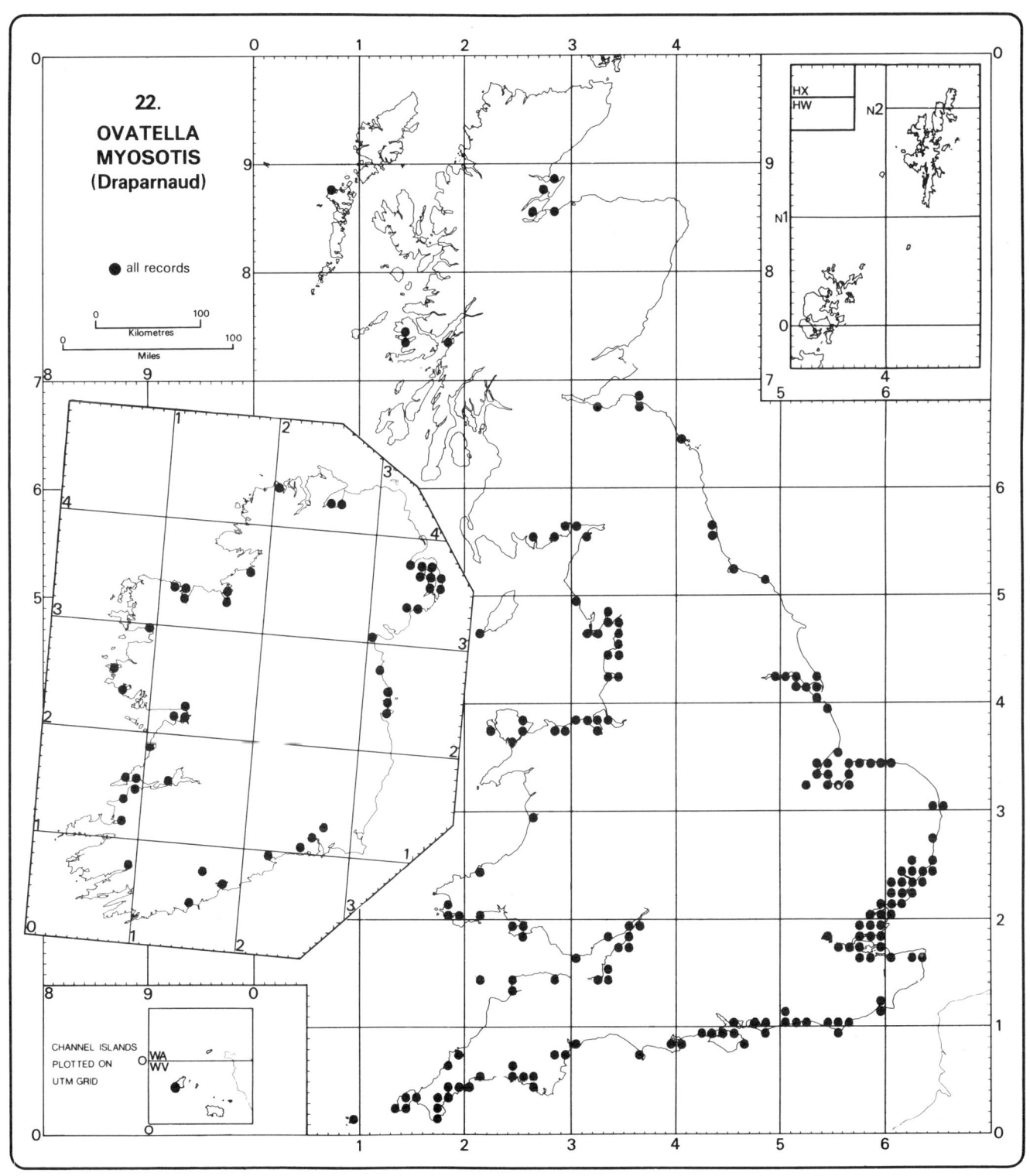

22.
OVATELLA
MYOSOTIS
(Draparnaud)

● all records

Kilometres

Miles

CHANNEL ISLANDS
PLOTTED ON
UTM GRID

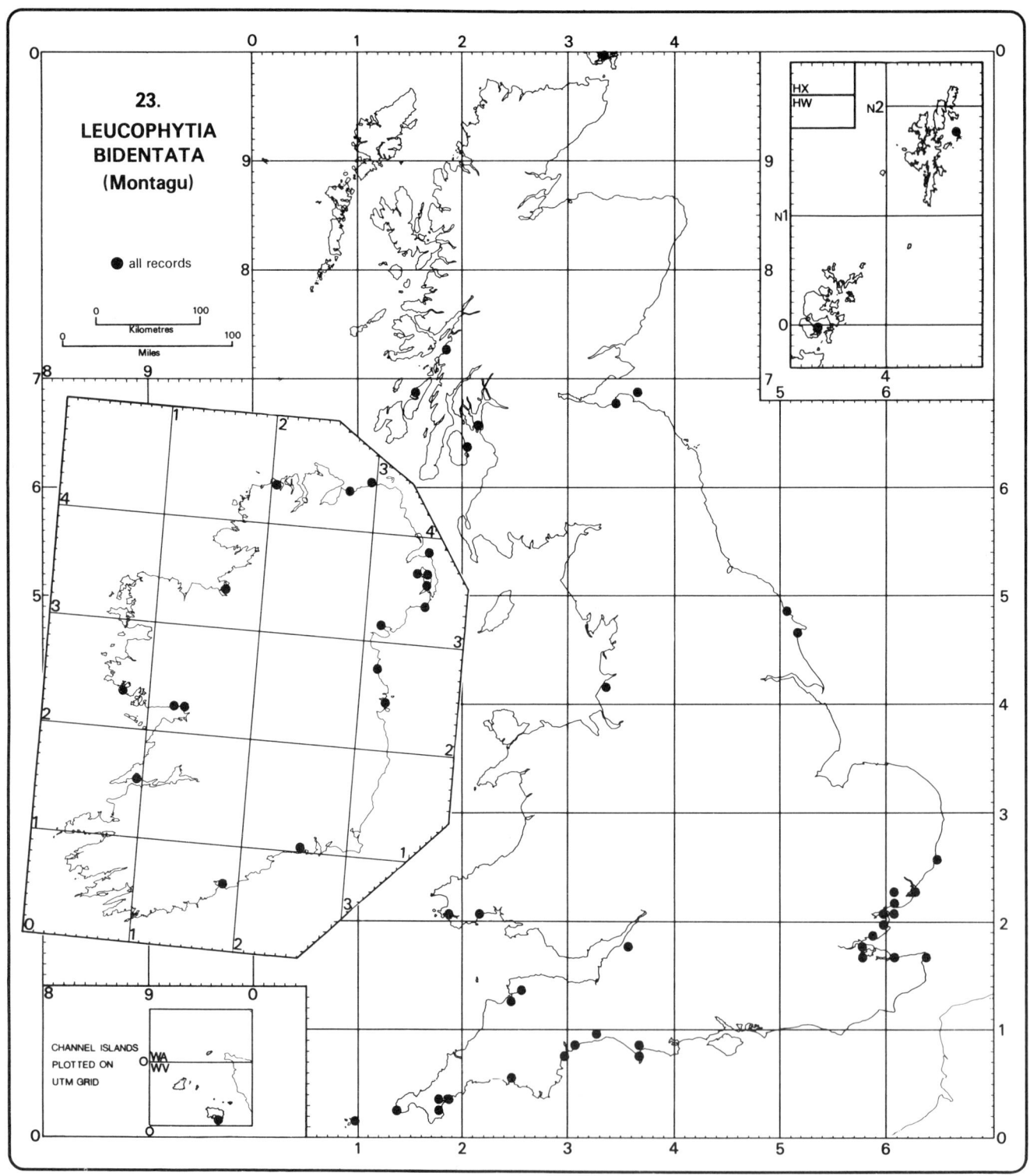

23.
LEUCOPHYTIA
BIDENTATA
(Montagu)

● all records

Kilometres

Miles

CHANNEL ISLANDS
PLOTTED ON
UTM GRID

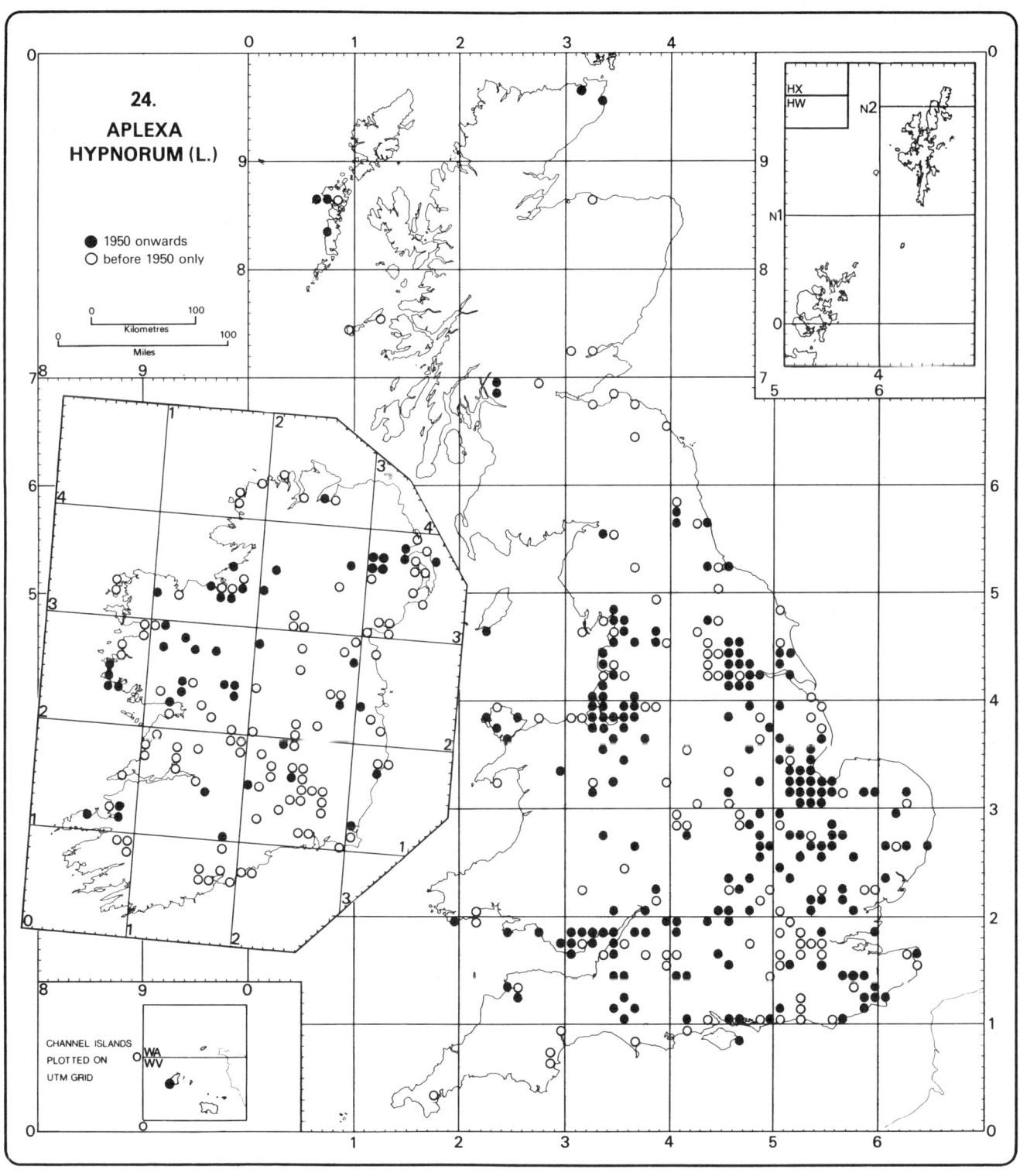

24.

APLEXA
HYPNORUM (L.)

● 1950 onwards
○ before 1950 only

Kilometres

Miles

CHANNEL ISLANDS
PLOTTED ON
UTM GRID

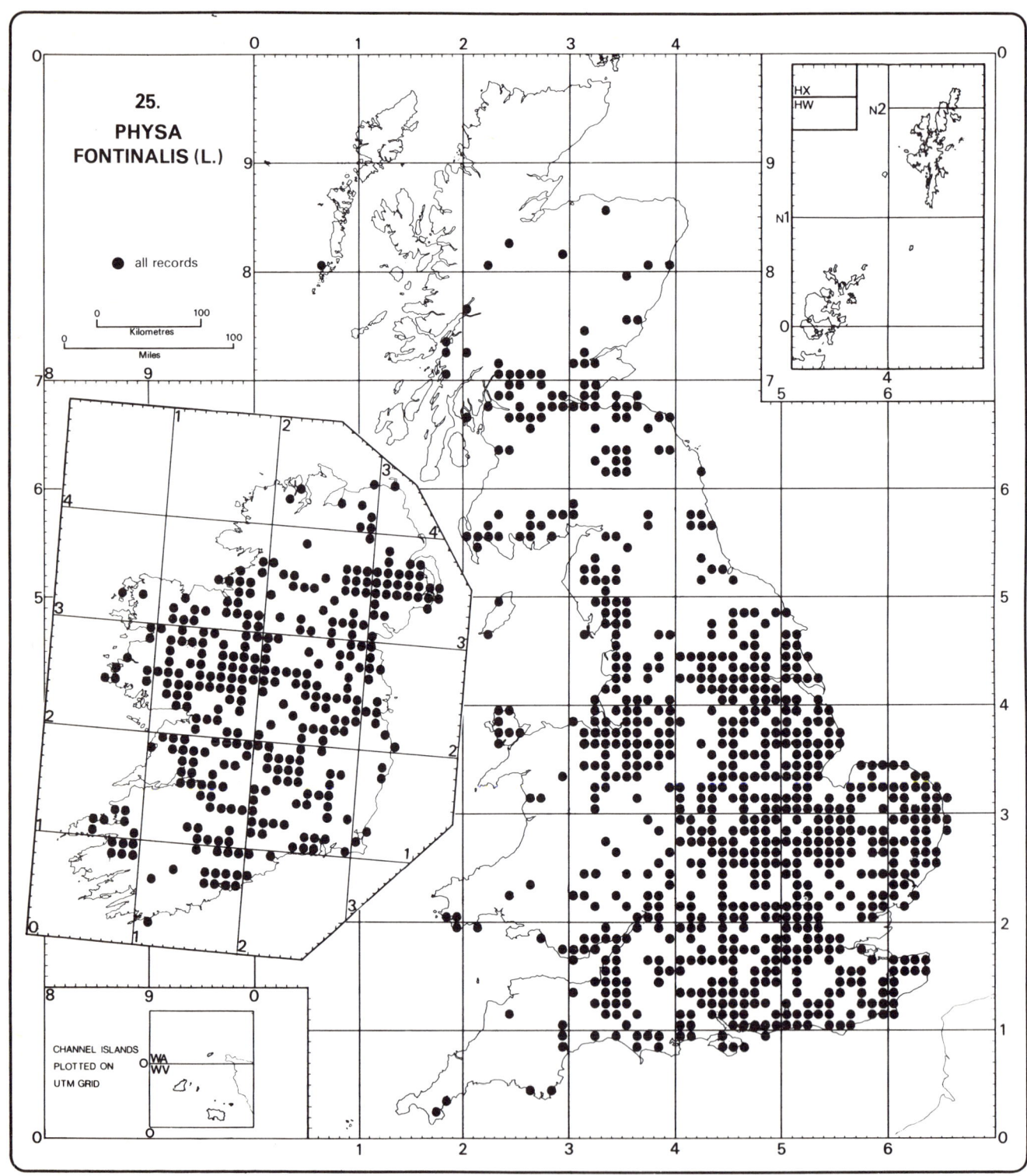

25.
PHYSA
FONTINALIS (L.)

● all records

0   100
Kilometres

0   100
Miles

CHANNEL ISLANDS
PLOTTED ON
UTM GRID

WA
WV

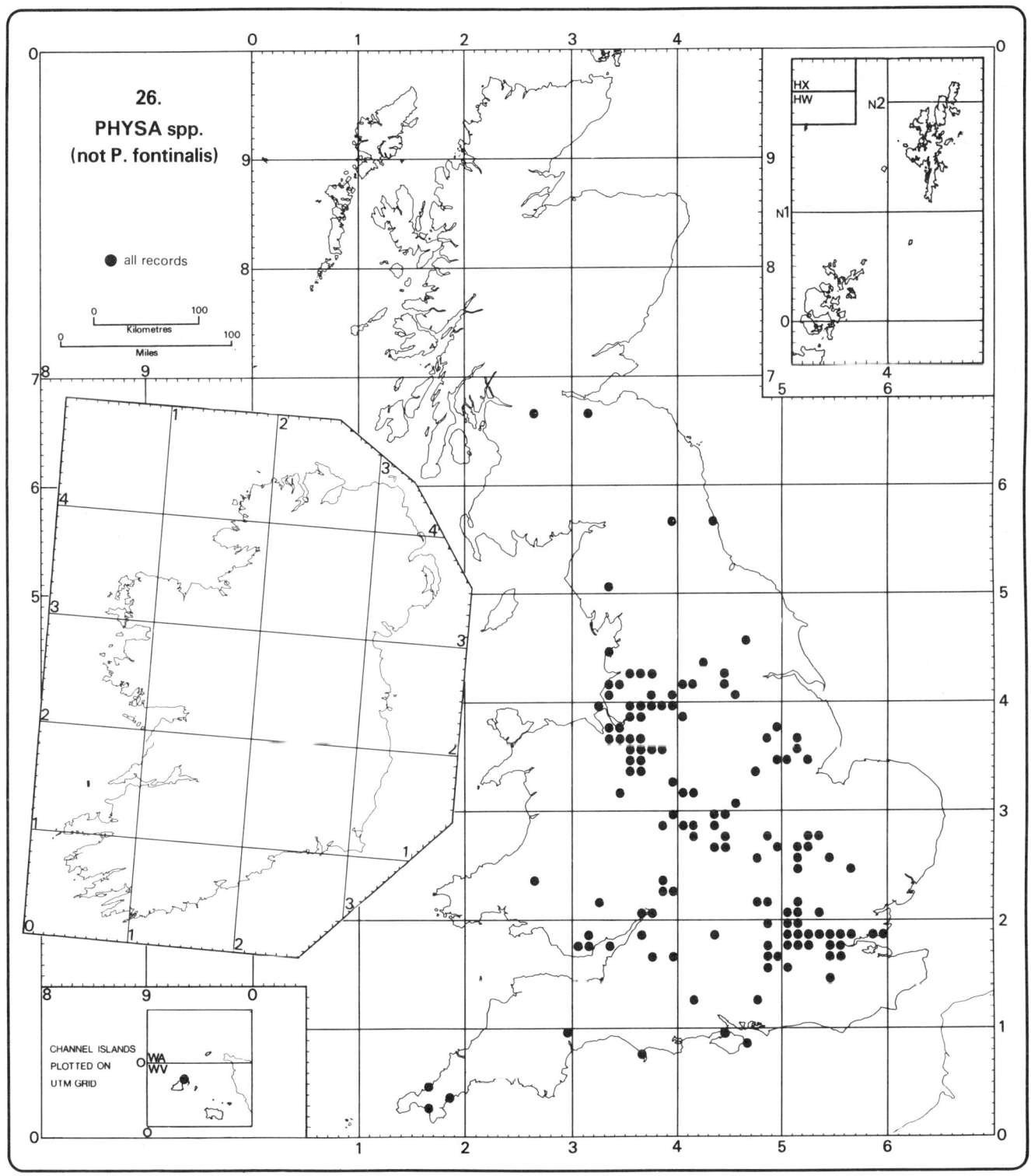

26.
PHYSA spp.
(not P. fontinalis)

● all records

0   100
Kilometres
0   100
Miles

CHANNEL ISLANDS
PLOTTED ON
UTM GRID

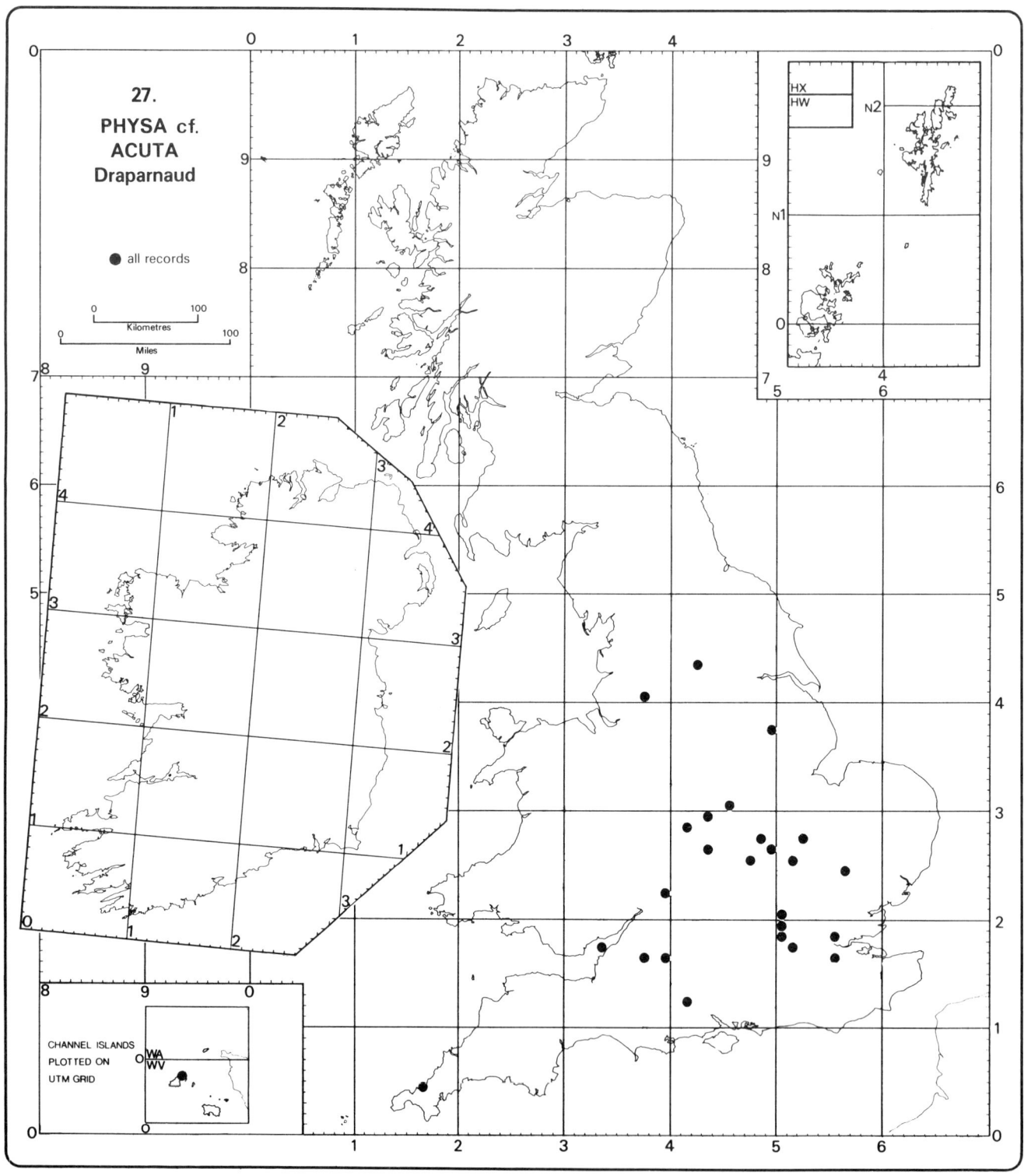

27.

PHYSA cf.
ACUTA
Draparnaud

● all records

Kilometres

Miles

CHANNEL ISLANDS
PLOTTED ON
UTM GRID

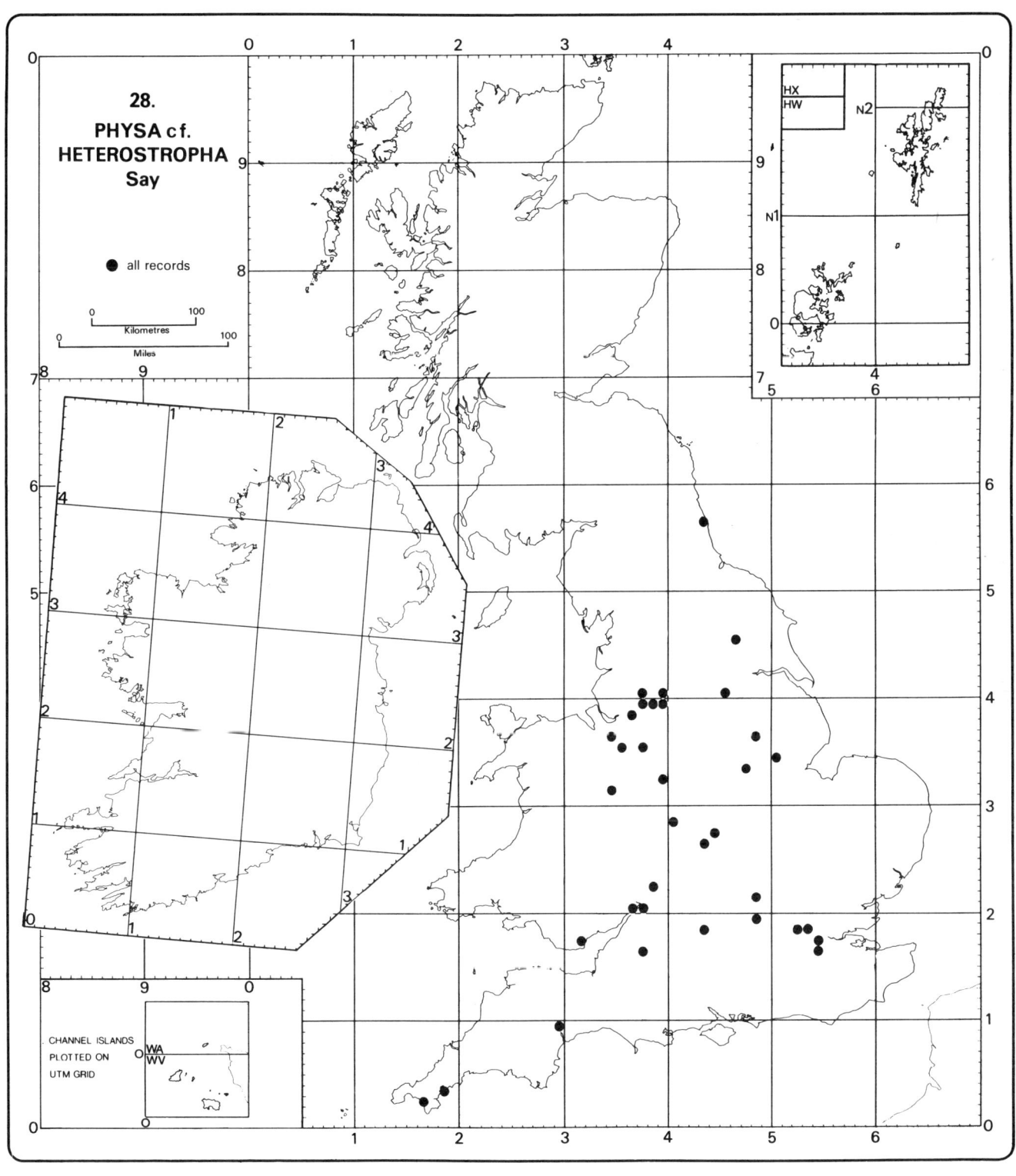

28.
PHYSA c f.
HETEROSTROPHA
Say

● all records

Kilometres

Miles

CHANNEL ISLANDS
PLOTTED ON
UTM GRID

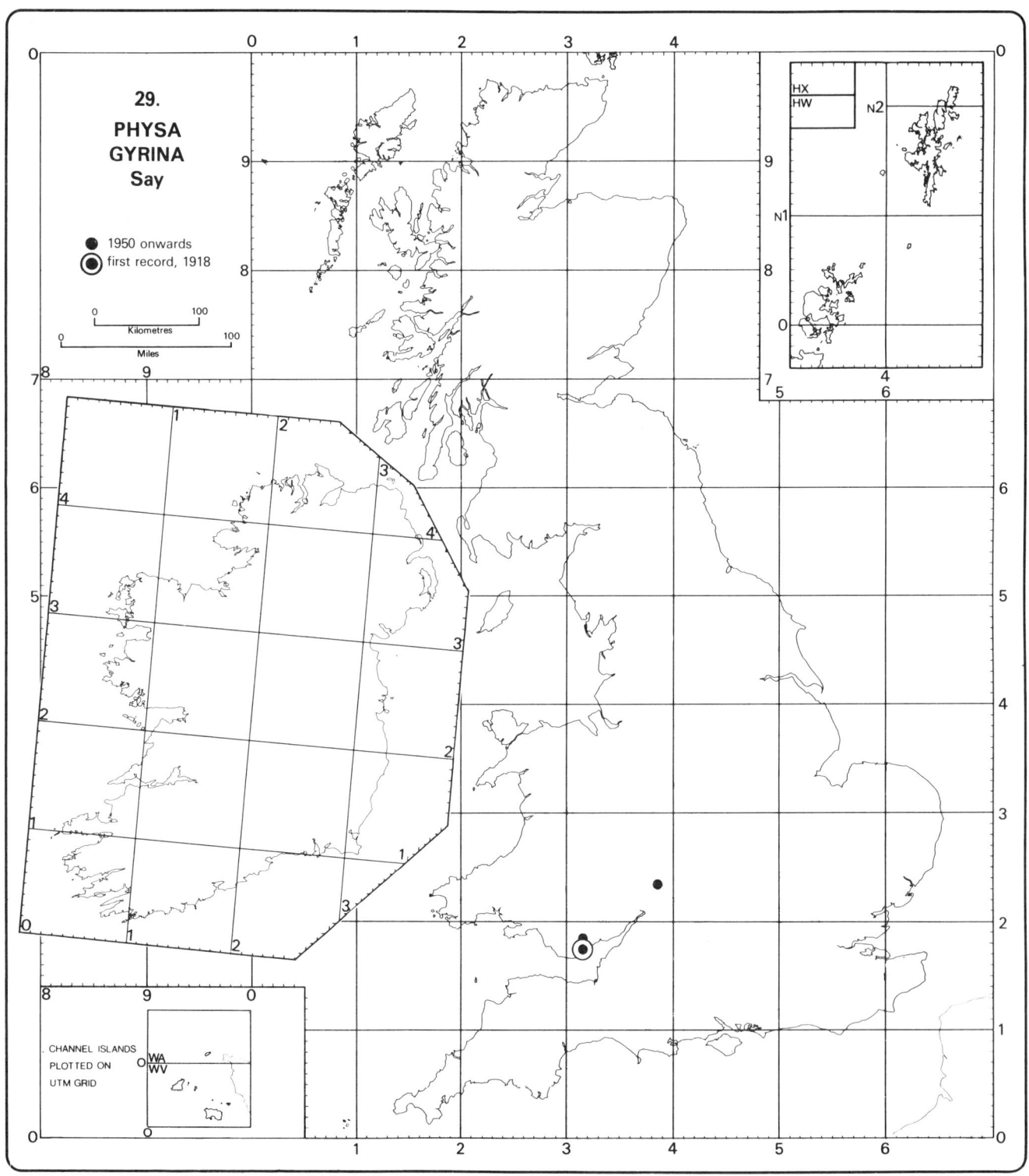

29.
PHYSA
GYRINA
Say

● 1950 onwards
◉ first record, 1918

0   100
Kilometres
0   100
Miles

CHANNEL ISLANDS
PLOTTED ON
UTM GRID

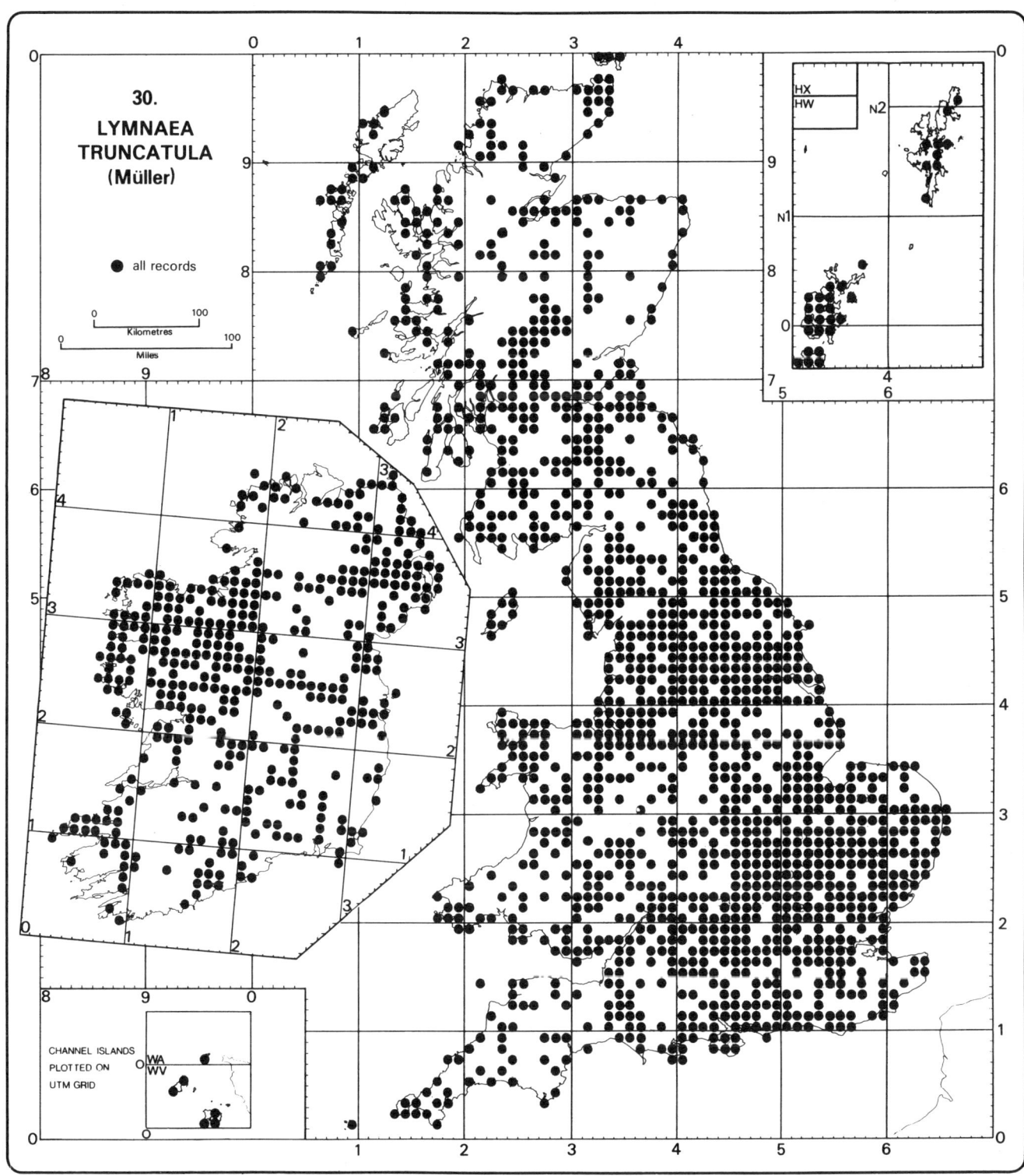

30.
LYMNAEA
TRUNCATULA
(Müller)

● all records

Kilometres

Miles

CHANNEL ISLANDS
PLOTTED ON
UTM GRID

WA
WV

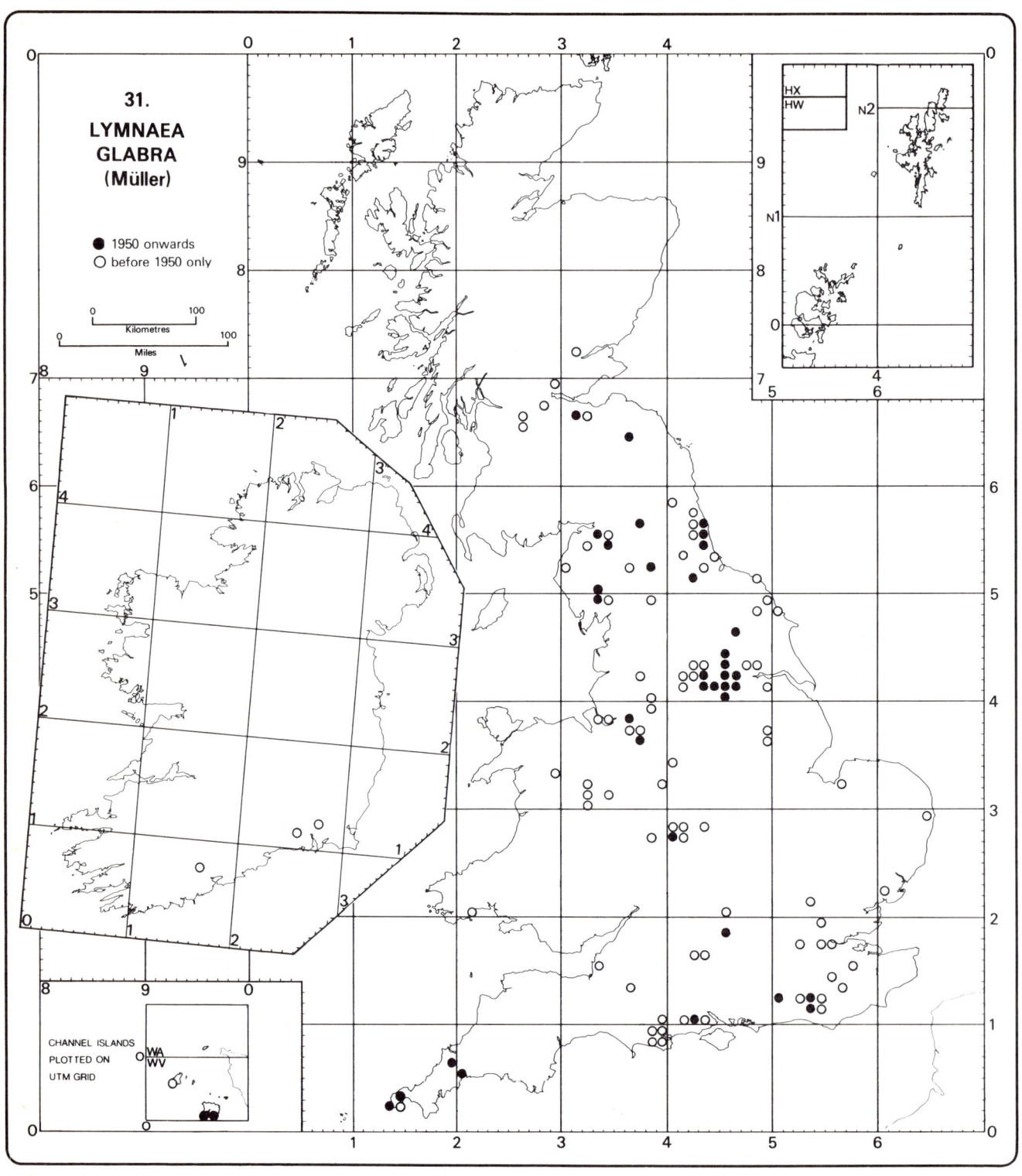

31.
LYMNAEA
GLABRA
(Müller)

● 1950 onwards
○ before 1950 only

Kilometres
Miles

CHANNEL ISLANDS
PLOTTED ON
UTM GRID

WA
WV

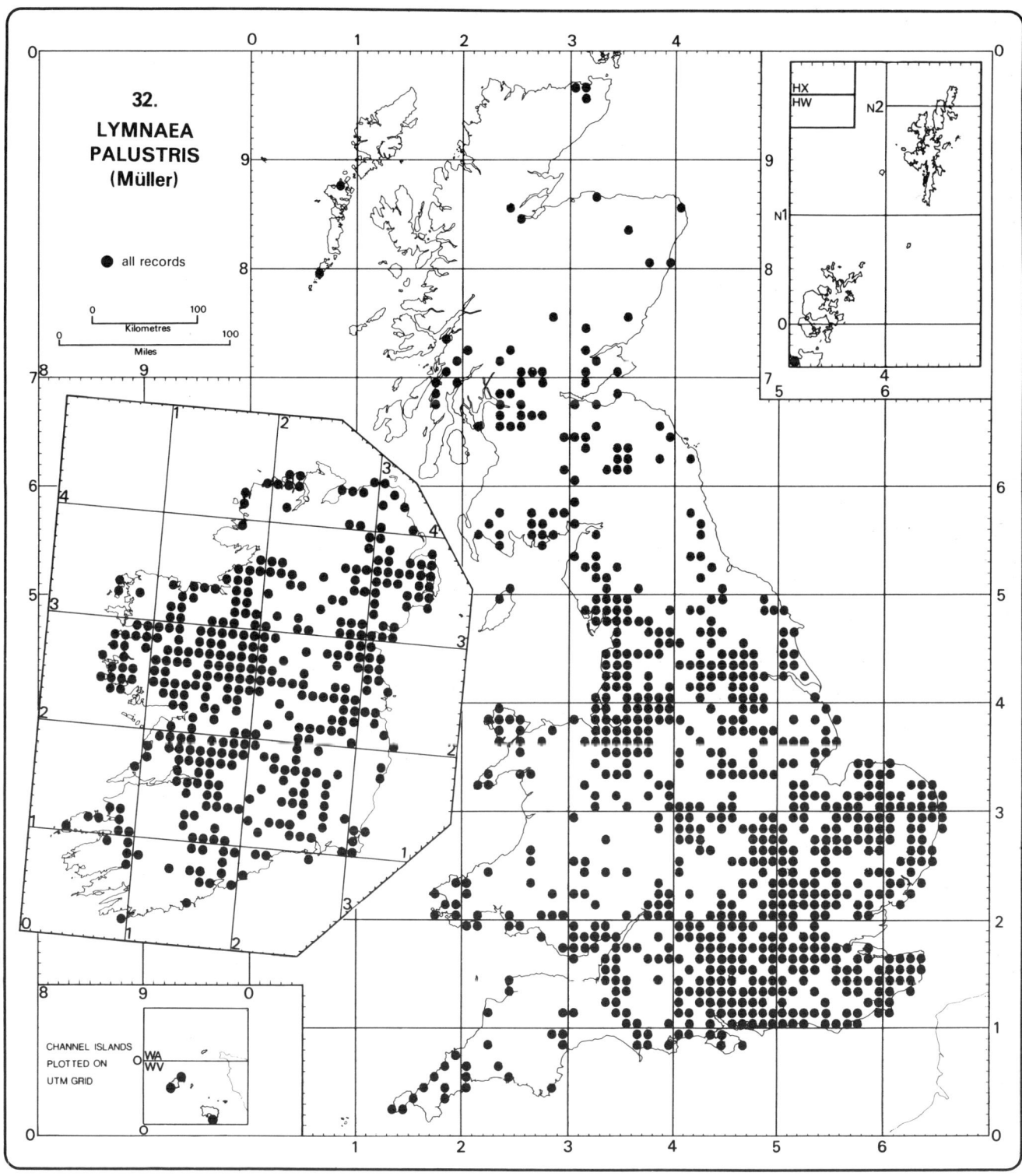

32.
LYMNAEA
PALUSTRIS
(Müller)

● all records

Kilometres

Miles

CHANNEL ISLANDS
PLOTTED ON
UTM GRID

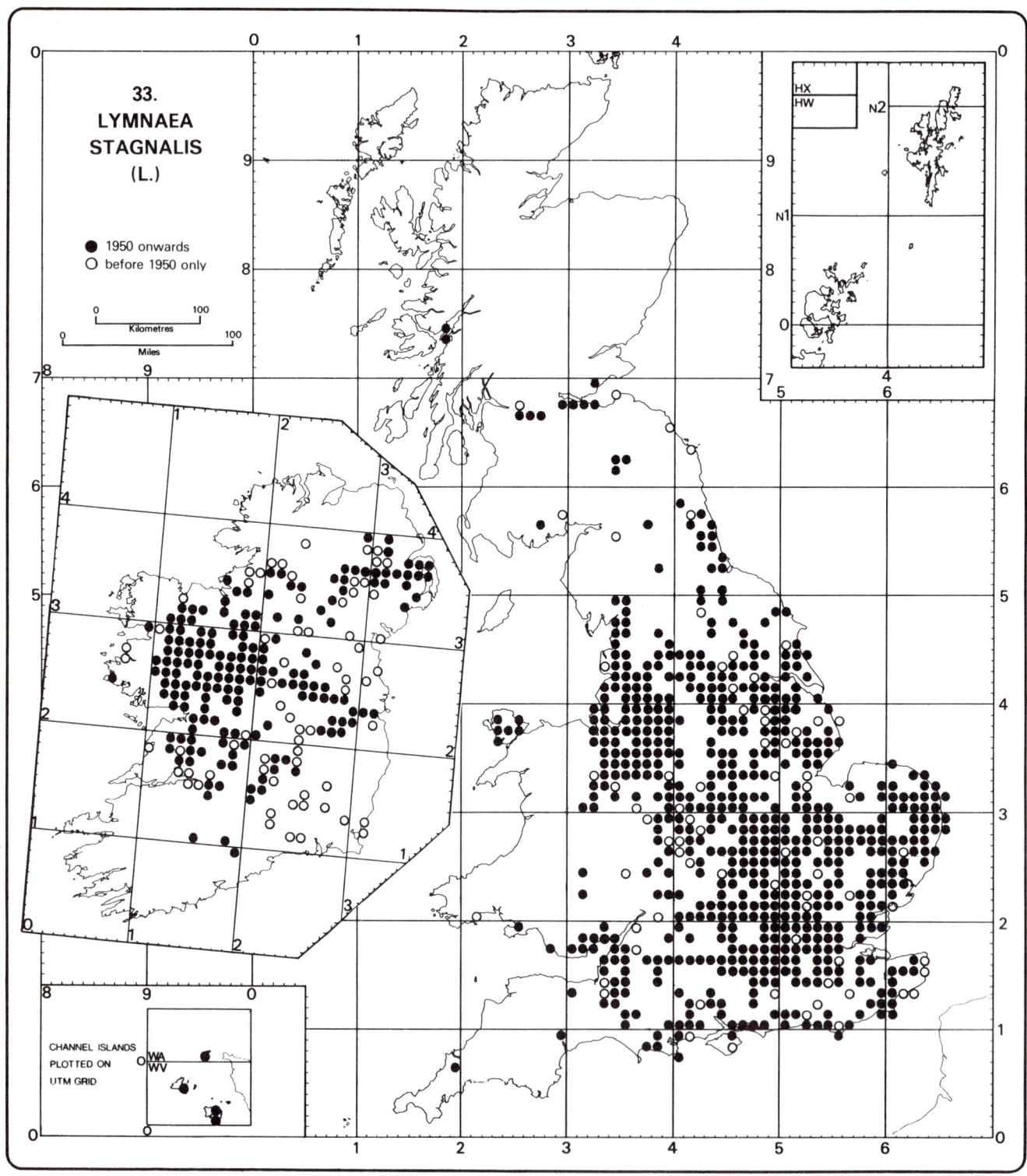

33.
LYMNAEA
STAGNALIS
(L.)

● 1950 onwards
○ before 1950 only

0
100
Kilometres
0
100
Miles

CHANNEL ISLANDS
PLOTTED ON
UTM GRID

WA
WV

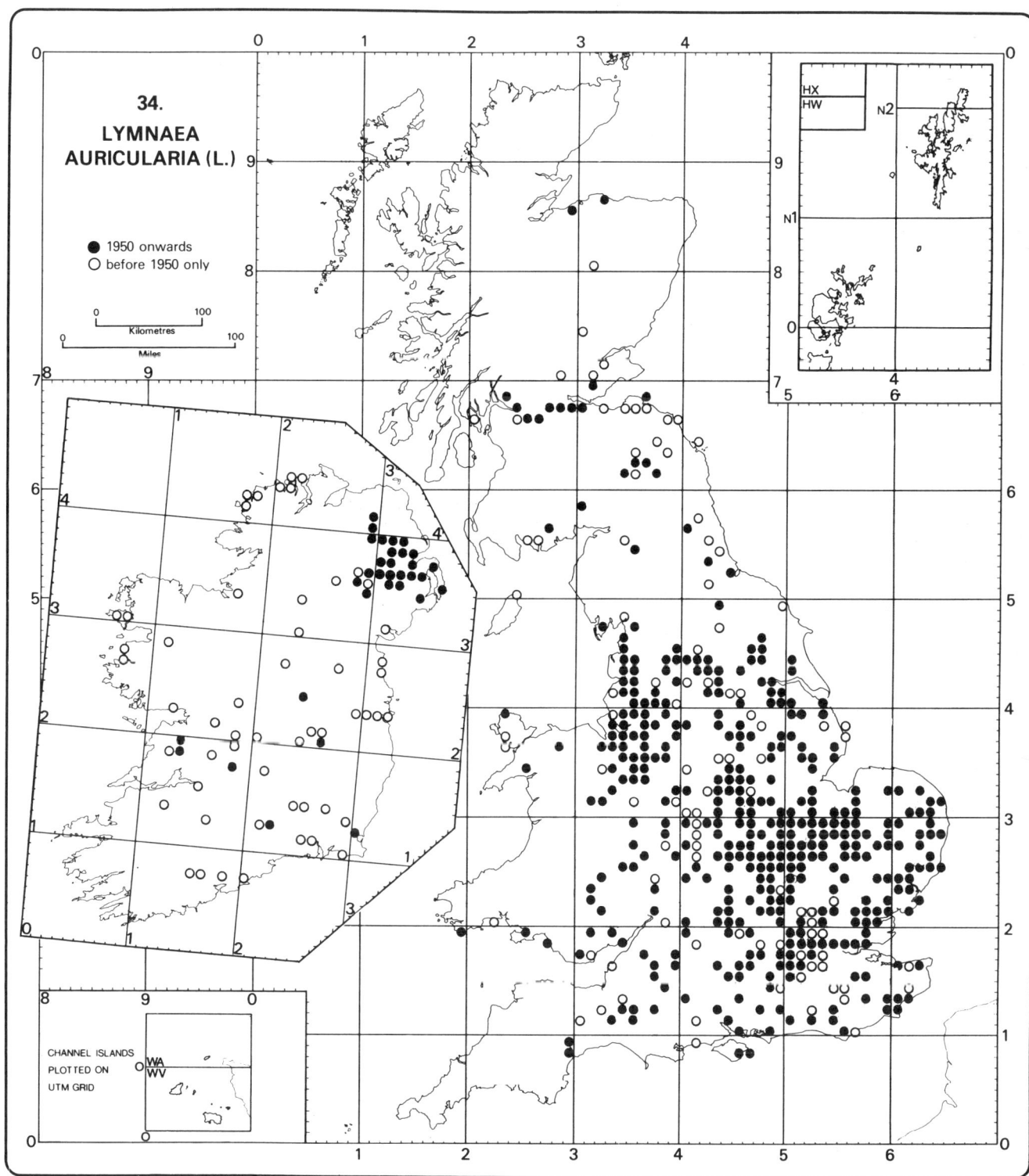

34.

LYMNAEA
AURICULARIA (L.)

● 1950 onwards
○ before 1950 only

Kilometres

Miles

CHANNEL ISLANDS
PLOTTED ON
UTM GRID

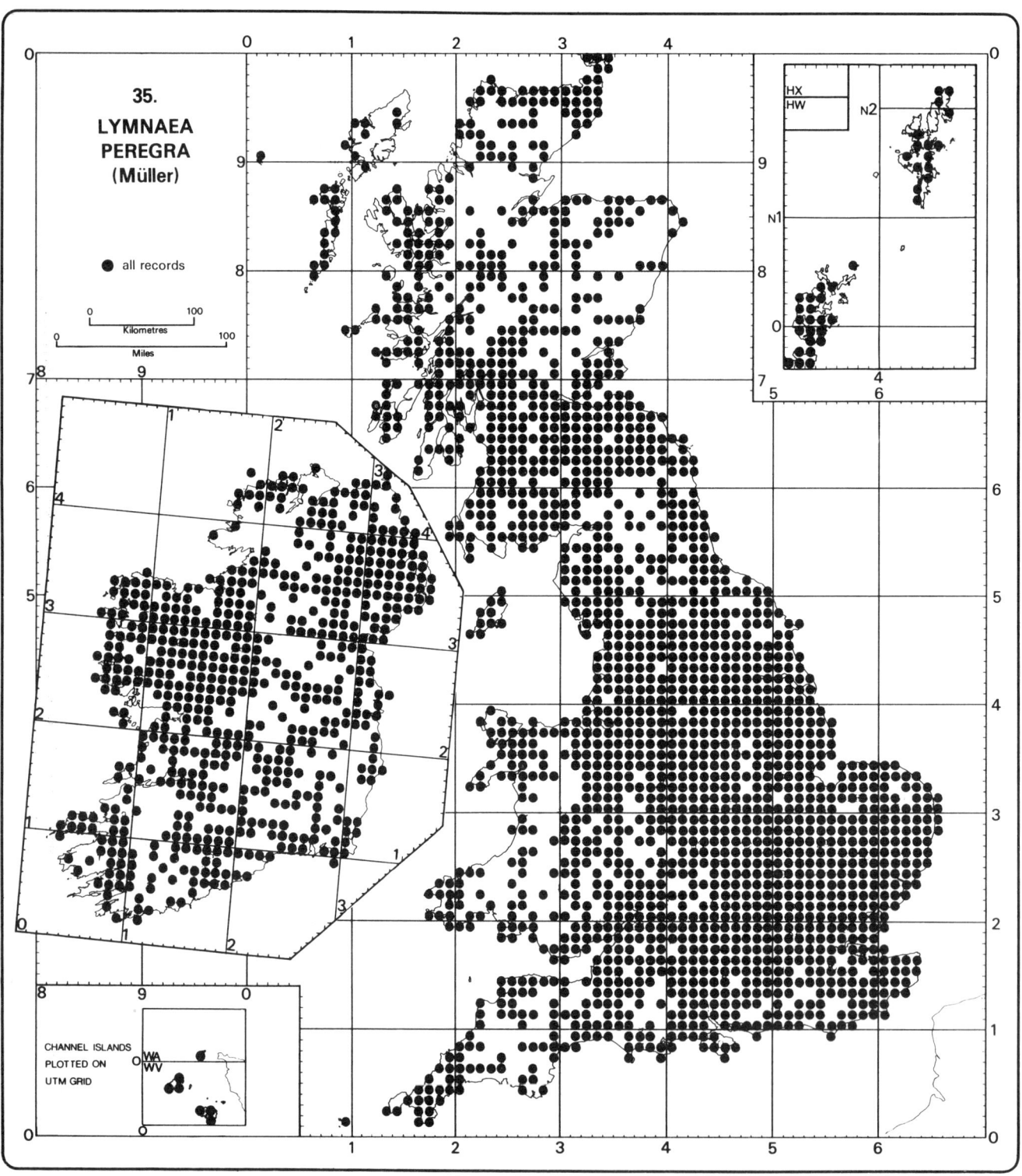

35.
LYMNAEA
PEREGRA
(Müller)

● all records

Kilometres

Miles

CHANNEL ISLANDS
PLOTTED ON
UTM GRID

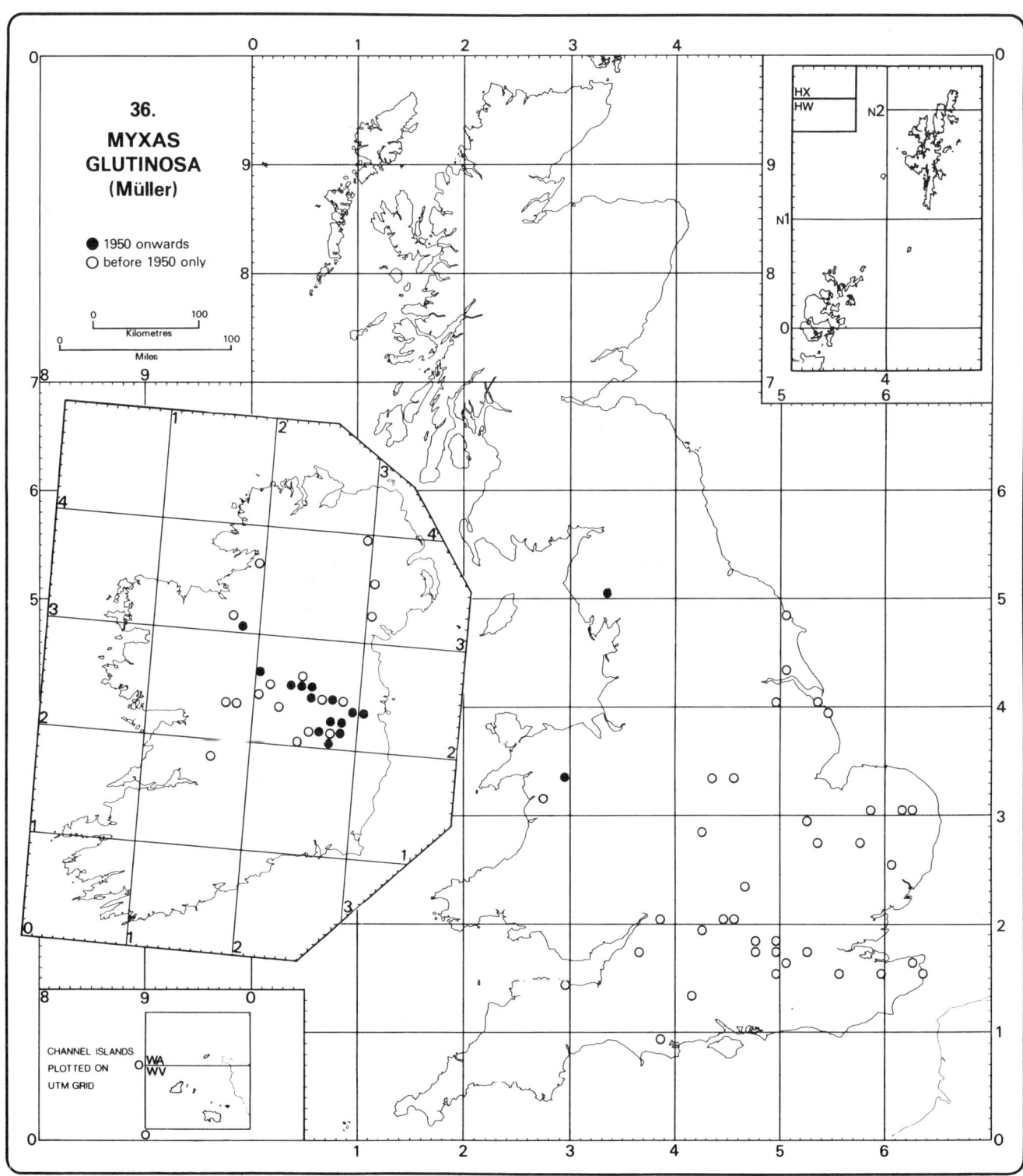

36.

MYXAS
GLUTINOSA
(Müller)

● 1950 onwards
○ before 1950 only

Kilometres

Miles

CHANNEL ISLANDS
PLOTTED ON
UTM GRID

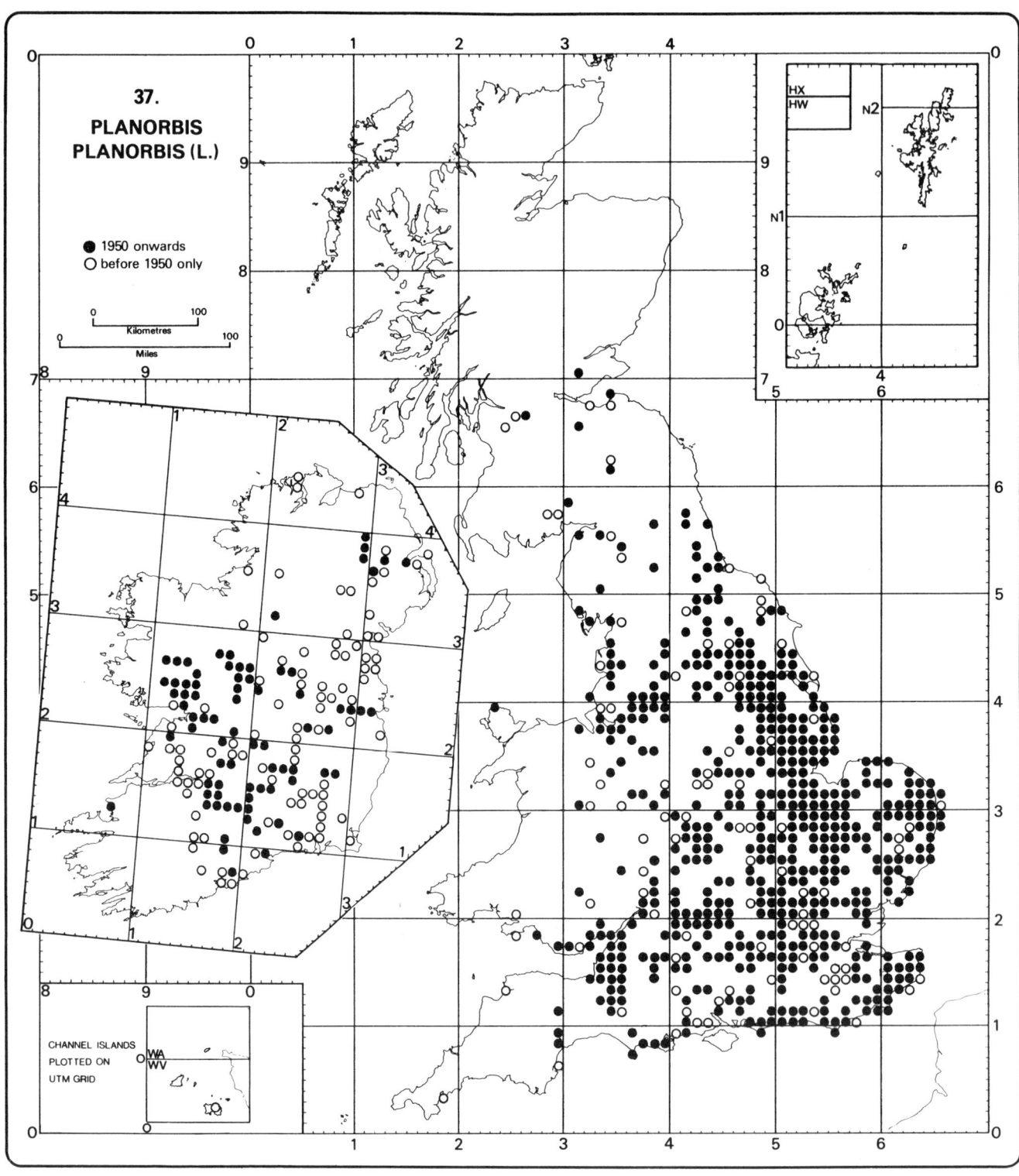

37.
PLANORBIS
PLANORBIS (L.)

● 1950 onwards
○ before 1950 only

0   100
Kilometres
0   100
Miles

CHANNEL ISLANDS
PLOTTED ON
UTM GRID

WA
WV

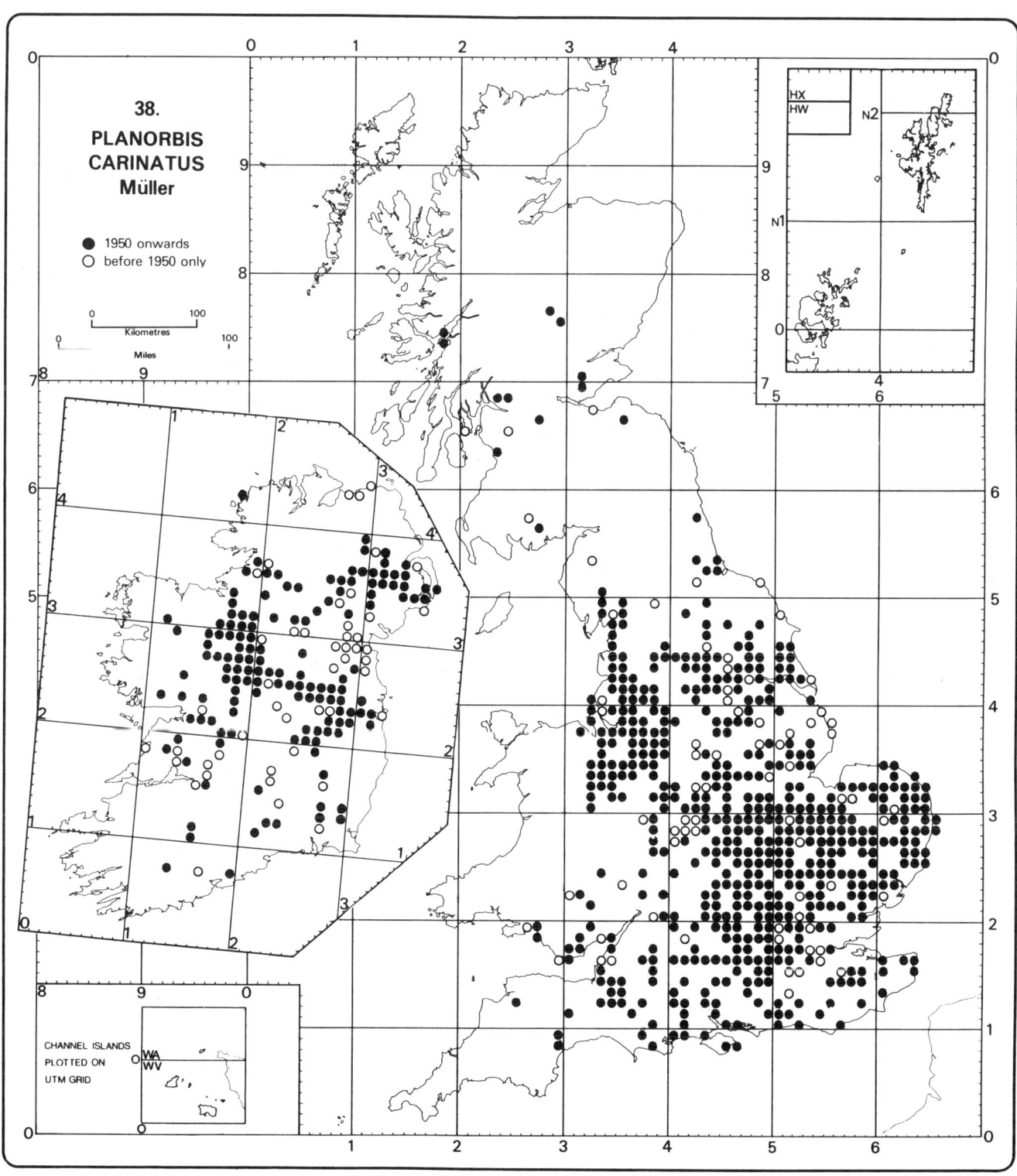

38.
PLANORBIS
CARINATUS
Müller

● 1950 onwards
○ before 1950 only

Kilometres

Miles

CHANNEL ISLANDS
PLOTTED ON
UTM GRID

WA
WV

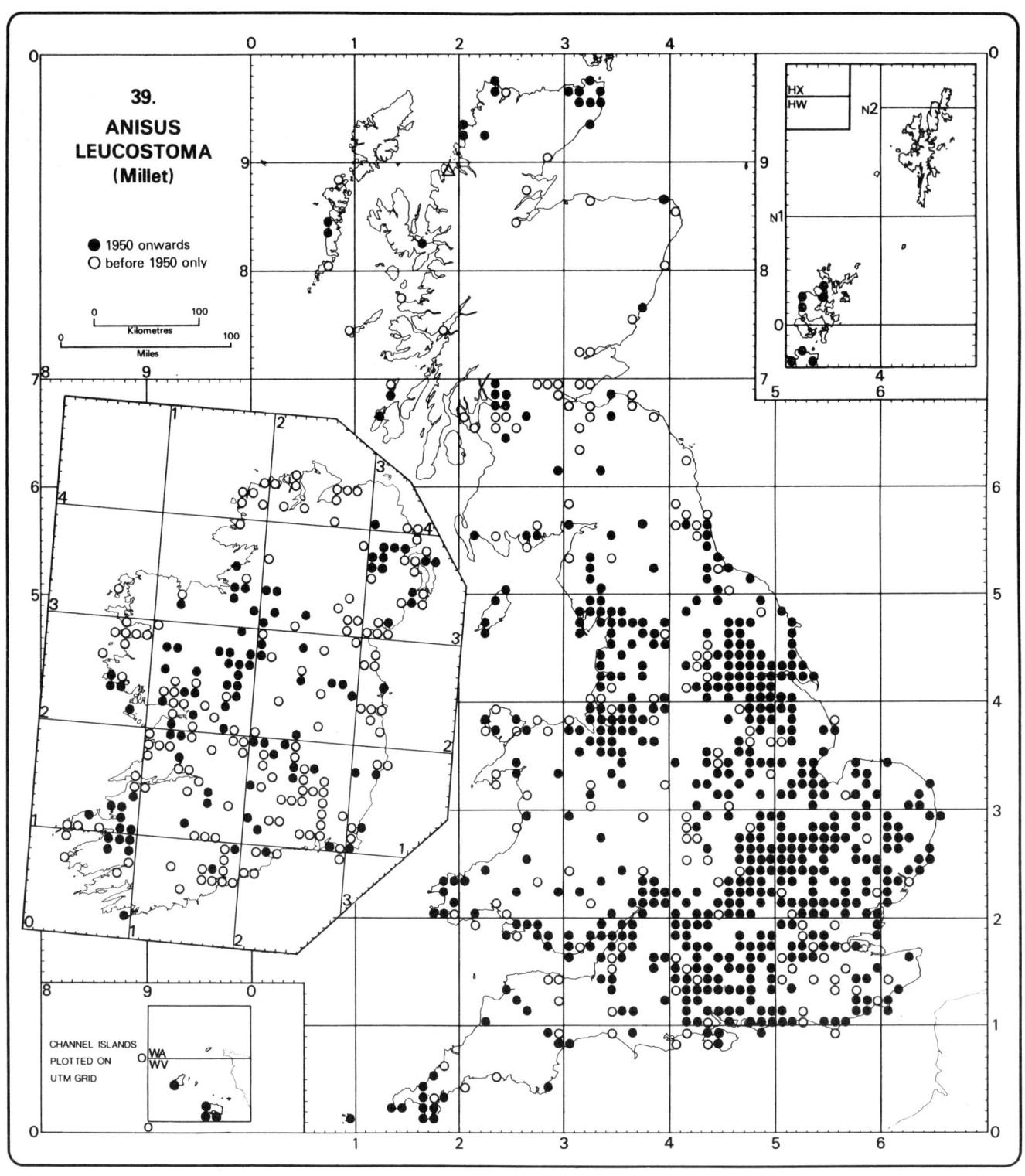

39.
ANISUS
LEUCOSTOMA
(Millet)

● 1950 onwards
○ before 1950 only

Kilometres
Miles

CHANNEL ISLANDS
PLOTTED ON
UTM GRID

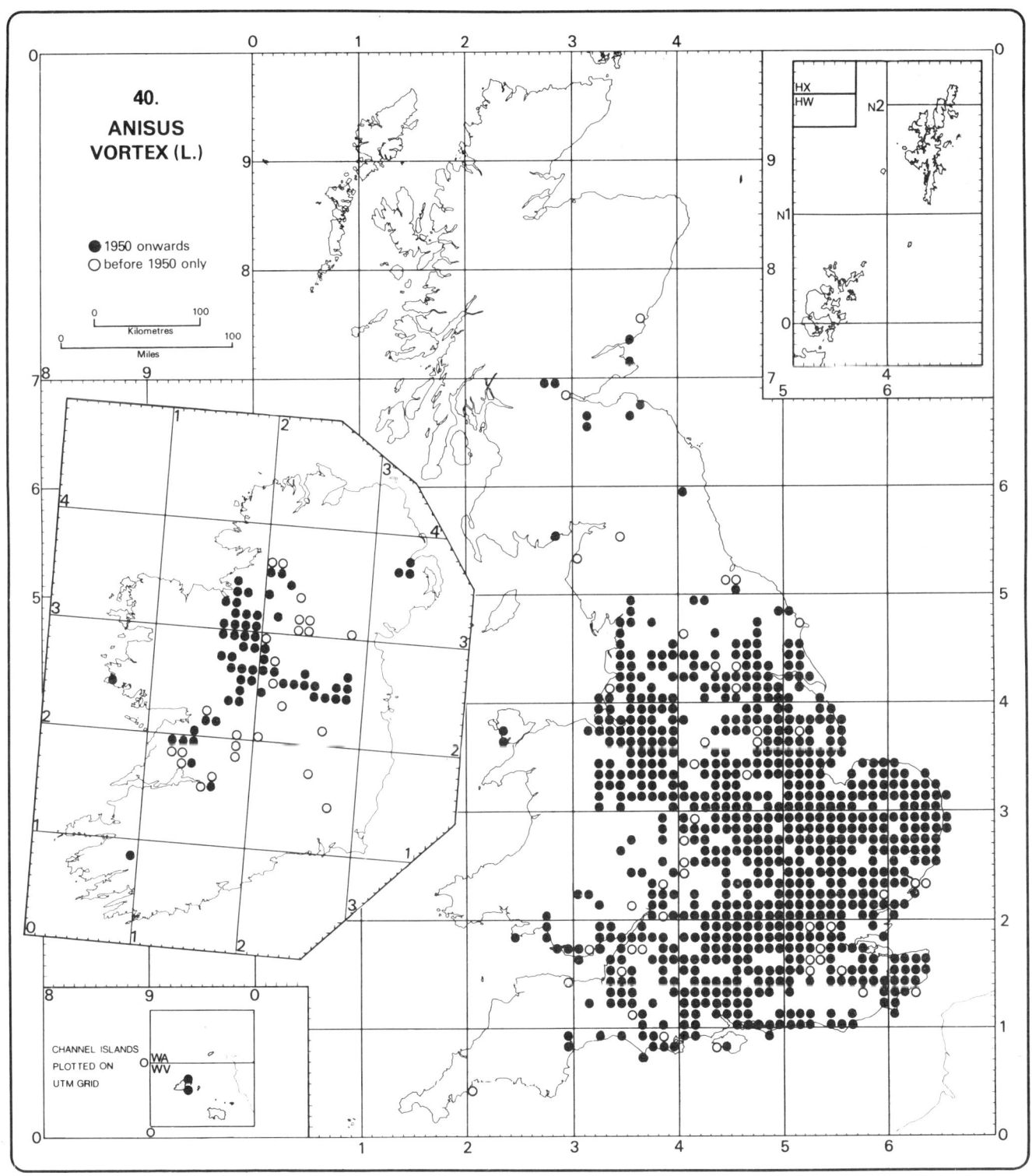

40.
ANISUS
VORTEX (L.)

● 1950 onwards
○ before 1950 only

0   100
Kilometres
0   100
Miles

CHANNEL ISLANDS
PLOTTED ON
UTM GRID

WA
WV

HX
HW

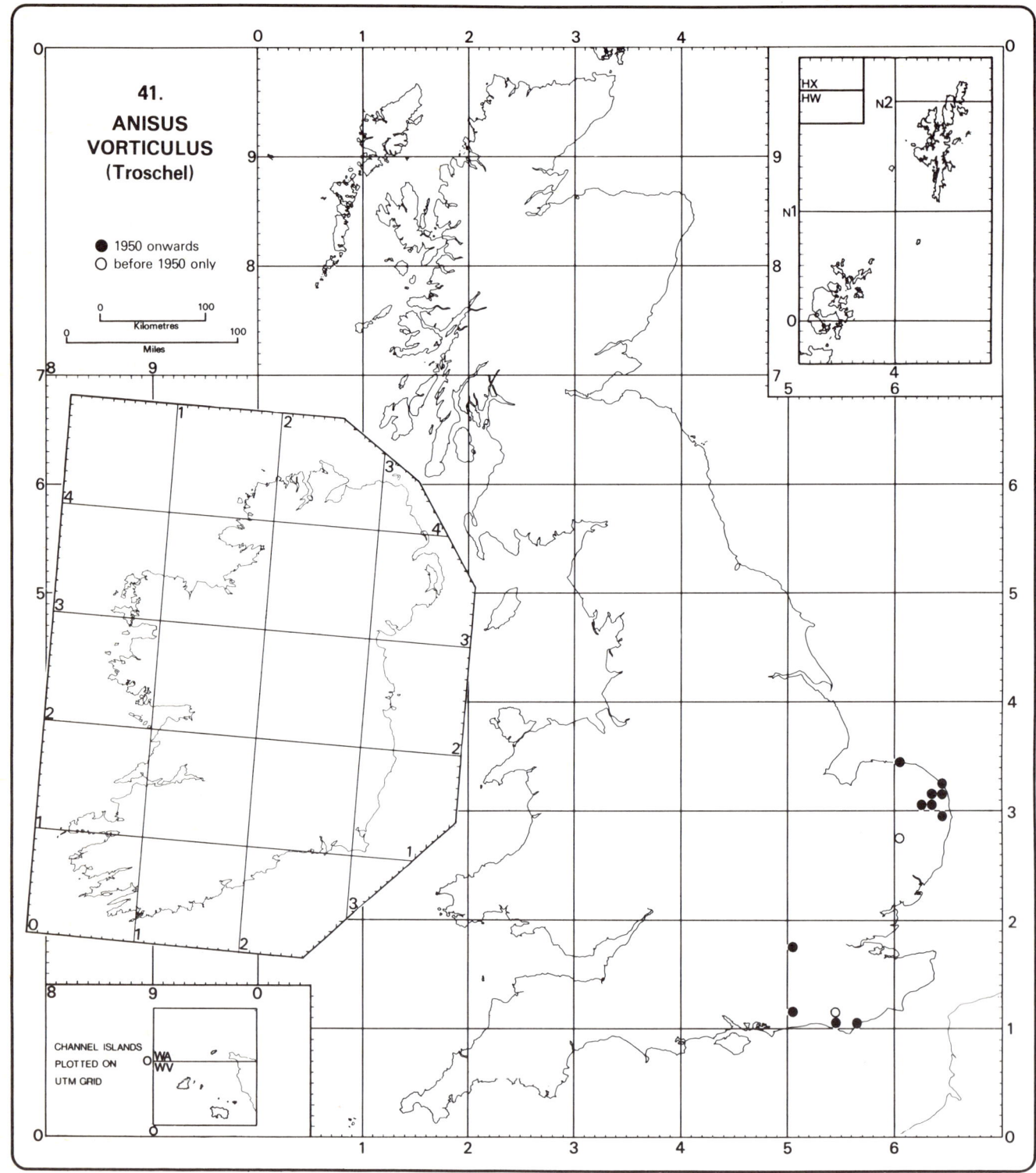

41.
ANISUS
VORTICULUS
(Troschel)

● 1950 onwards
○ before 1950 only

Kilometres
Miles

CHANNEL ISLANDS
PLOTTED ON
UTM GRID

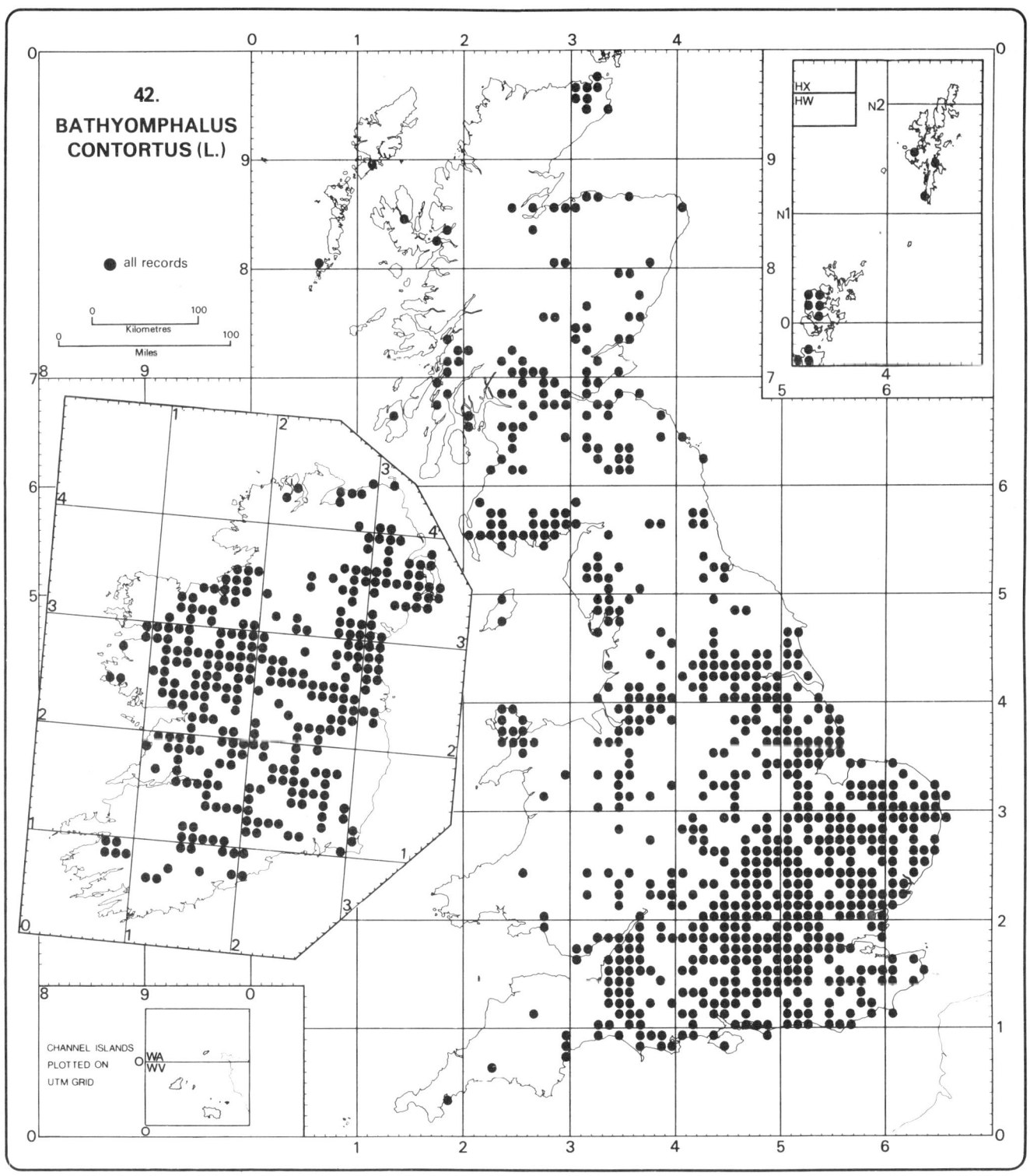

42.
BATHYOMPHALUS
CONTORTUS (L.)

● all records

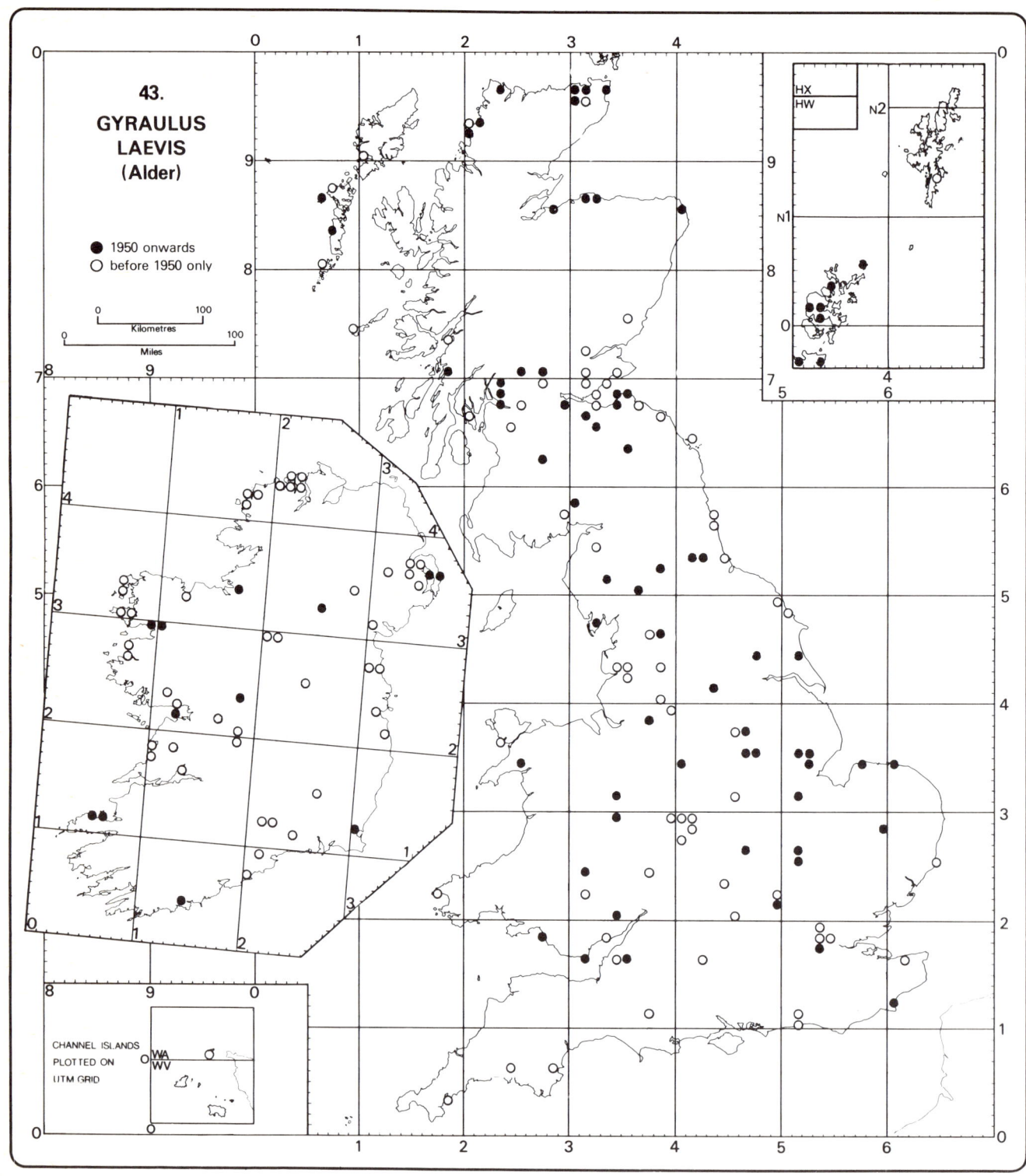

43.
GYRAULUS
LAEVIS
(Alder)

● 1950 onwards
○ before 1950 only

0   100
Kilometres
0   100
Miles

CHANNEL ISLANDS
PLOTTED ON
UTM GRID

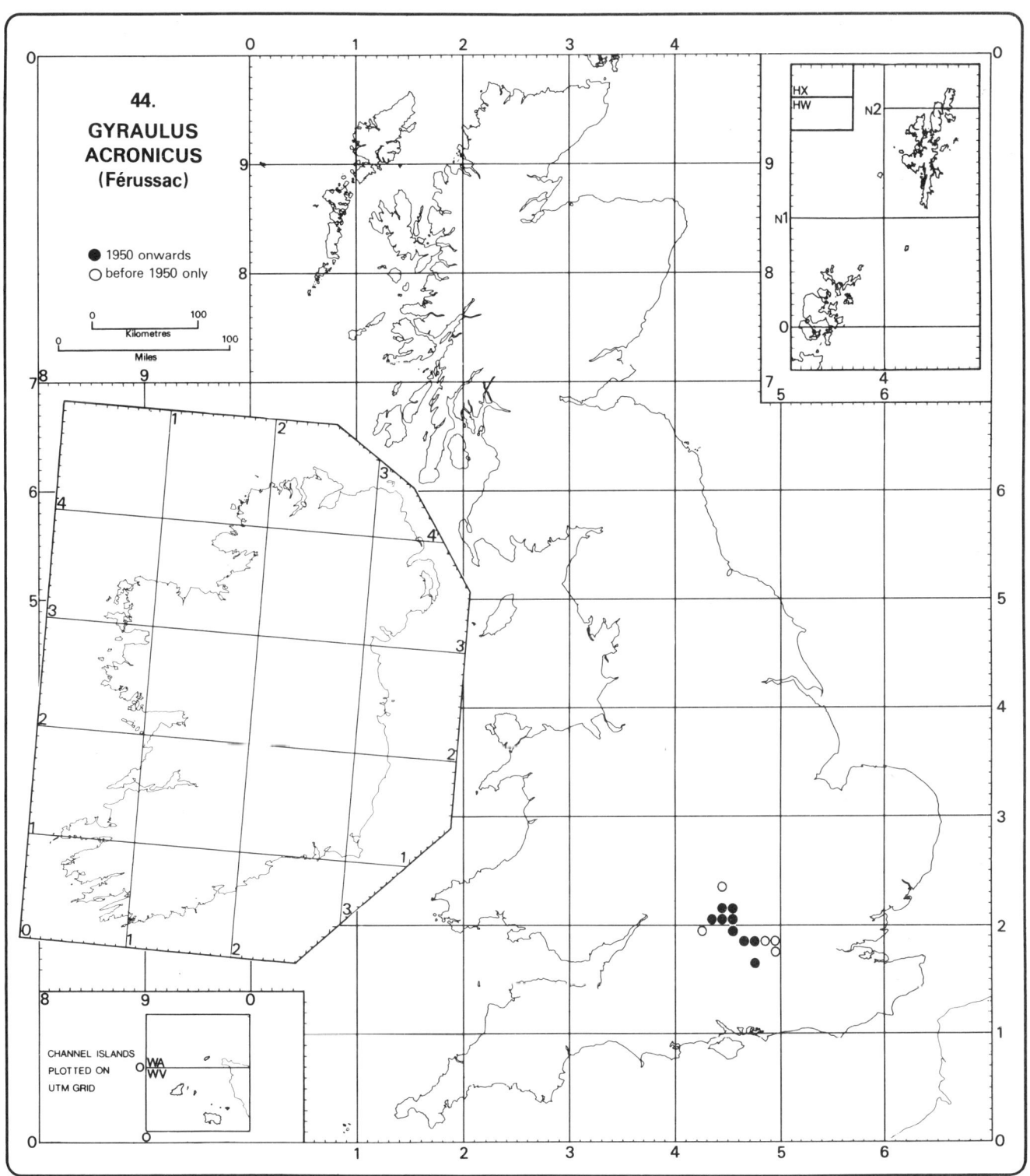

44.

**GYRAULUS
ACRONICUS
(Férussac)**

● 1950 onwards
○ before 1950 only

0   100
Kilometres
0   100
Miles

CHANNEL ISLANDS
PLOTTED ON
UTM GRID

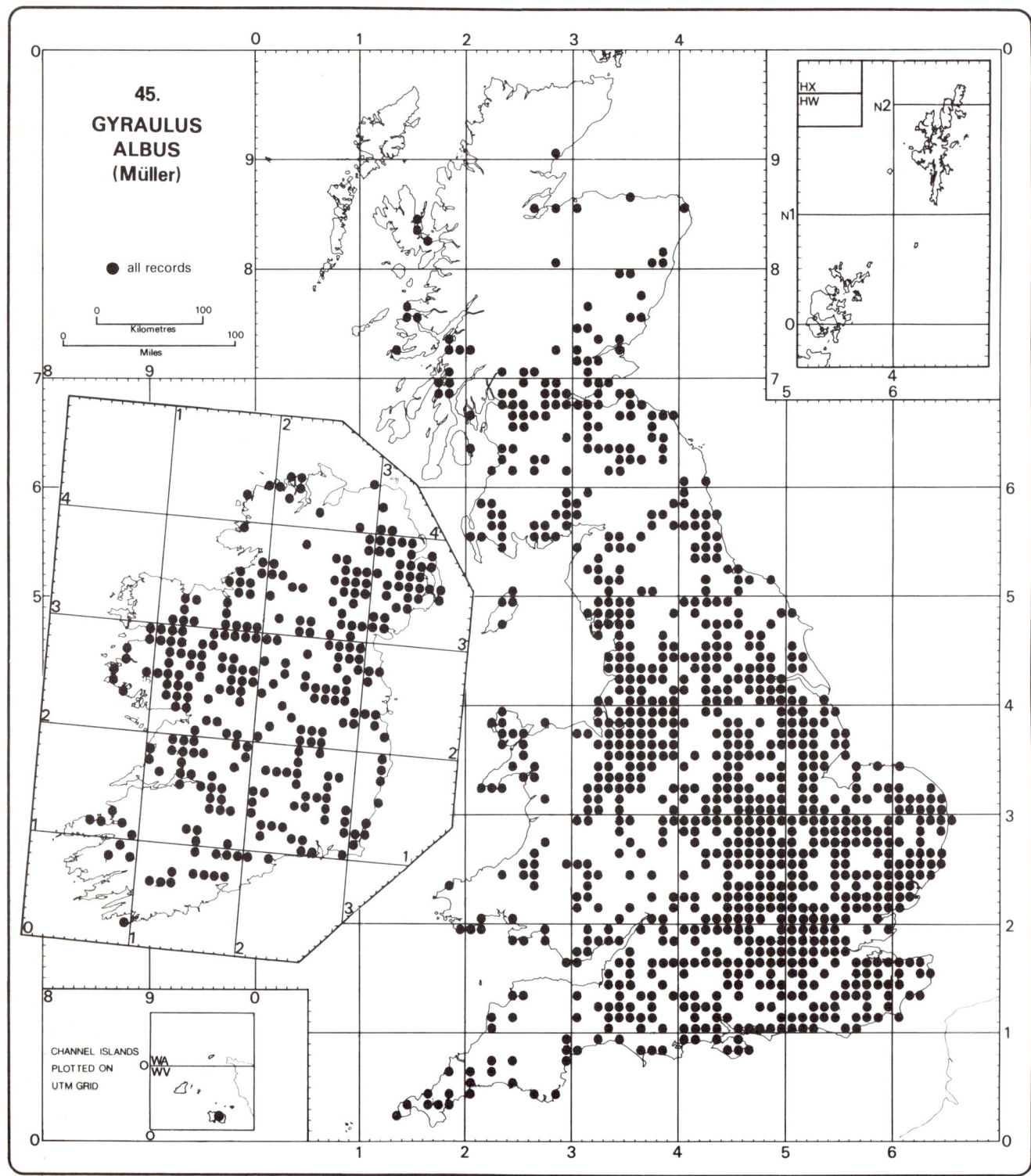

45.
GYRAULUS
ALBUS
(Müller)

● all records

Kilometres

Miles

CHANNEL ISLANDS
PLOTTED ON
UTM GRID

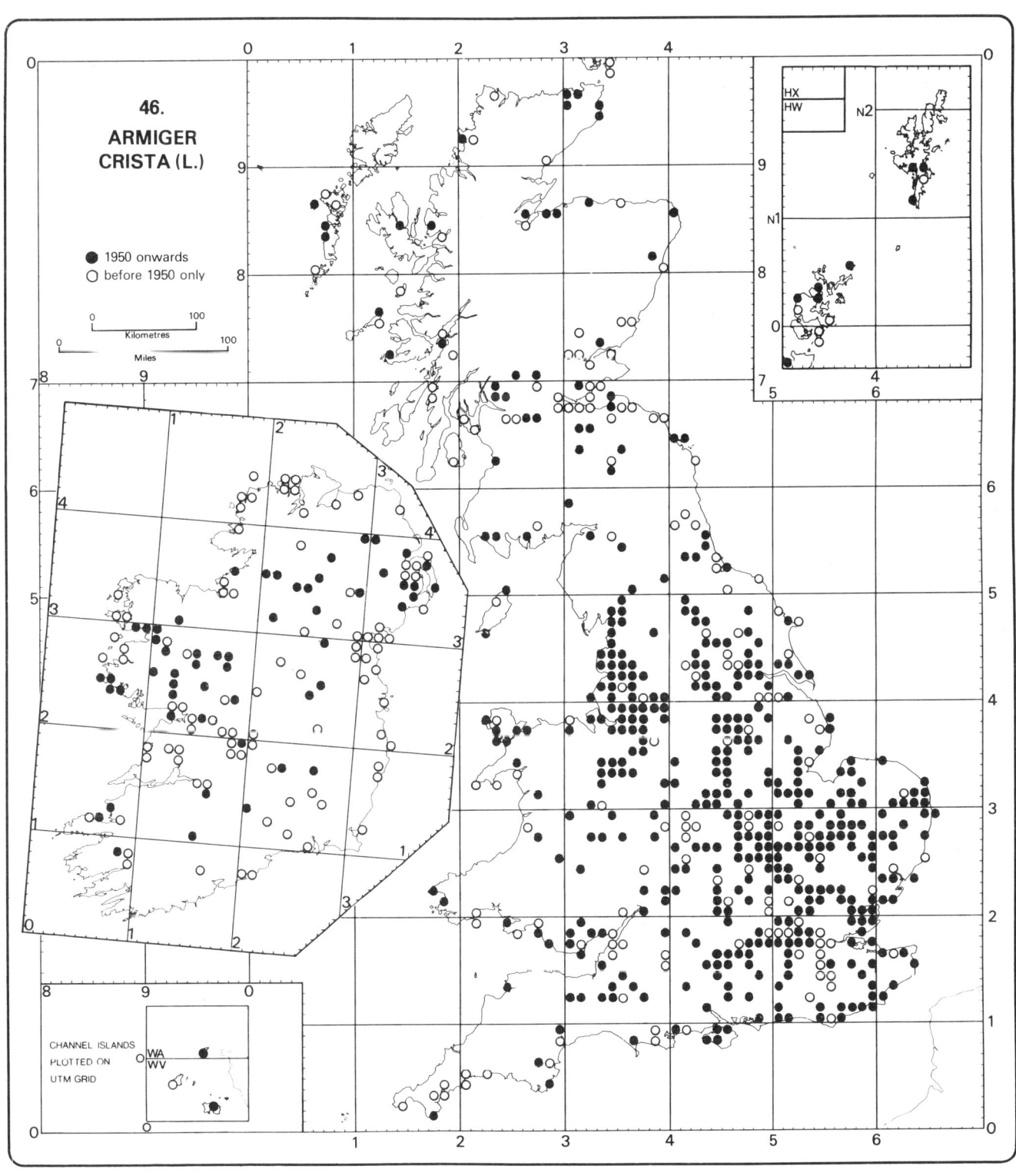

46.
ARMIGER
CRISTA (L.)

● 1950 onwards
○ before 1950 only

Kilometres
Miles

HX
HW
N2
N1

CHANNEL ISLANDS
PLOTTED ON
UTM GRID

WA
WV

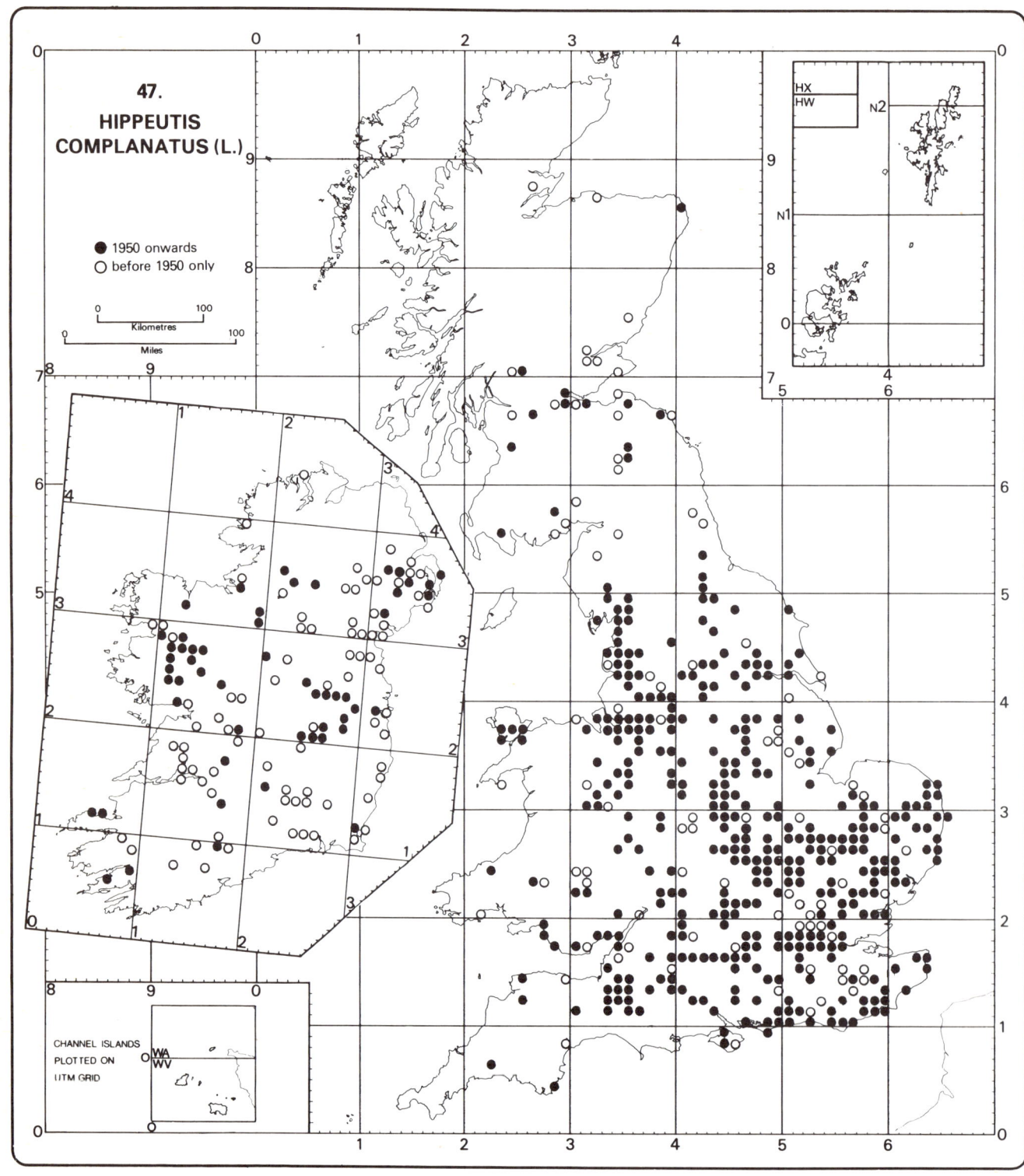

47.

HIPPEUTIS
COMPLANATUS (L.)

● 1950 onwards
○ before 1950 only

0 — 100
Kilometres
0 — 100
Miles

CHANNEL ISLANDS
PLOTTED ON
UTM GRID

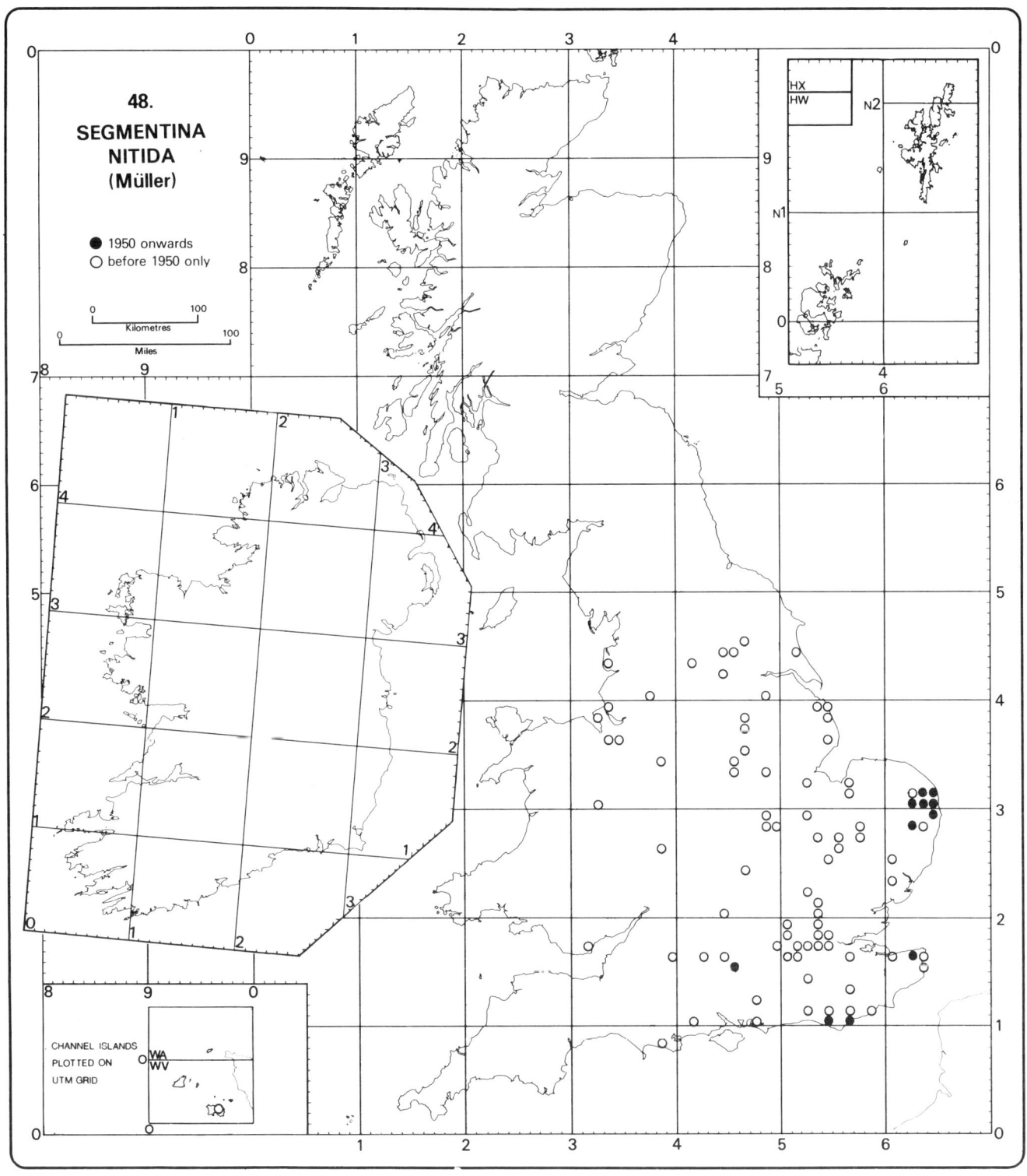

48.
SEGMENTINA
NITIDA
(Müller)

● 1950 onwards
○ before 1950 only

Kilometres
Miles

CHANNEL ISLANDS
PLOTTED ON
UTM GRID

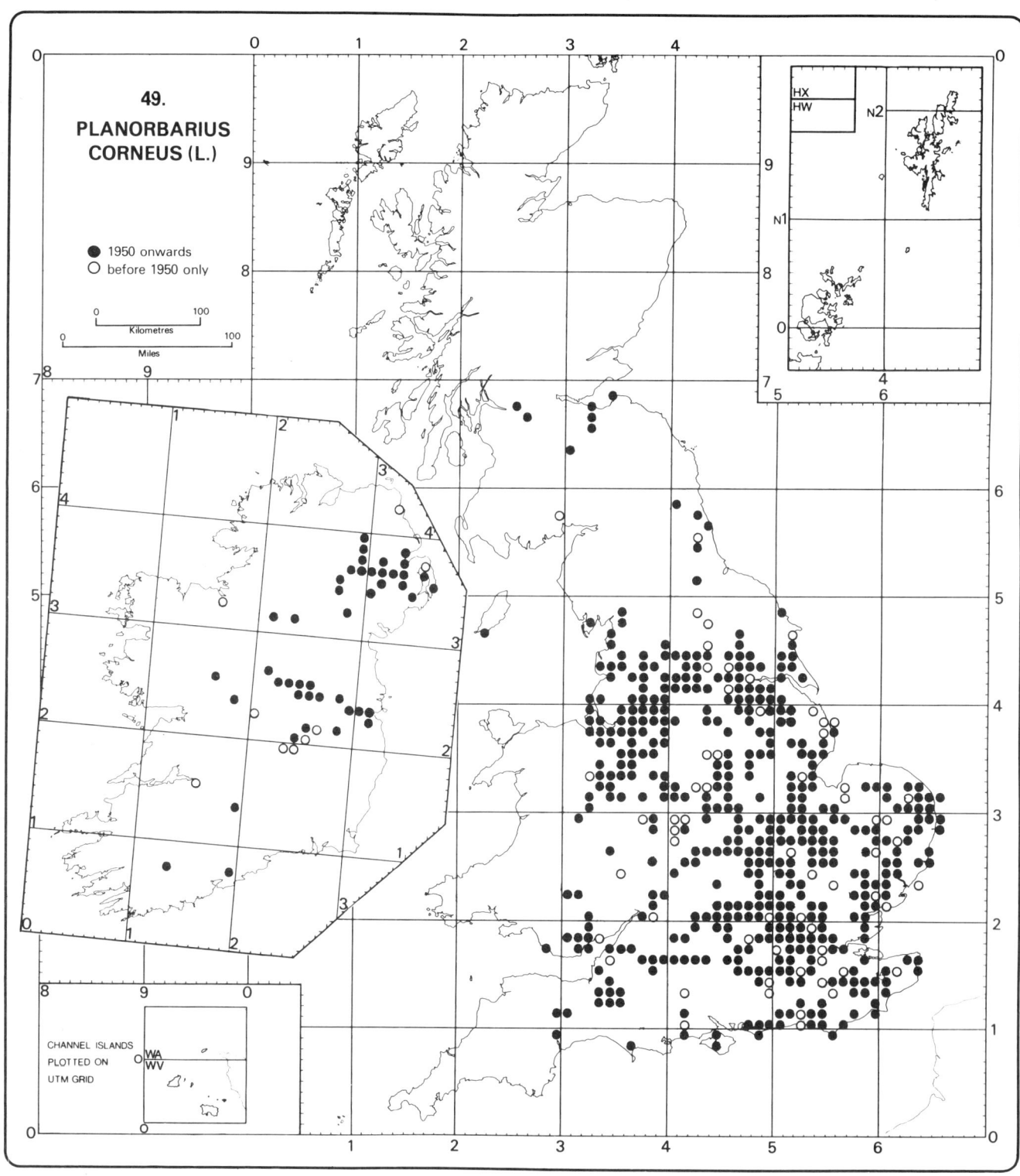

49.
PLANORBARIUS
CORNEUS (L.)

● 1950 onwards
○ before 1950 only

0   100
Kilometres
0   100
Miles

CHANNEL ISLANDS
PLOTTED ON
UTM GRID

WA
WV

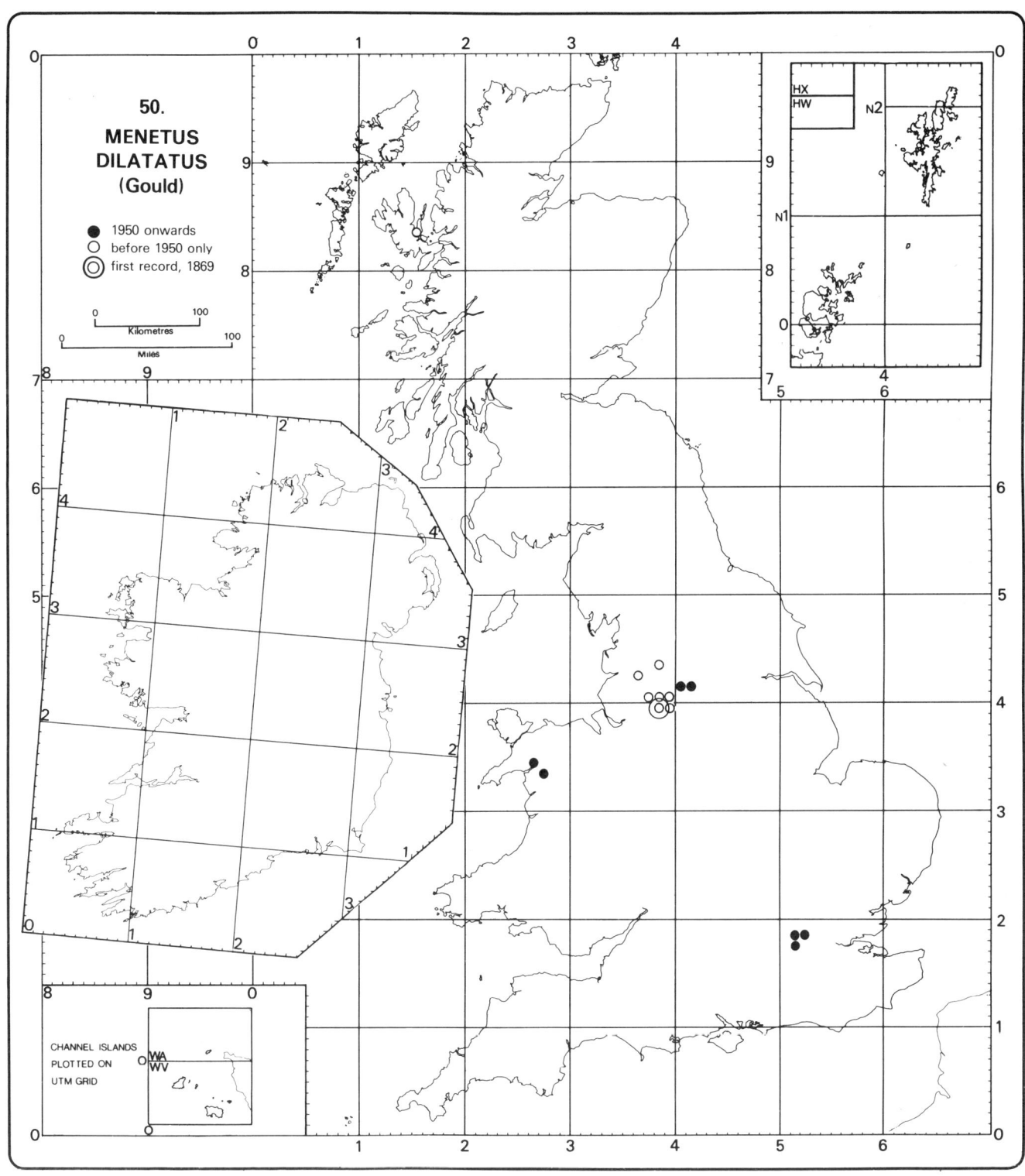

50.

**MENETUS
DILATATUS
(Gould)**

● 1950 onwards
○ before 1950 only
◎ first record, 1869

Kilometres

Miles

HX
HW

N2

N1

CHANNEL ISLANDS
PLOTTED ON
UTM GRID

WA
WV

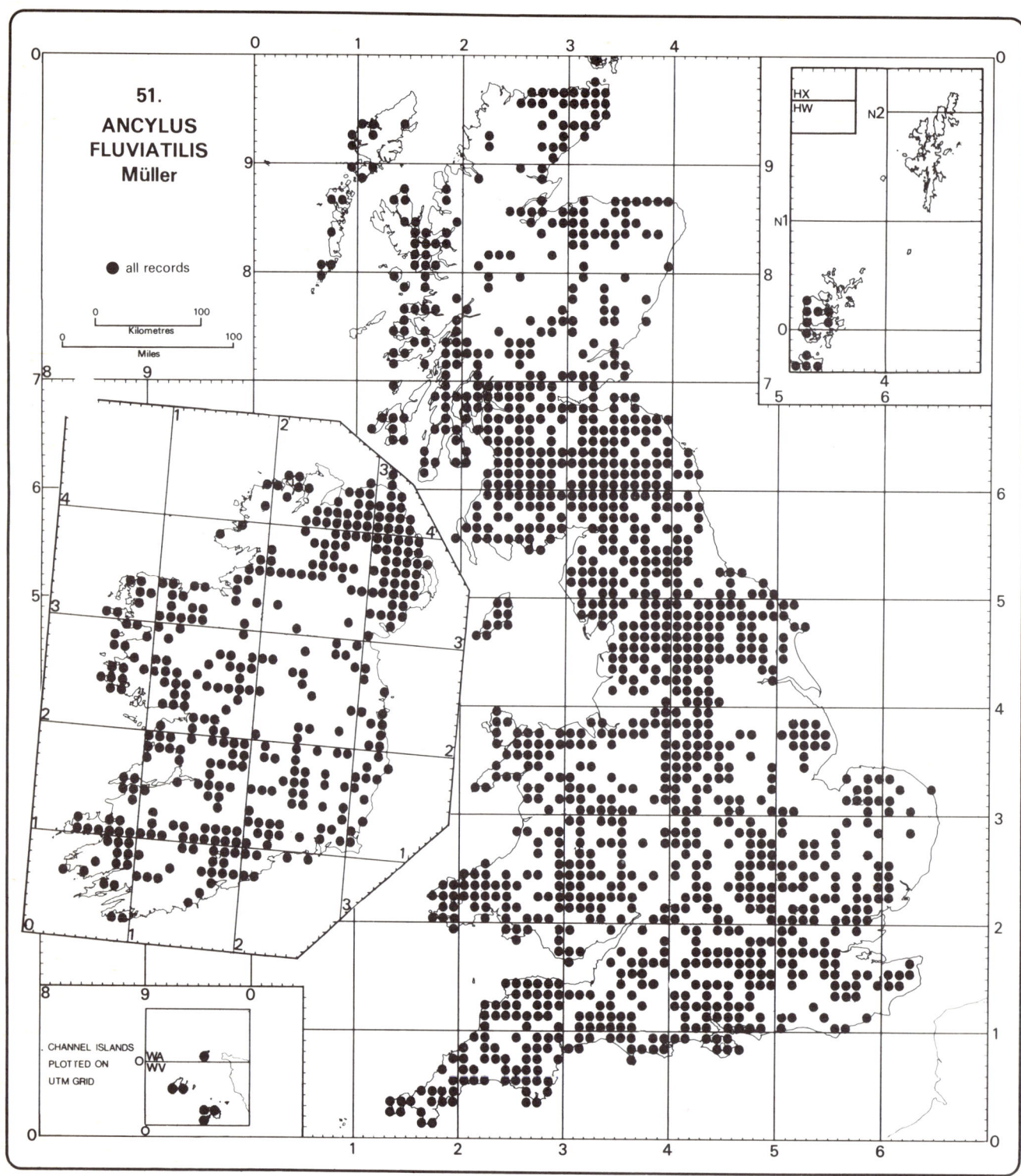

51.
ANCYLUS
FLUVIATILIS
Müller

● all records

CHANNEL ISLANDS
PLOTTED ON
UTM GRID

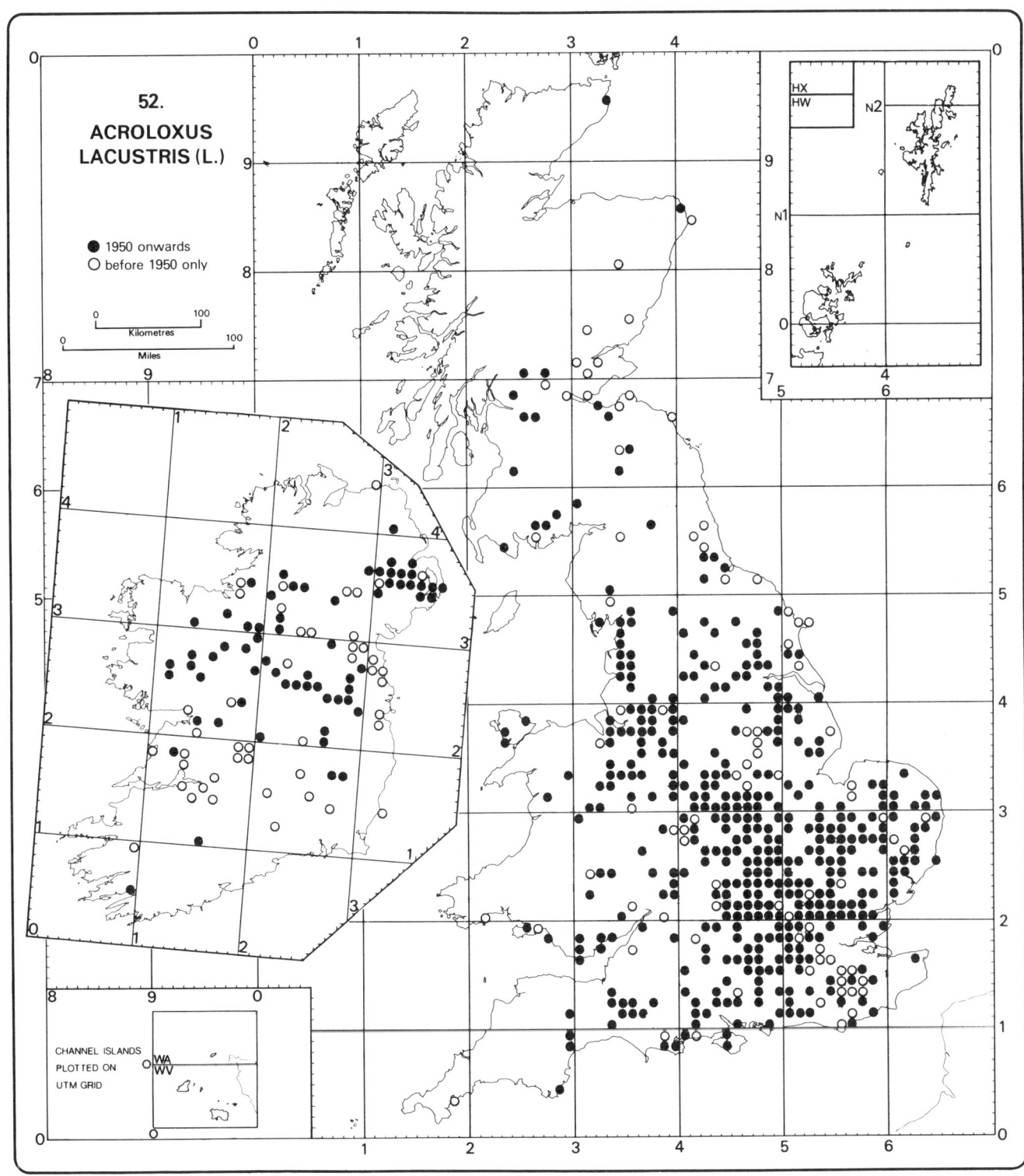

52.
ACROLOXUS
LACUSTRIS (L.)

● 1950 onwards
○ before 1950 only

0 _____ 100
Kilometres
0 _____ 100
Miles

CHANNEL ISLANDS
PLOTTED ON
UTM GRID

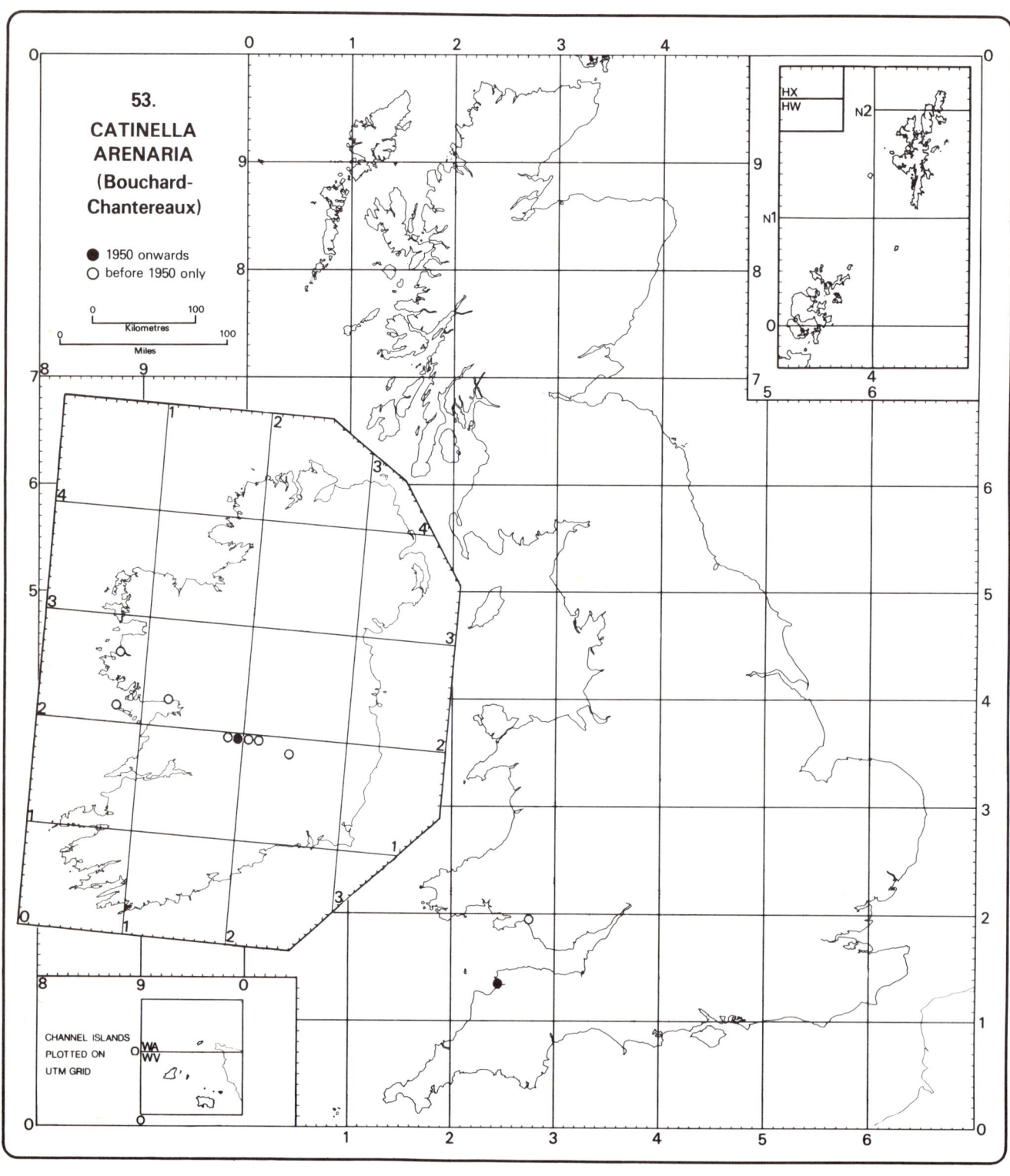

53.
CATINELLA
ARENARIA
(Bouchard-
Chantereaux)

● 1950 onwards
○ before 1950 only

Kilometres
Miles

HX
HW

CHANNEL ISLANDS
PLOTTED ON
UTM GRID

WA
WV

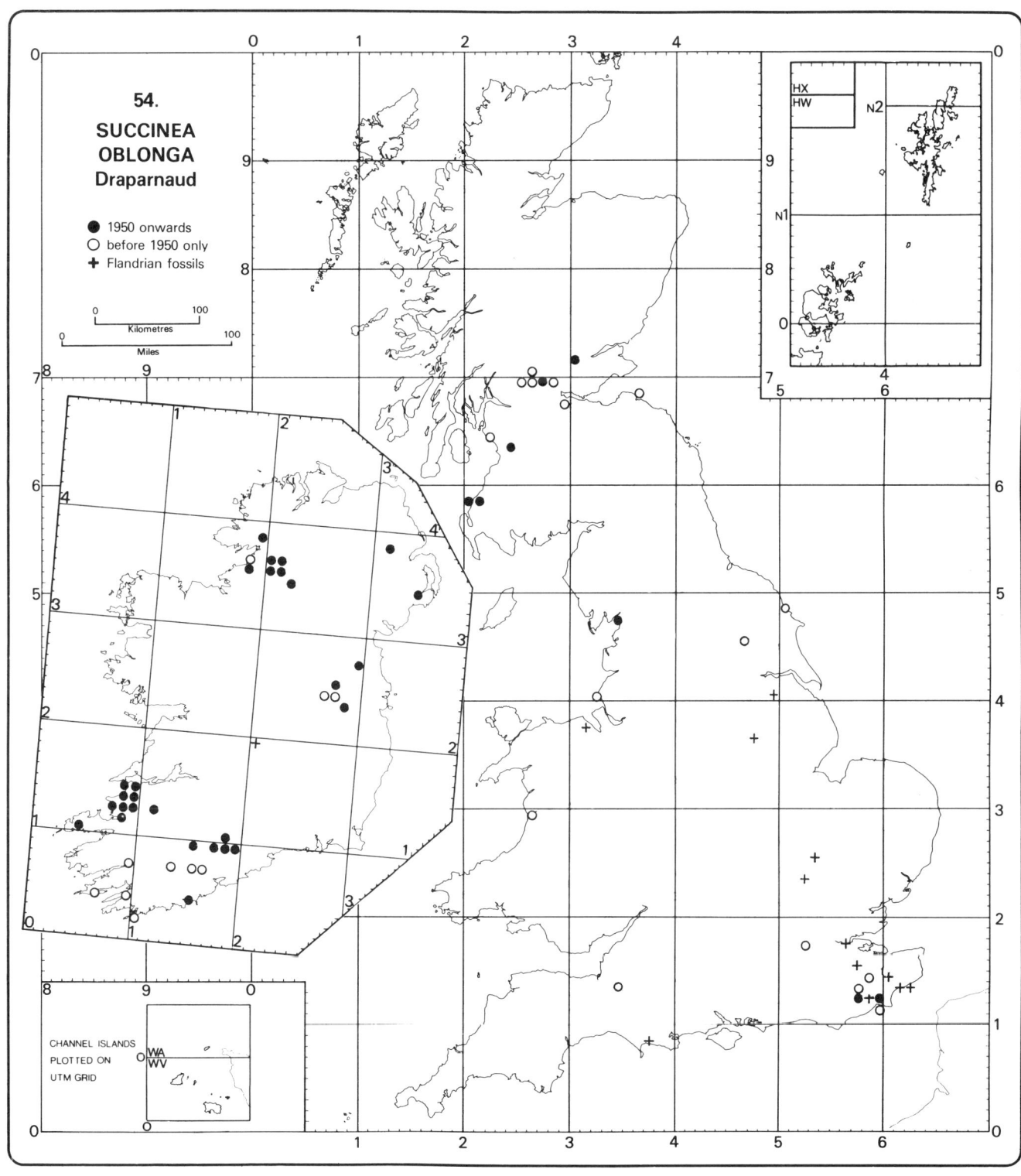

54.
SUCCINEA
OBLONGA
Draparnaud

● 1950 onwards
○ before 1950 only
+ Flandrian fossils

Kilometres
Miles

CHANNEL ISLANDS
PLOTTED ON
UTM GRID

WA
WV

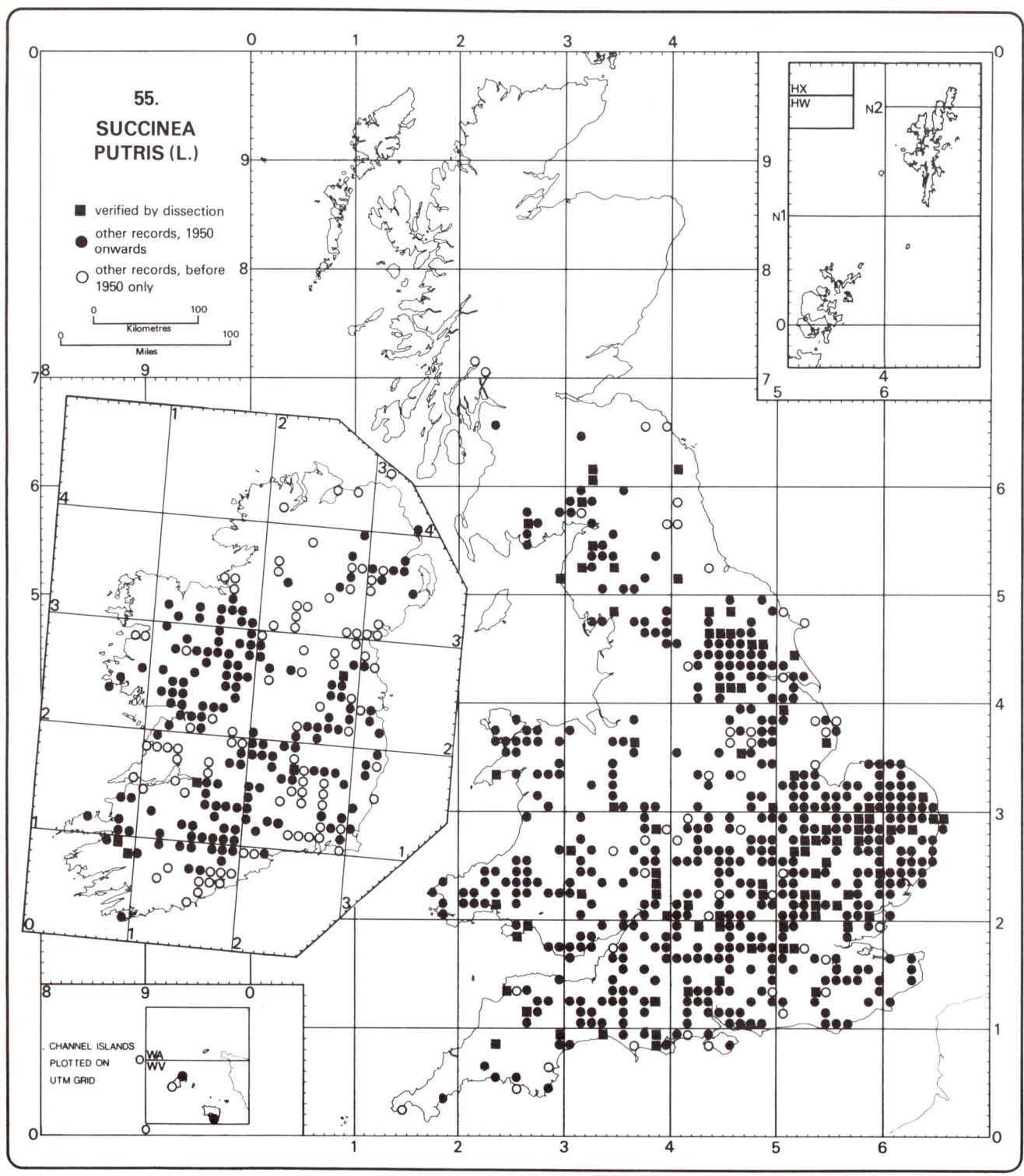

55.

SUCCINEA
PUTRIS (L.)

■ verified by dissection

● other records, 1950 onwards

○ other records, before 1950 only

0   100
Kilometres

0   100
Miles

CHANNEL ISLANDS
PLOTTED ON
UTM GRID

HX
HW

N2

N1

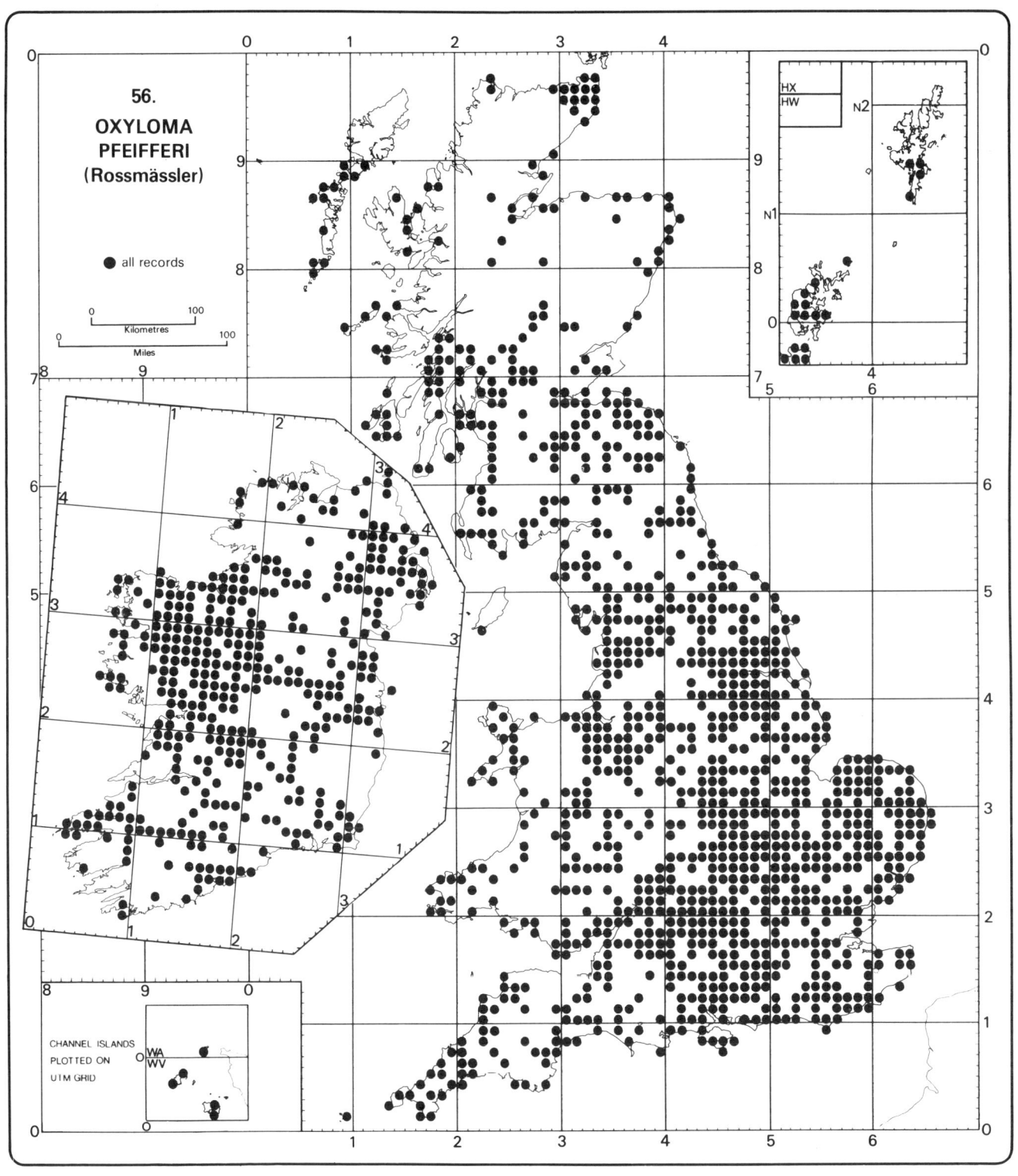

56.
OXYLOMA
PFEIFFERI
(Rossmässler)

● all records

0 _____ 100
Kilometres
0 _____ 100
Miles

CHANNEL ISLANDS
PLOTTED ON
UTM GRID

WA
WV

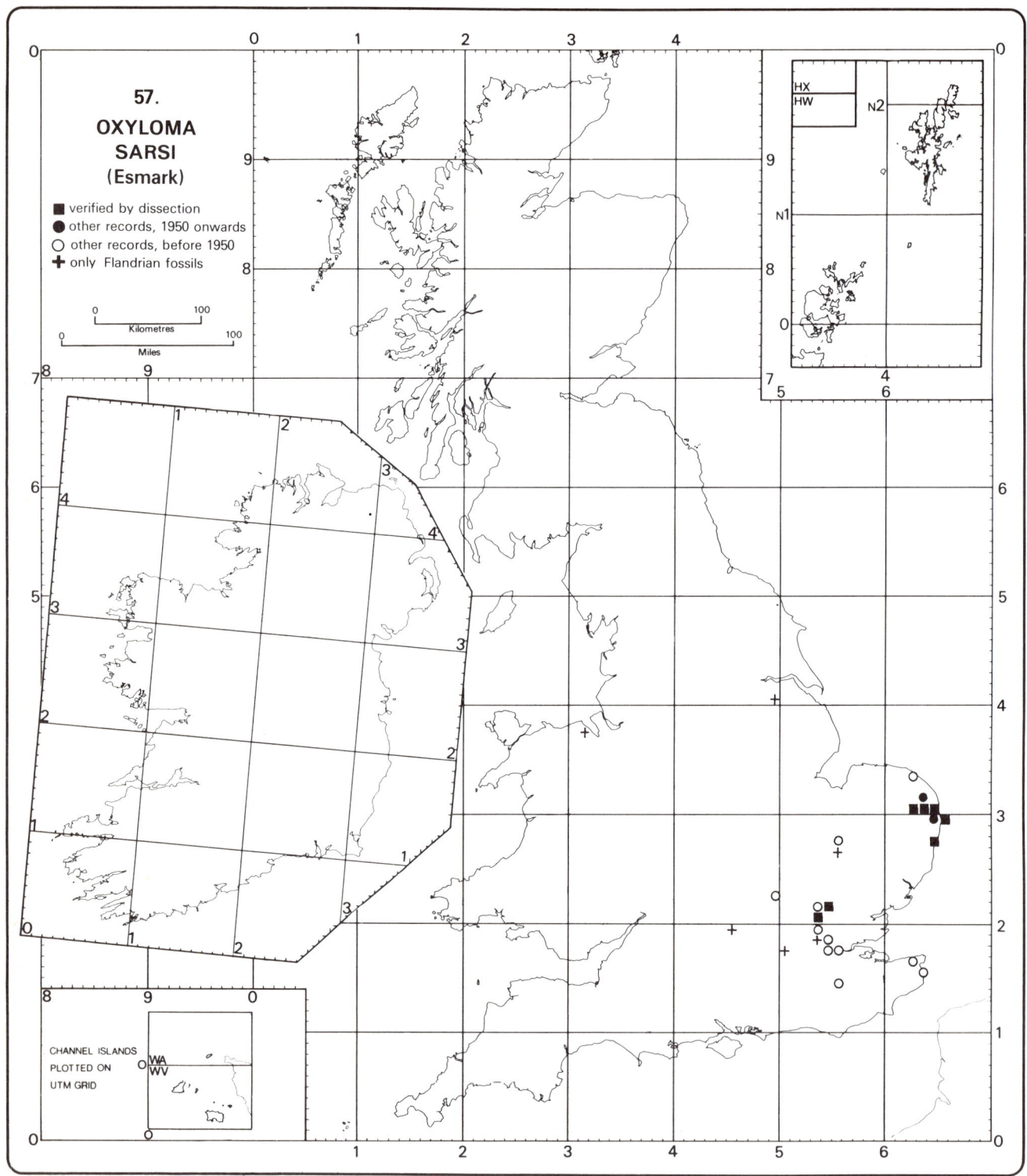

57.

OXYLOMA
SARSI
(Esmark)

■ verified by dissection
● other records, 1950 onwards
○ other records, before 1950
✛ only Flandrian fossils

Kilometres
Miles

CHANNEL ISLANDS
PLOTTED ON
UTM GRID

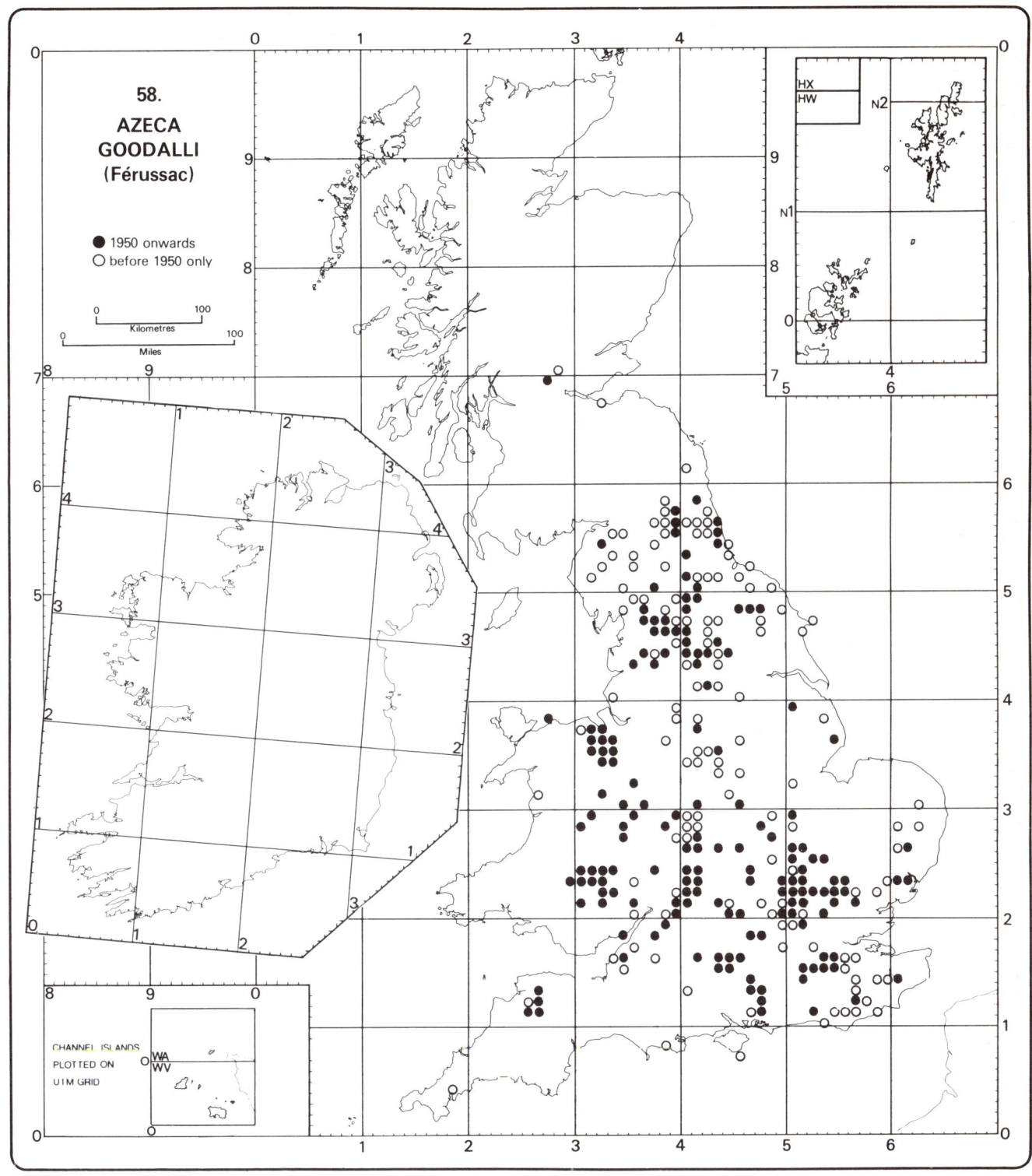

58.

AZECA
GOODALLI
(Férussac)

● 1950 onwards
○ before 1950 only

Kilometres

Miles

HX
HW

N2

N1

CHANNEL ISLANDS
PLOTTED ON
UTM GRID

WA
WV

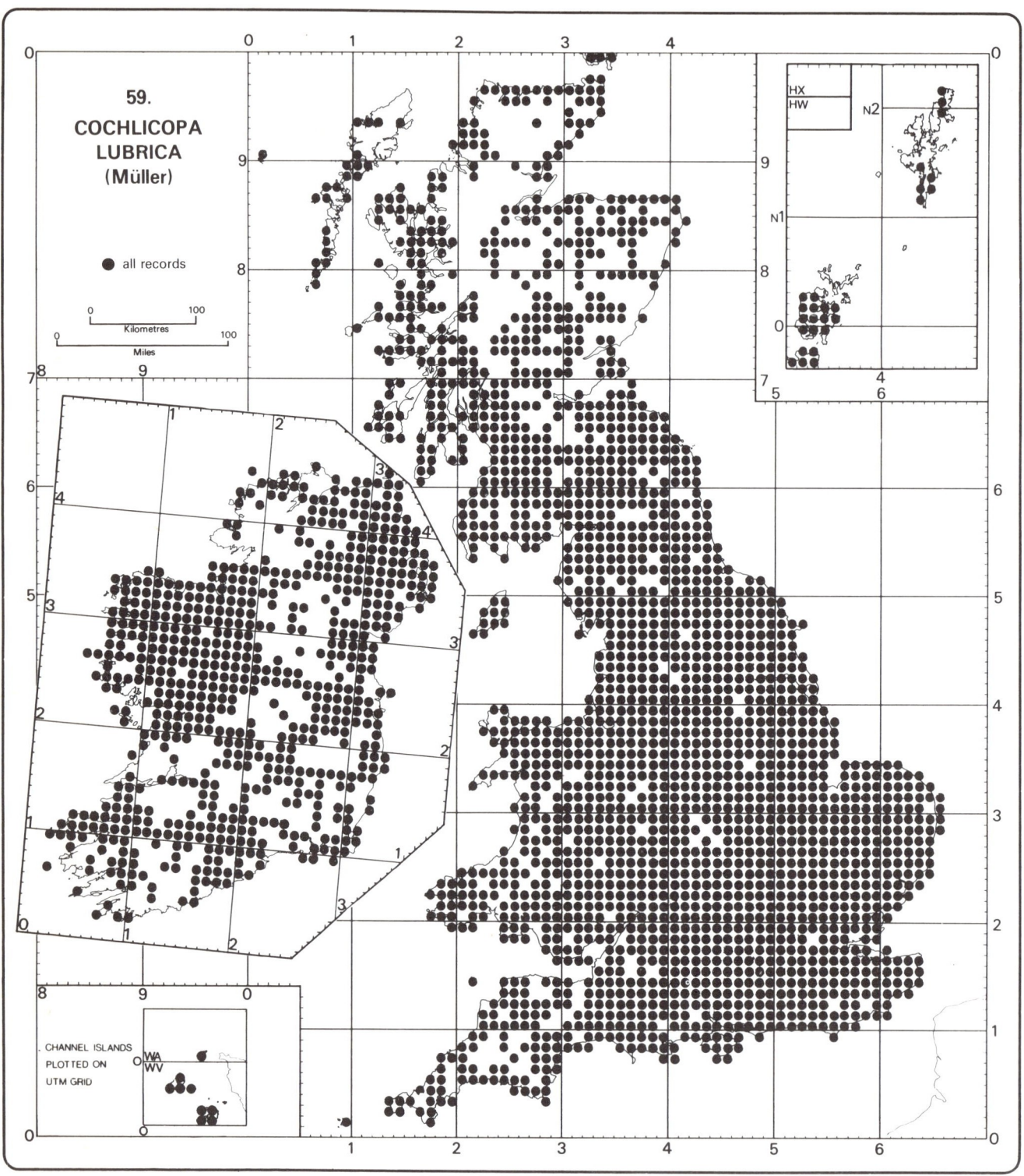

59.
COCHLICOPA
LUBRICA
(Müller)

● all records

Kilometres
Miles

CHANNEL ISLANDS
PLOTTED ON
UTM GRID

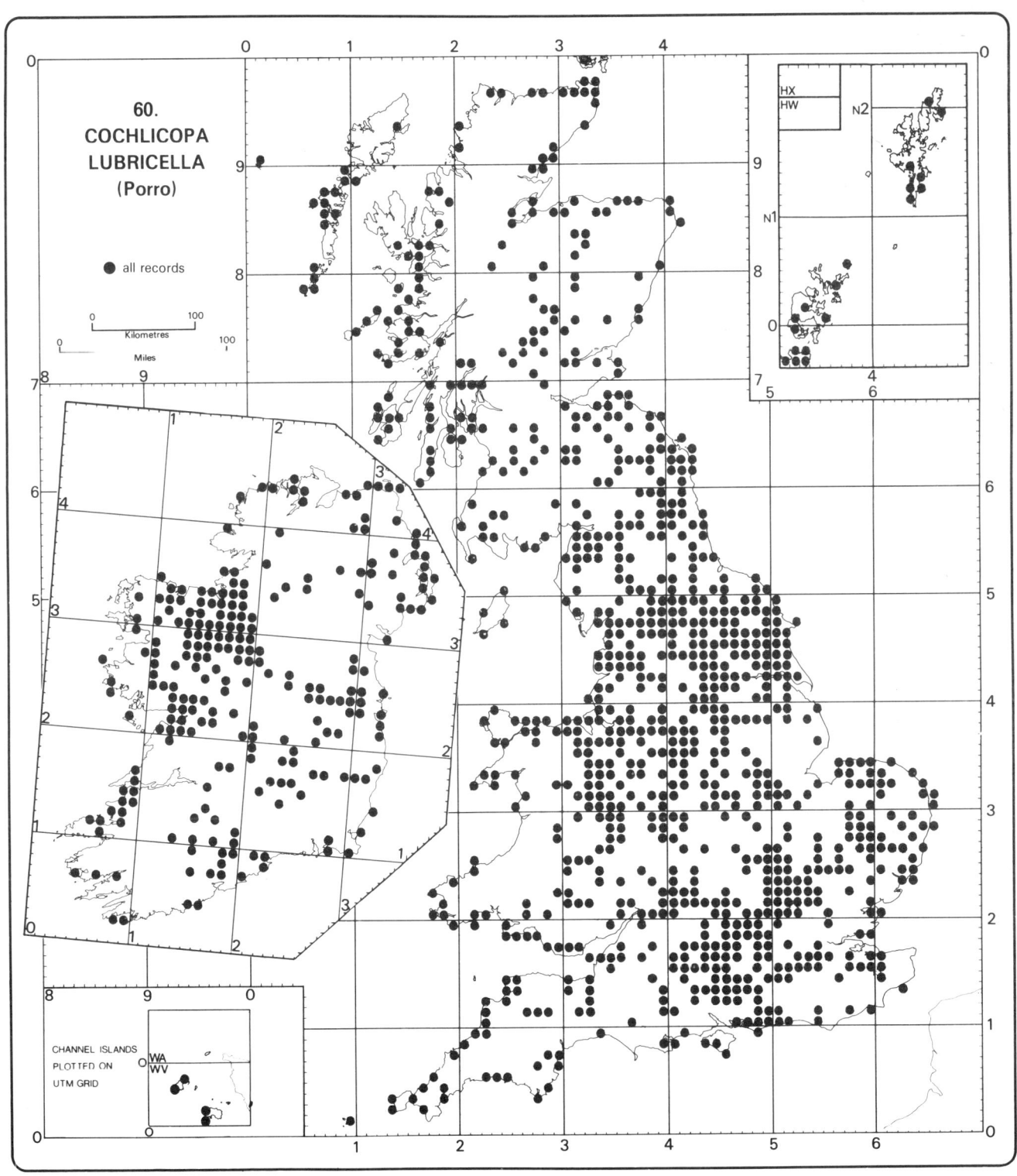

**60.
COCHLICOPA
LUBRICELLA
(Porro)**

● all records

Kilometres

Miles

CHANNEL ISLANDS
PLOTTED ON
UTM GRID

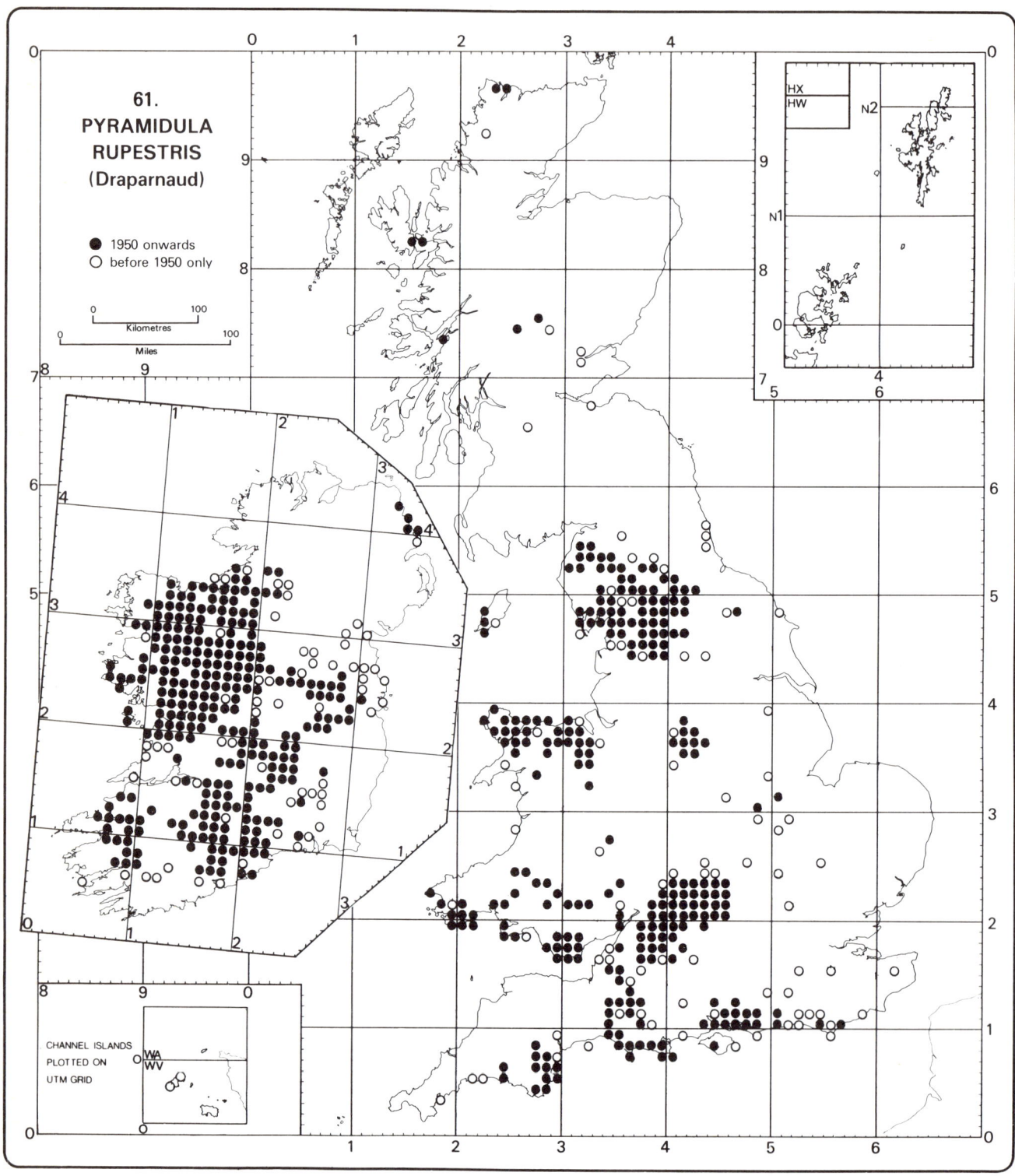

61.
PYRAMIDULA
RUPESTRIS
(Draparnaud)

● 1950 onwards
○ before 1950 only

0   100
Kilometres
0   100
Miles

CHANNEL ISLANDS
PLOTTED ON
UTM GRID

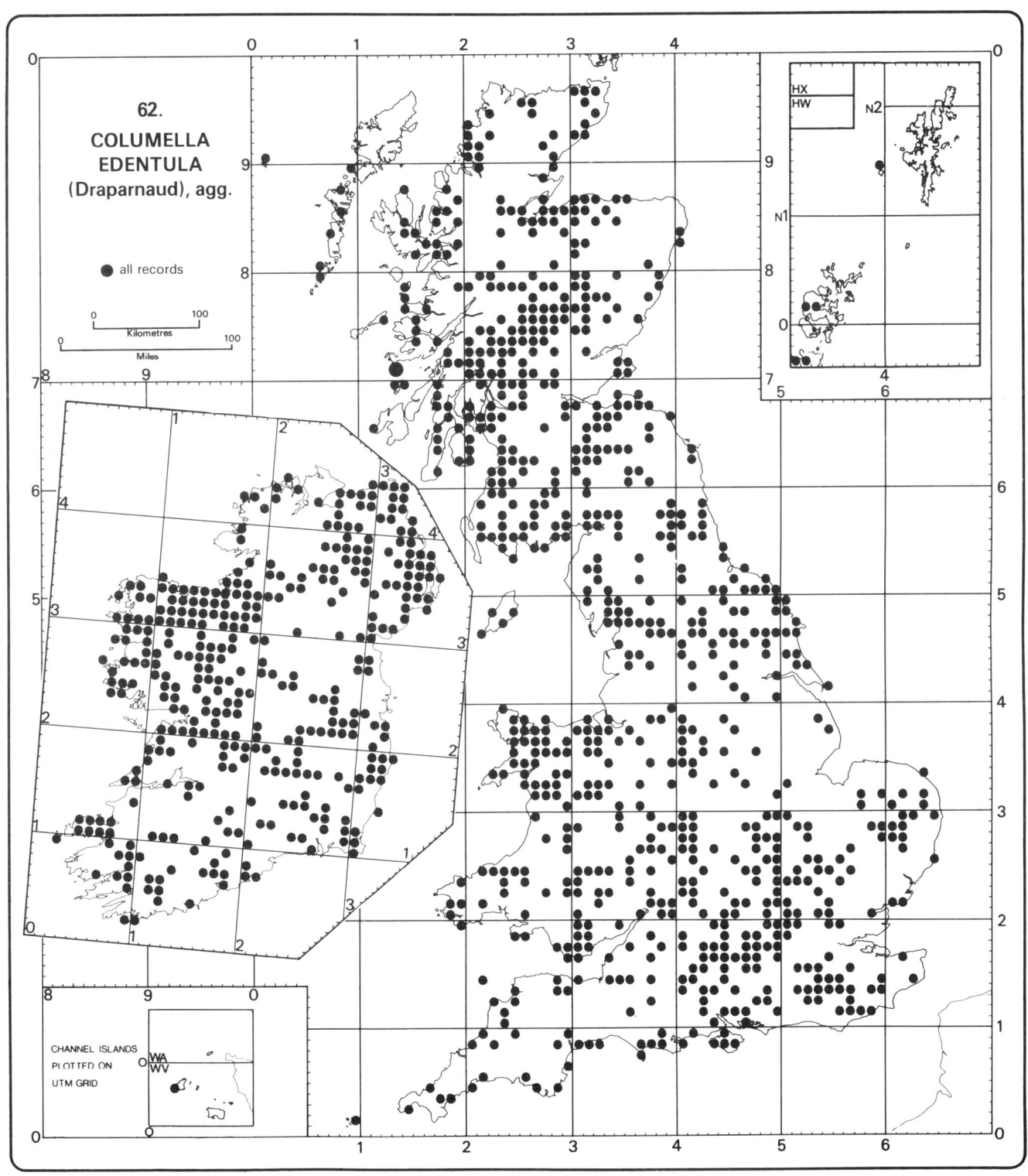

62.
COLUMELLA
EDENTULA
(Draparnaud), agg.

● all records

Kilometres

Miles

CHANNEL ISLANDS
PLOTTED ON
UTM GRID

WA
WV

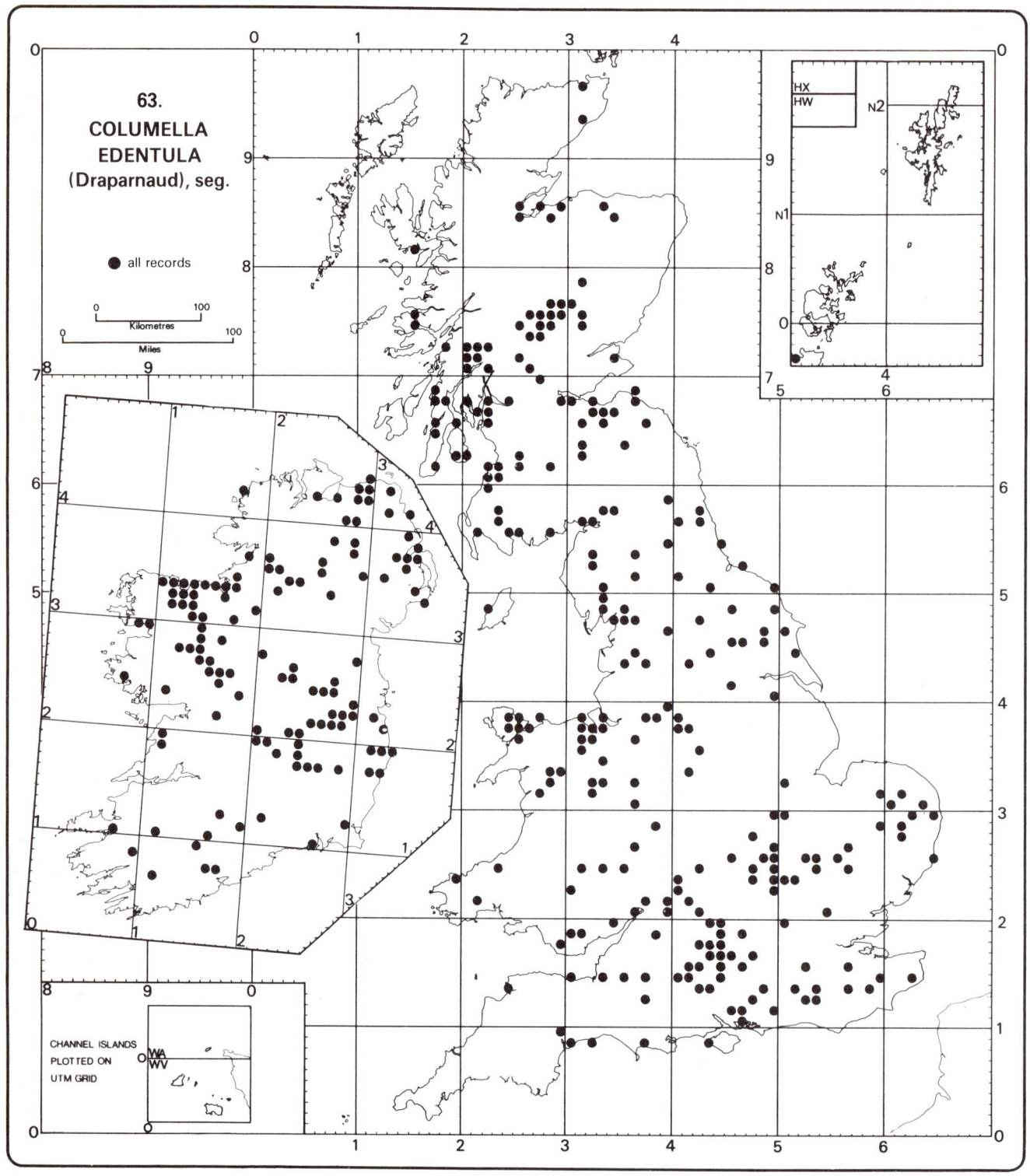

63.
COLUMELLA
EDENTULA
(Draparnaud), seg.

● all records

Kilometres

Miles

CHANNEL ISLANDS
PLOTTED ON
UTM GRID

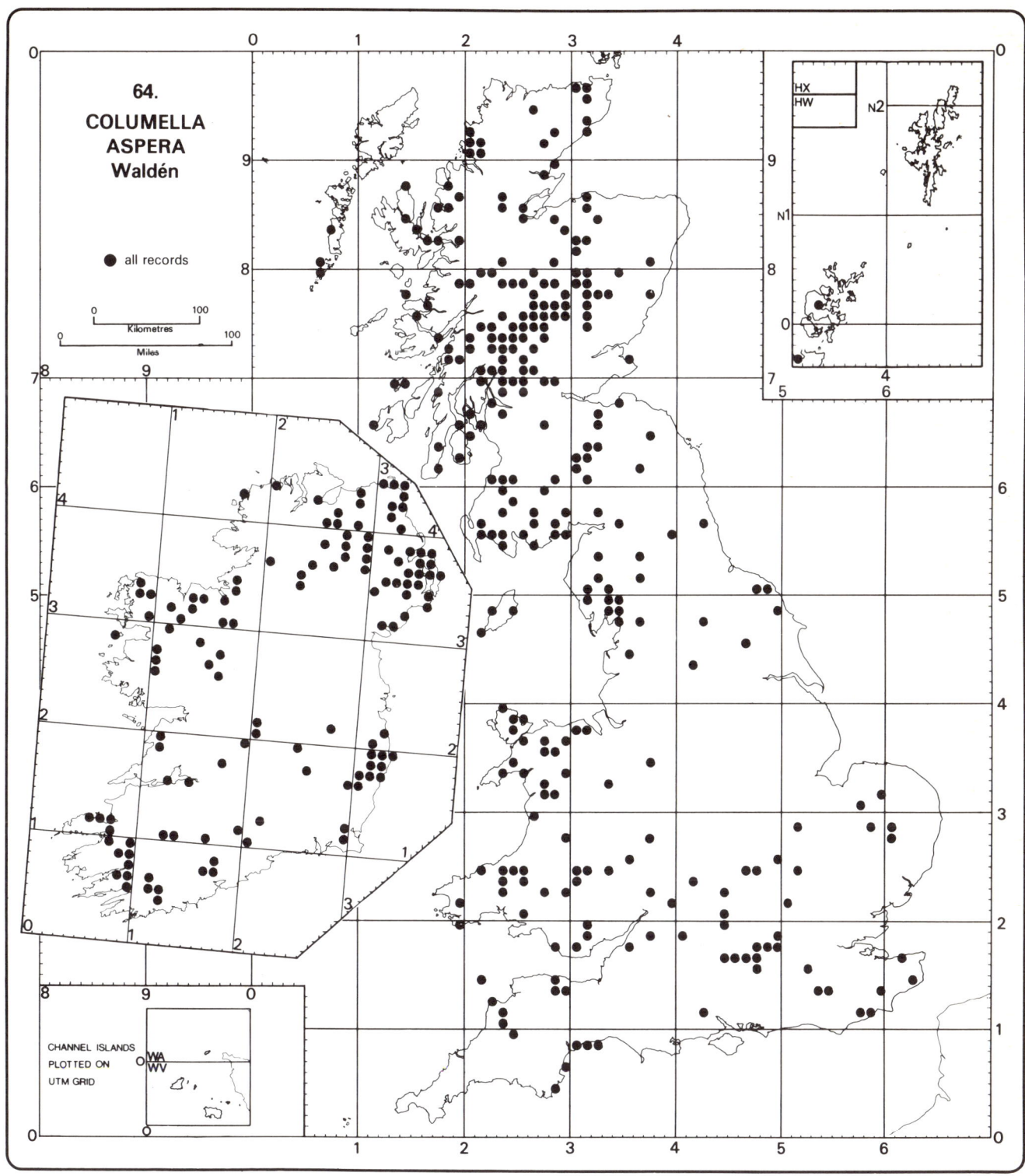

64.
COLUMELLA
ASPERA
Waldén

● all records

0 — Kilometres — 100

0 — Miles — 100

HX
HW
N2

CHANNEL ISLANDS
PLOTTED ON
UTM GRID

WA
WV

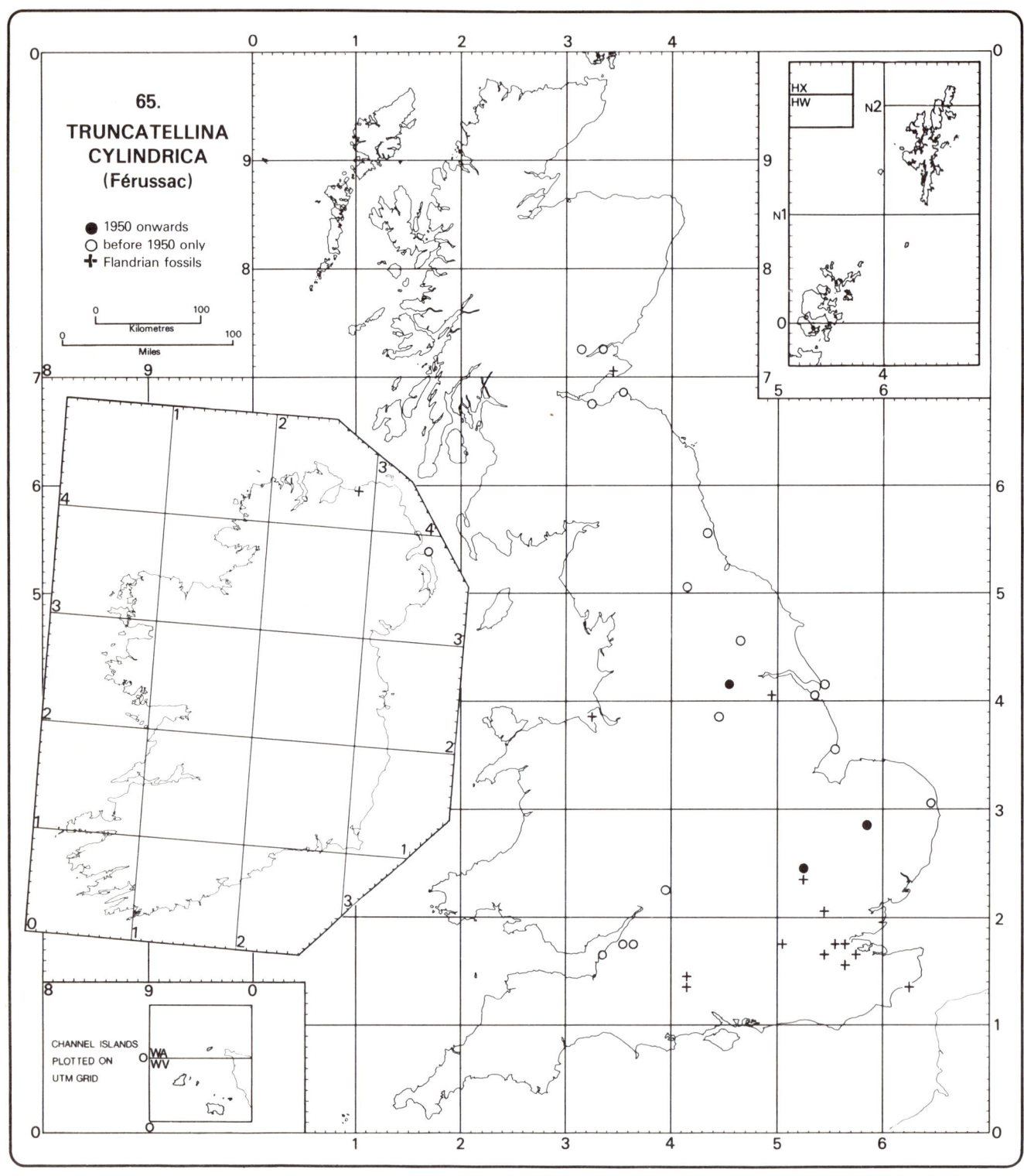

65.
TRUNCATELLINA
CYLINDRICA
(Férussac)

● 1950 onwards
○ before 1950 only
✛ Flandrian fossils

0 — 100
Kilometres
0 — 100
Miles

CHANNEL ISLANDS
PLOTTED ON
UTM GRID

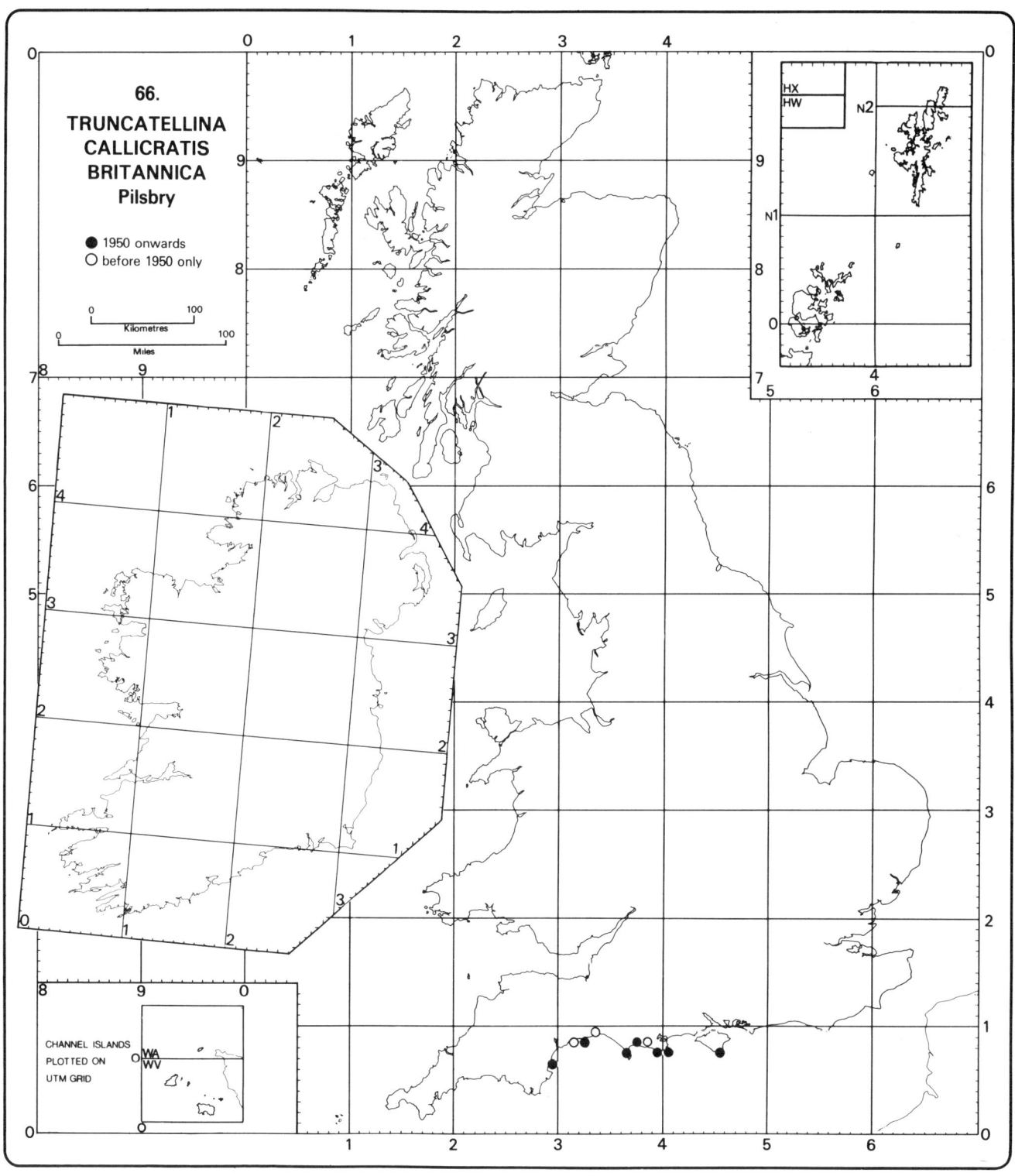

66.
TRUNCATELLINA
CALLICRATIS
BRITANNICA
Pilsbry

● 1950 onwards
○ before 1950 only

Kilometres
Miles

HX
HW

N2

N1

CHANNEL ISLANDS
PLOTTED ON
UTM GRID

WA
WV

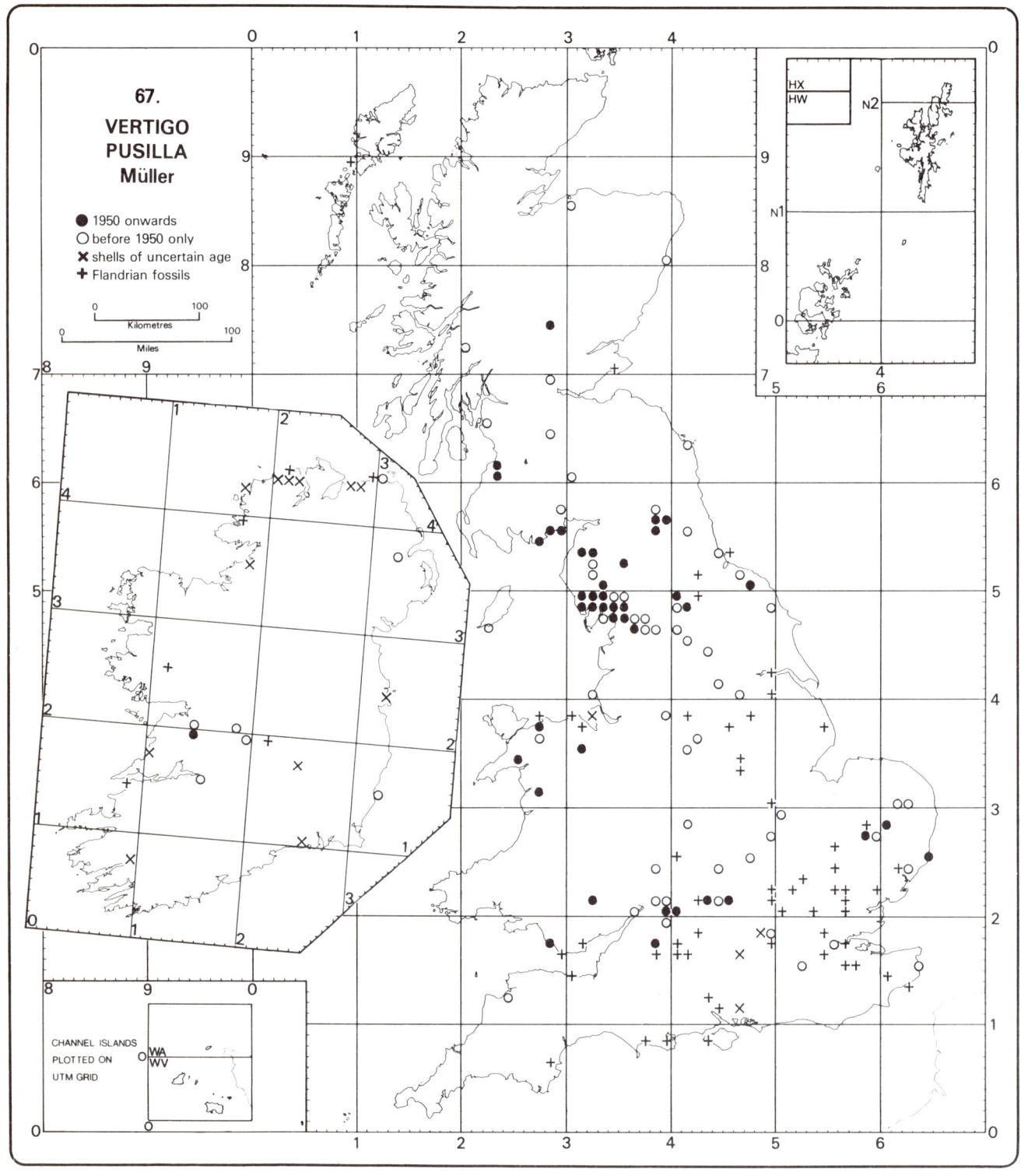

67.
VERTIGO
PUSILLA
Müller

● 1950 onwards
○ before 1950 only
✖ shells of uncertain age
✚ Flandrian fossils

Kilometres
Miles

CHANNEL ISLANDS
PLOTTED ON
UTM GRID

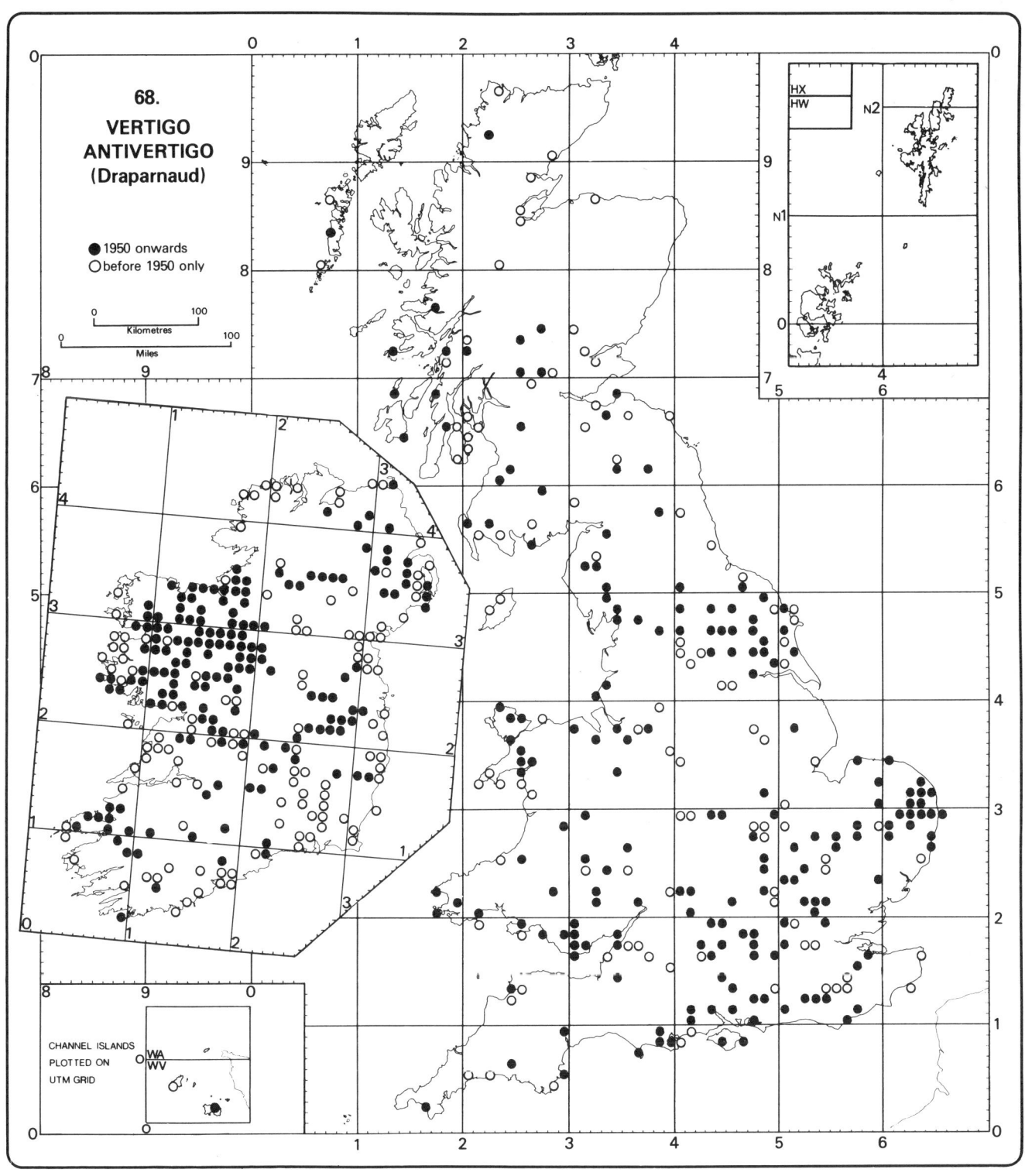

68.
VERTIGO
ANTIVERTIGO
(Draparnaud)

● 1950 onwards
○ before 1950 only

0 _____ 100
Kilometres
0 _____ 100
Miles

CHANNEL ISLANDS
PLOTTED ON
UTM GRID

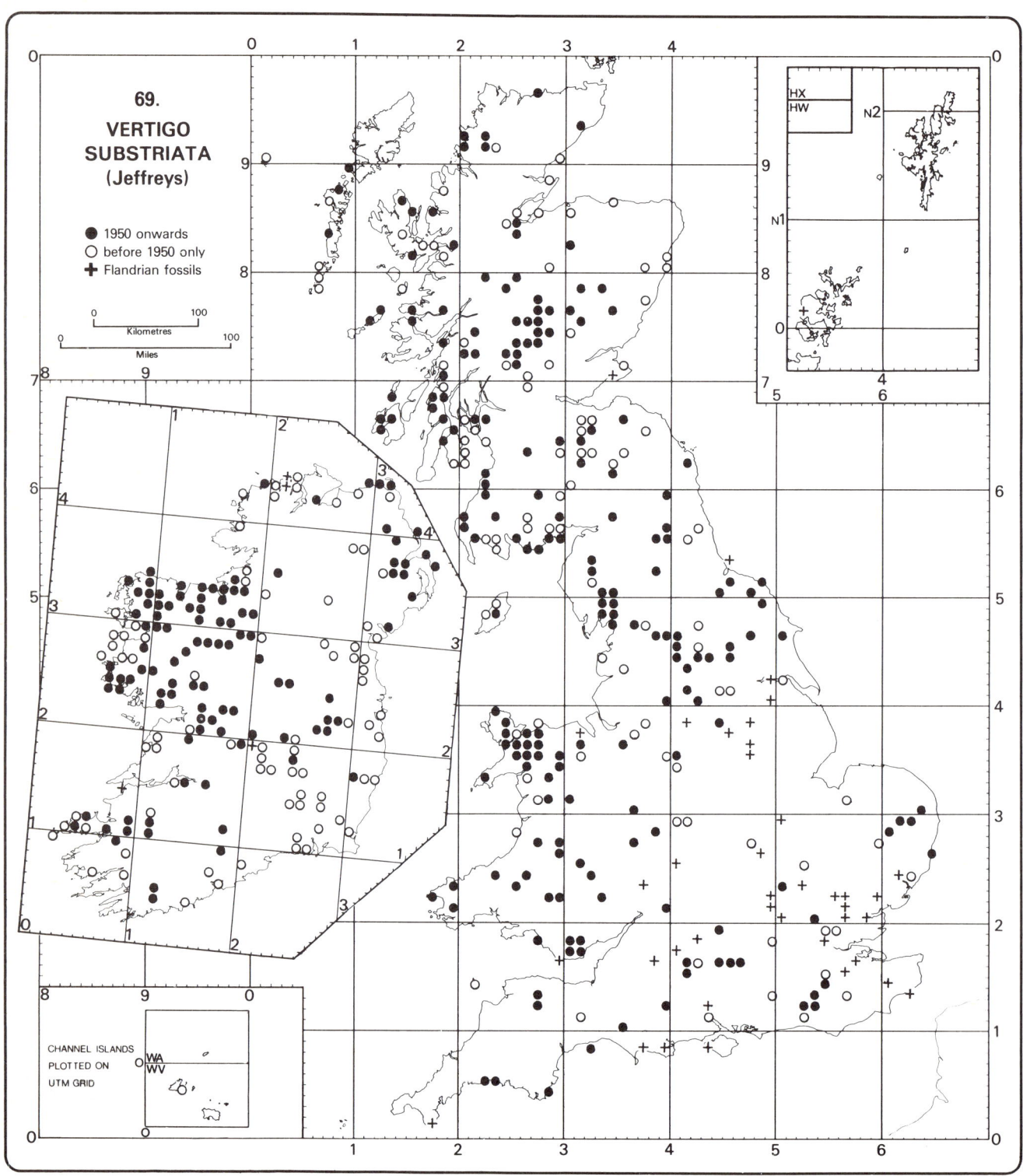

69.
VERTIGO
SUBSTRIATA
(Jeffreys)

● 1950 onwards
○ before 1950 only
+ Flandrian fossils

0 Kilometres 100
0 Miles 100

CHANNEL ISLANDS
PLOTTED ON
UTM GRID

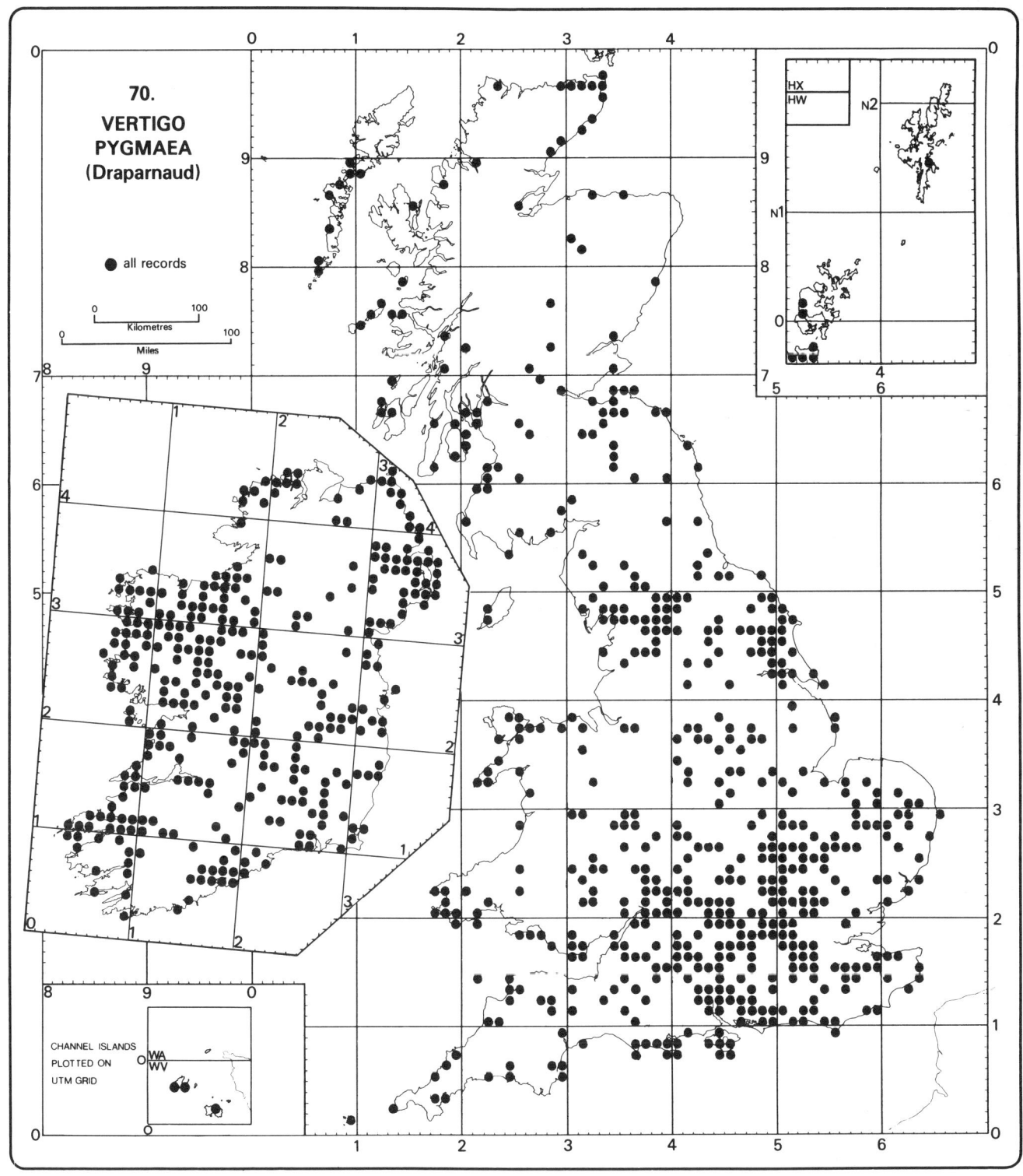

70.
VERTIGO
PYGMAEA
(Draparnaud)

● all records

0   Kilometres   100
0   Miles   100

CHANNEL ISLANDS
PLOTTED ON
UTM GRID

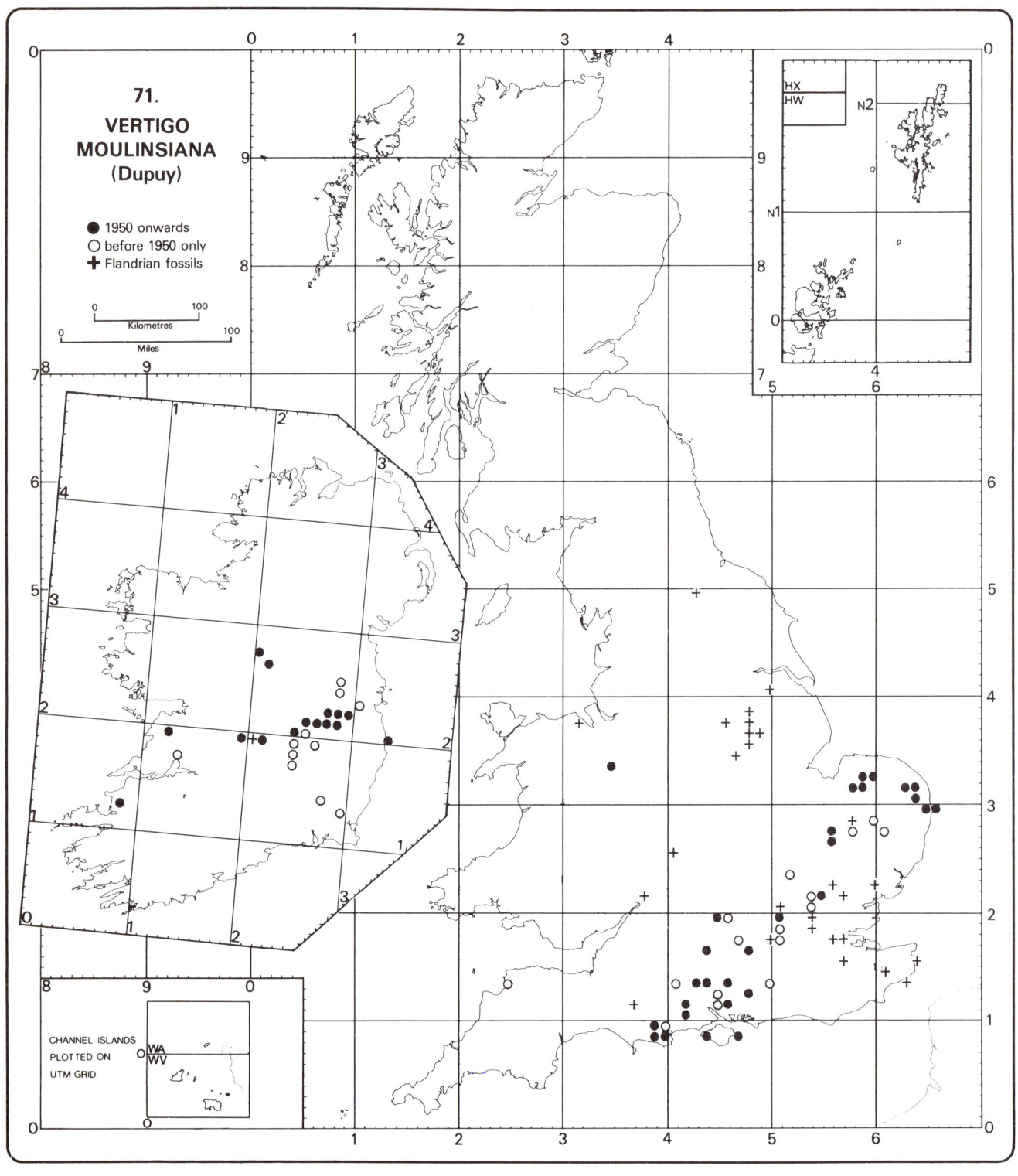

71.

VERTIGO
MOULINSIANA
(Dupuy)

● 1950 onwards
○ before 1950 only
✛ Flandrian fossils

0   100
Kilometres
0   100
Miles

CHANNEL ISLANDS
PLOTTED ON
UTM GRID

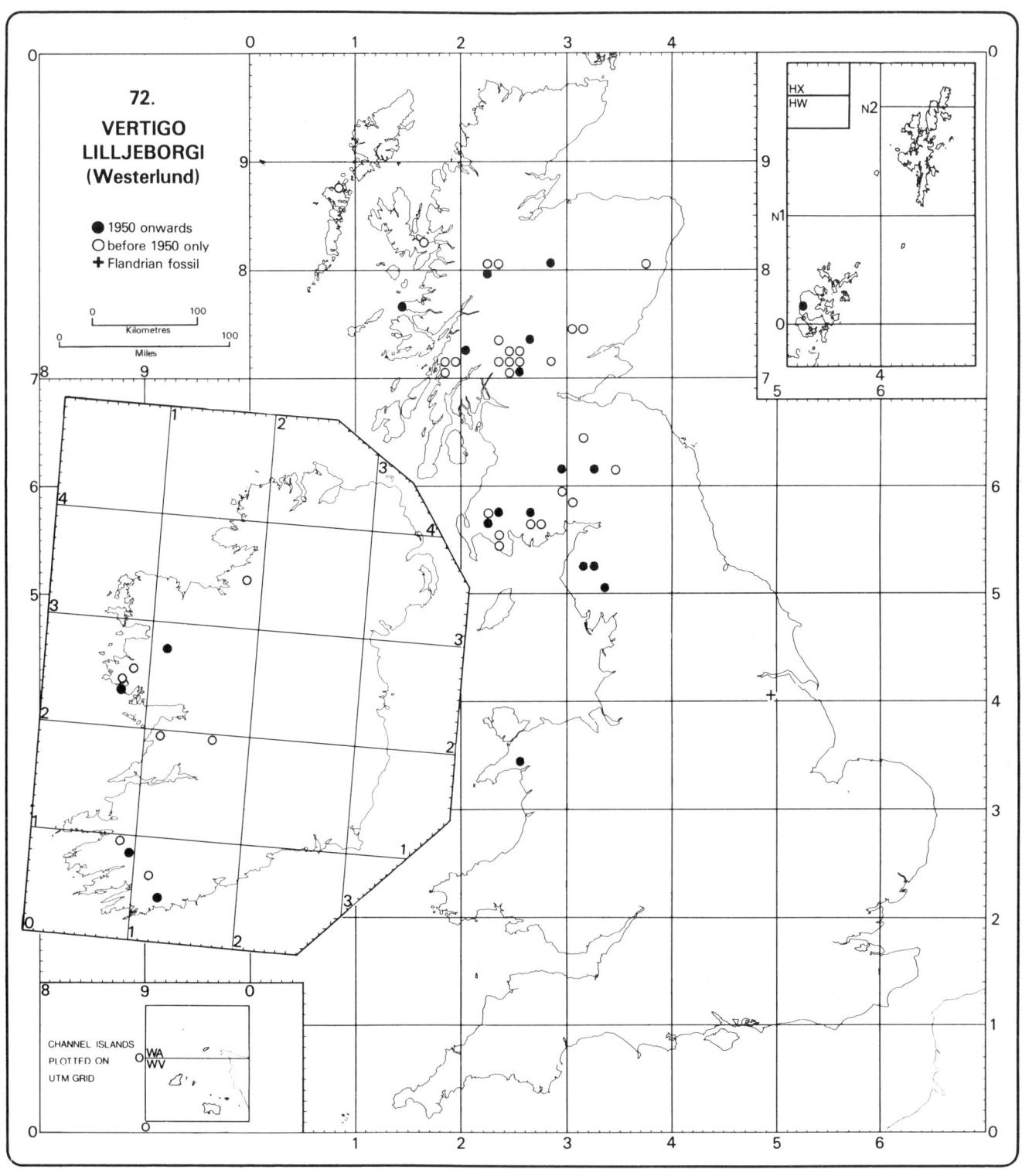

72.
VERTIGO
LILLJEBORGI
(Westerlund)

● 1950 onwards
○ before 1950 only
+ Flandrian fossil

Kilometres
Miles

CHANNEL ISLANDS
PLOTTED ON
UTM GRID

WA
WV

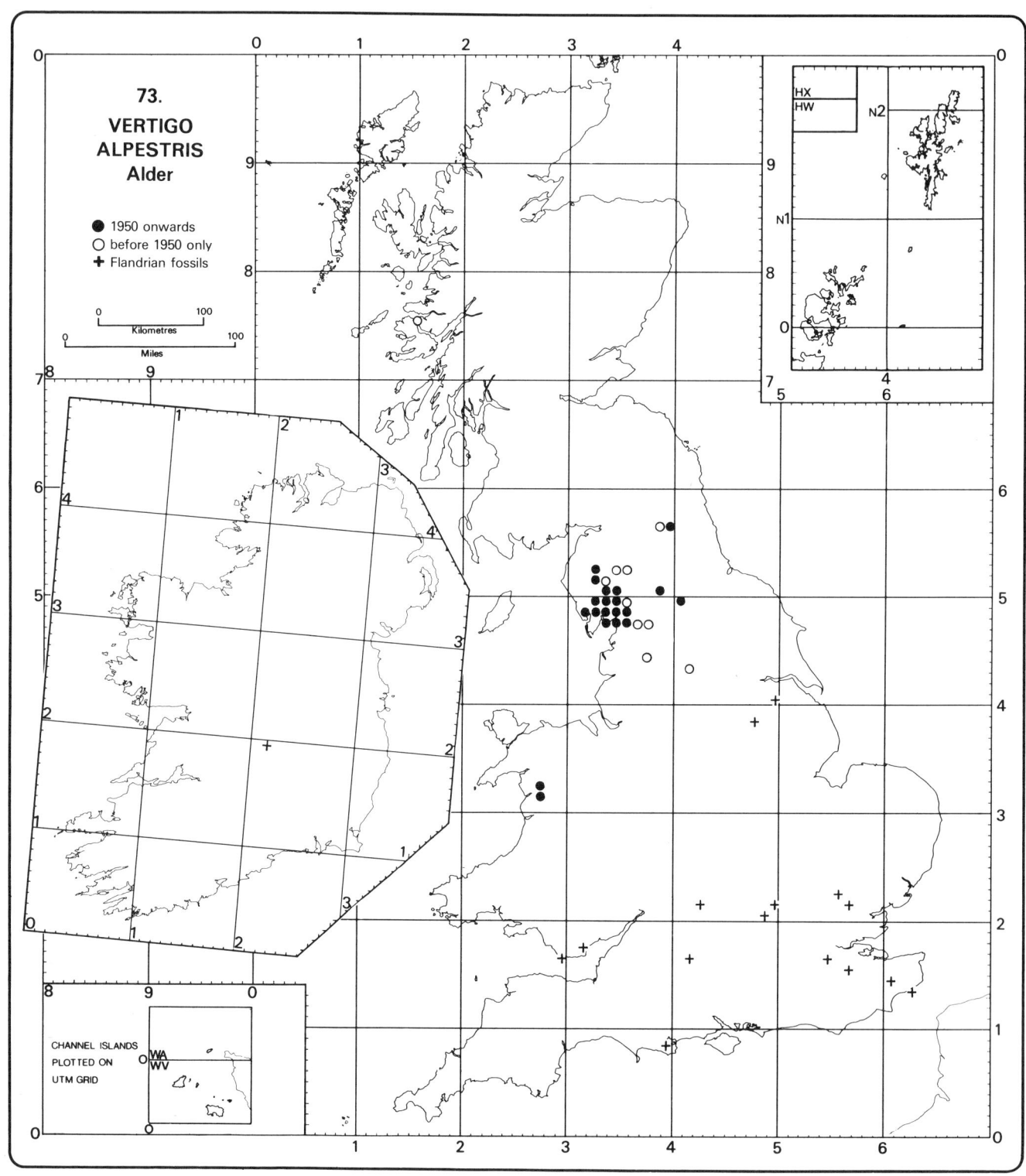

73.
VERTIGO
ALPESTRIS
Alder

● 1950 onwards
○ before 1950 only
+ Flandrian fossils

Kilometres
Miles

CHANNEL ISLANDS
PLOTTED ON
UTM GRID

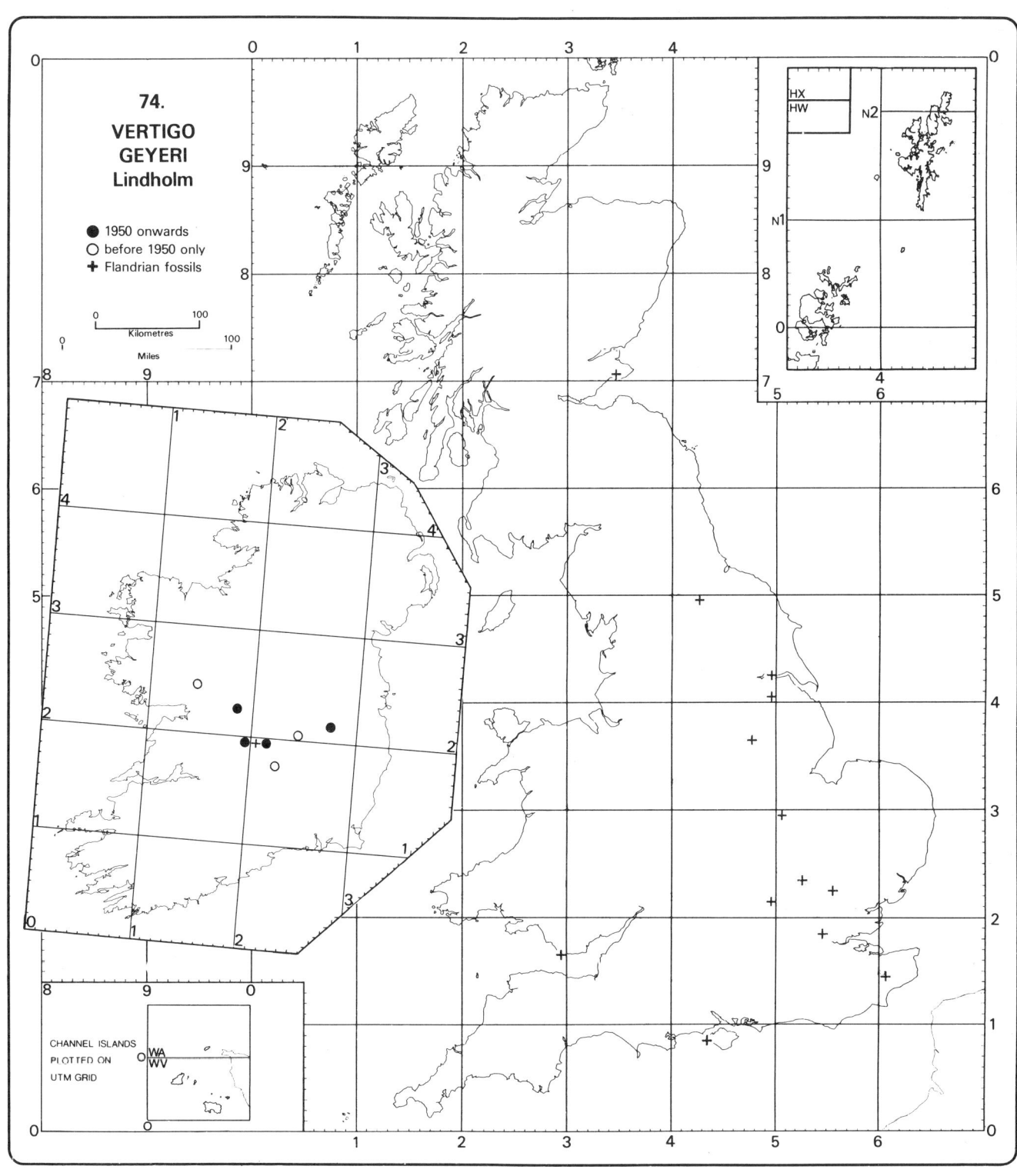

74.
VERTIGO
GEYERI
Lindholm

● 1950 onwards
○ before 1950 only
+ Flandrian fossils

Kilometres
Miles

CHANNEL ISLANDS
PLOTTED ON
UTM GRID

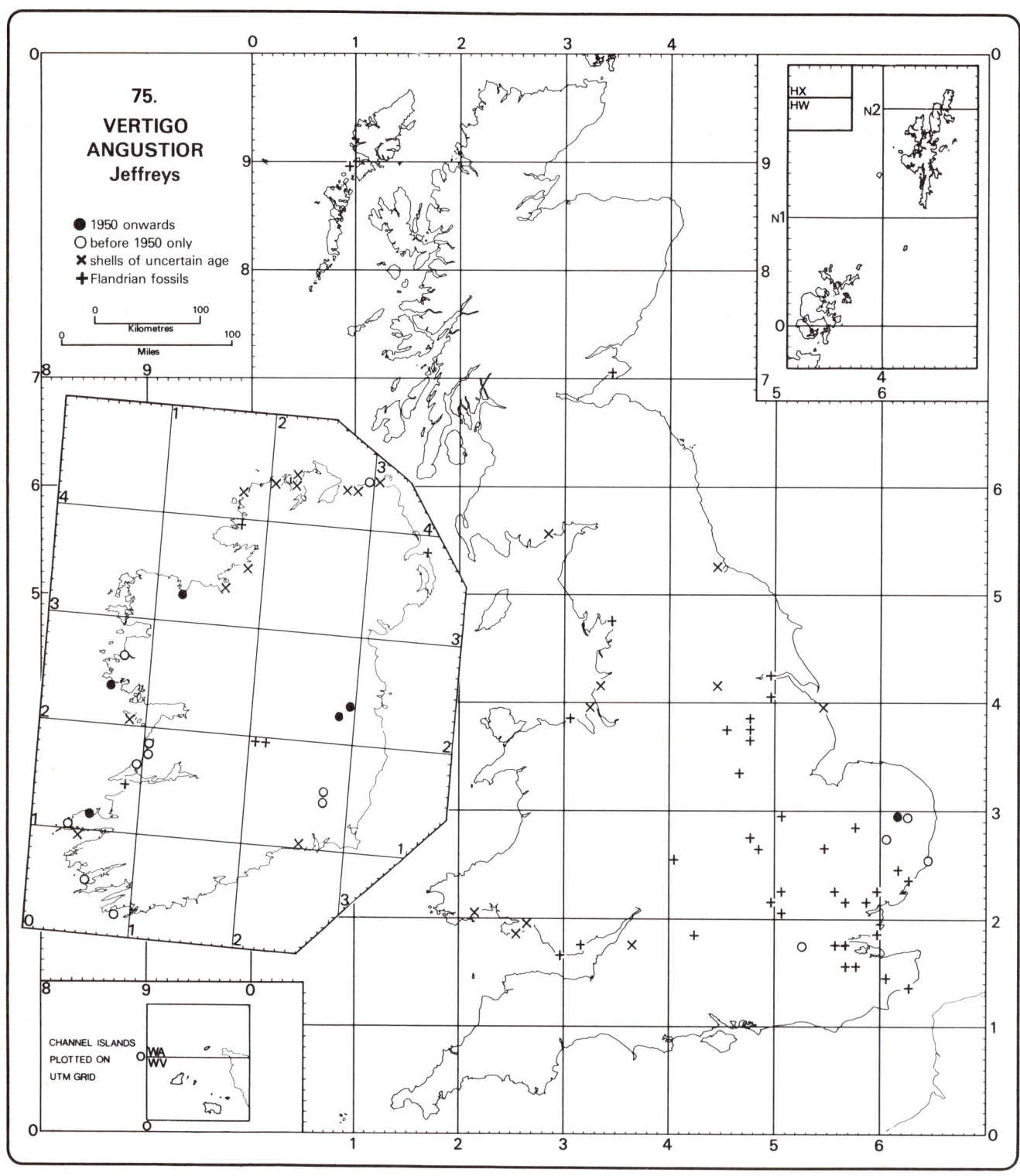

75.
VERTIGO
ANGUSTIOR
Jeffreys

● 1950 onwards
○ before 1950 only
✖ shells of uncertain age
✚ Flandrian fossils

Kilometres
Miles

CHANNEL ISLANDS
PLOTTED ON
UTM GRID

WA
WV

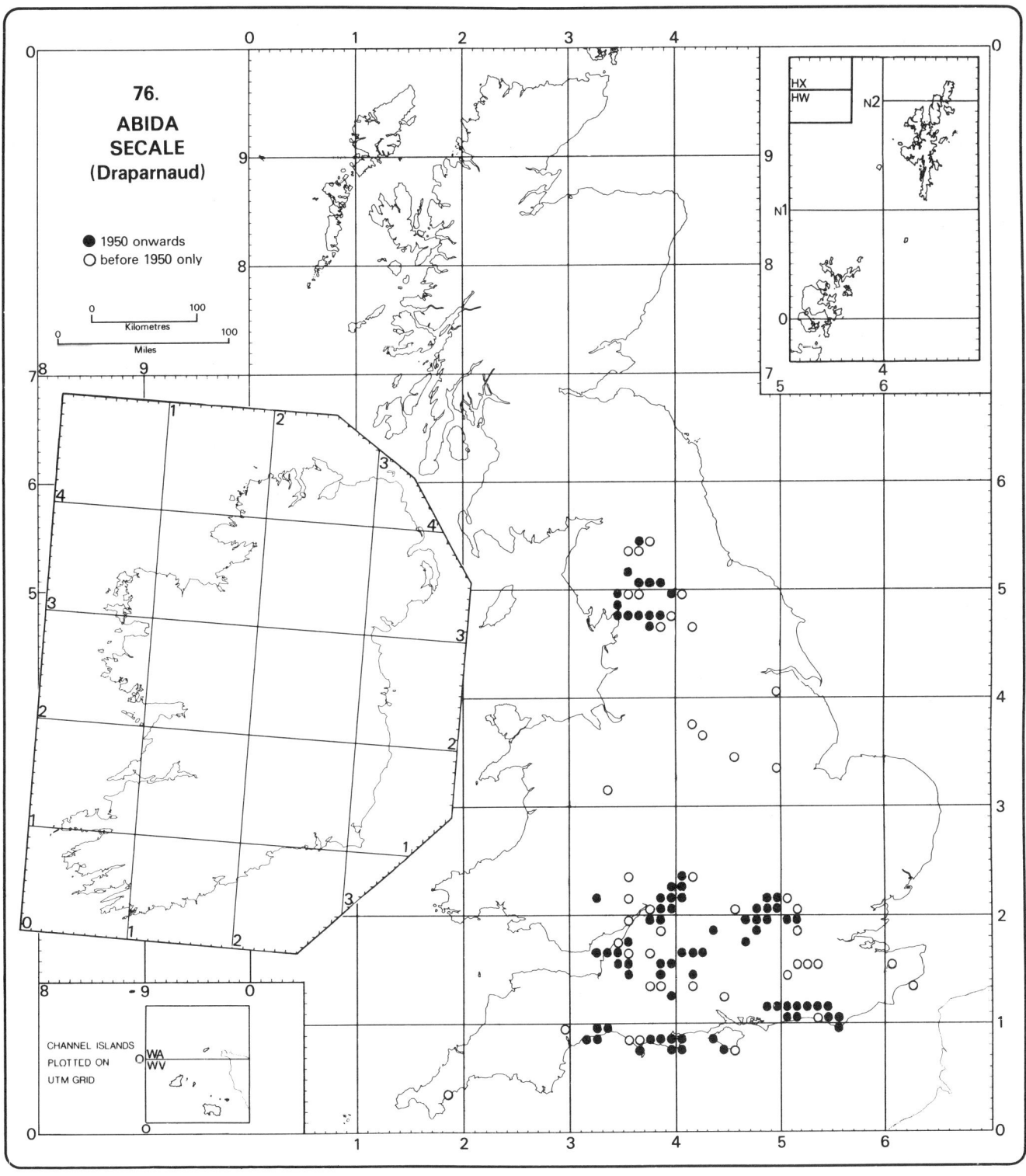

76.
ABIDA
SECALE
(Draparnaud)

● 1950 onwards
○ before 1950 only

0 _____ 100
Kilometres
0 _____ 100
Miles

HX
HW
N2
N1
0

CHANNEL ISLANDS
PLOTTED ON
UTM GRID
WA
WV

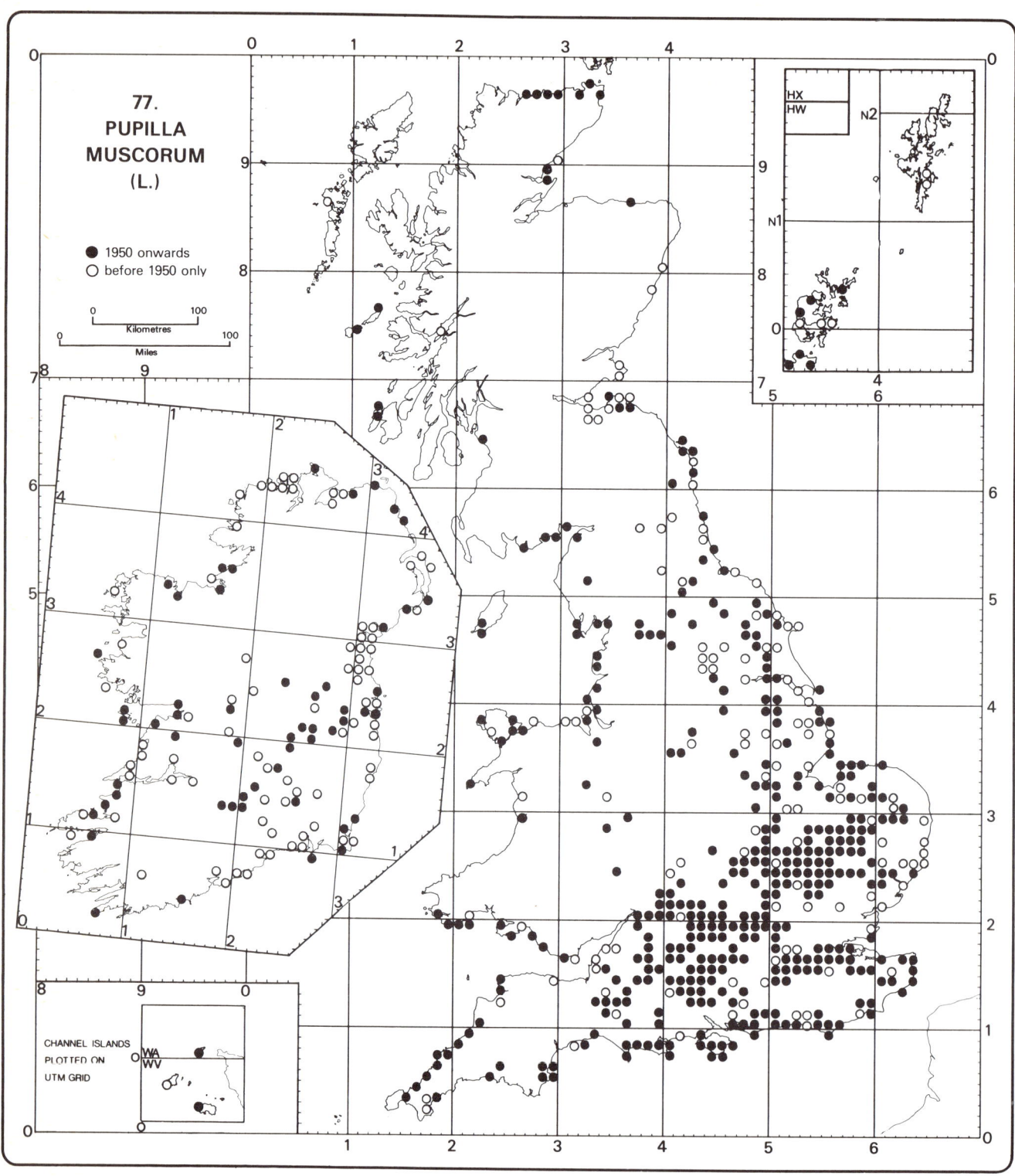

77.
PUPILLA
MUSCORUM
(L.)

● 1950 onwards
○ before 1950 only

0   100
Kilometres
0   100
Miles

HX
HW

N2

N1

CHANNEL ISLANDS
PLOTTED ON
UTM GRID

WA
WV

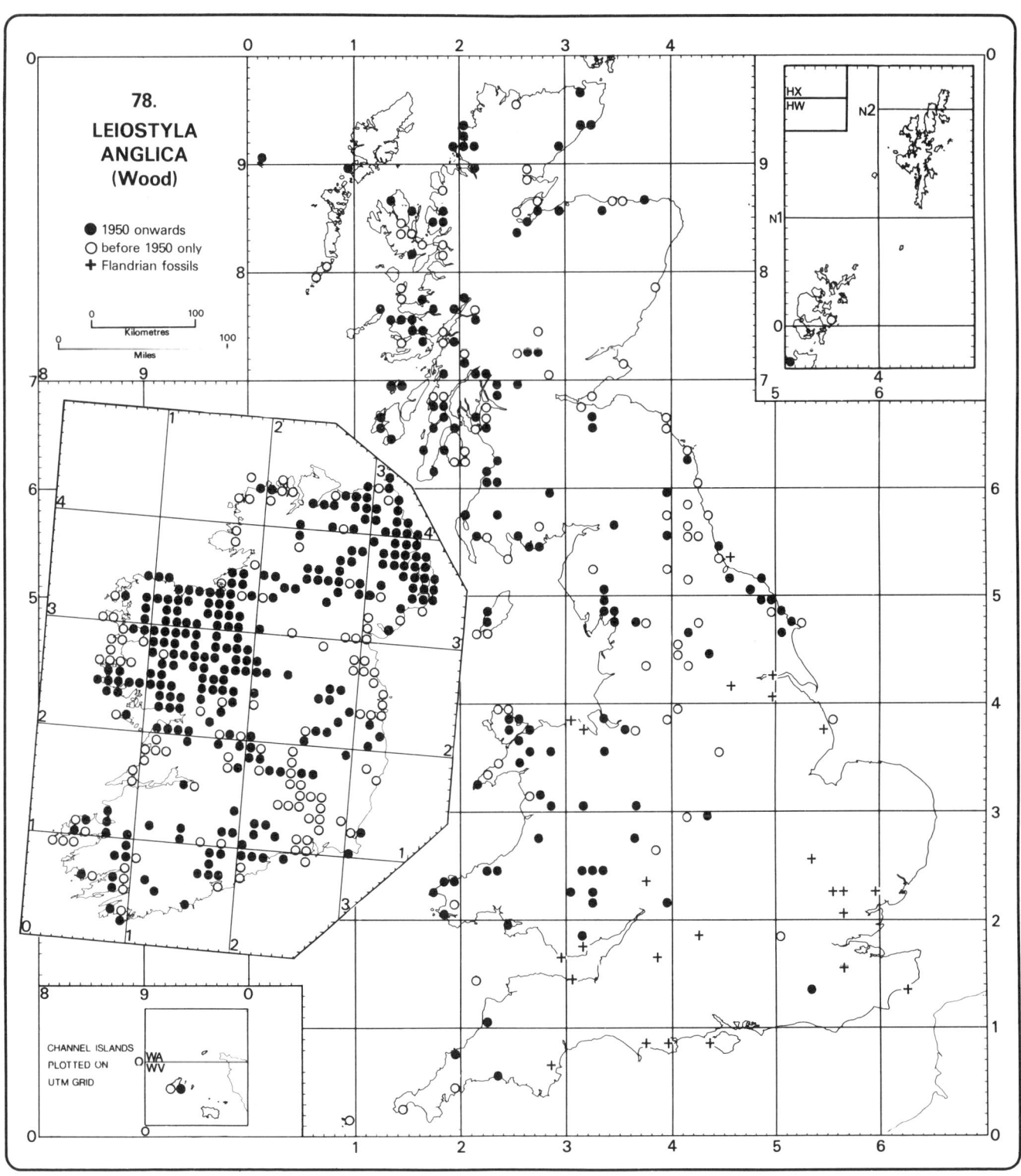

78.
LEIOSTYLA
ANGLICA
(Wood)

● 1950 onwards
○ before 1950 only
+ Flandrian fossils

Kilometres
Miles

CHANNEL ISLANDS
PLOTTED ON
UTM GRID

WA
WV

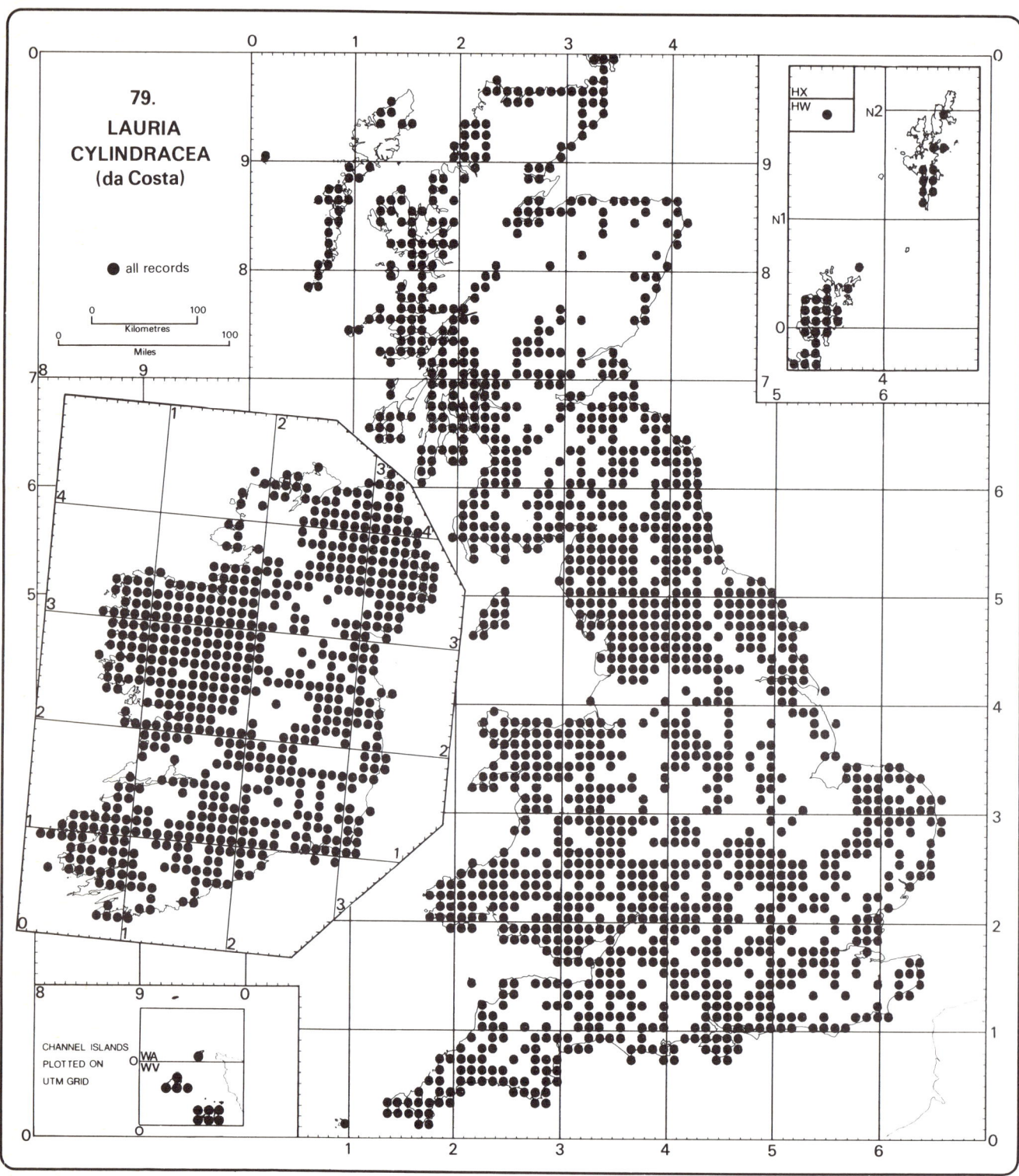

79.
LAURIA
CYLINDRACEA
(da Costa)

● all records

Kilometres
0 — 100

Miles
0 — 100

CHANNEL ISLANDS
PLOTTED ON
UTM GRID

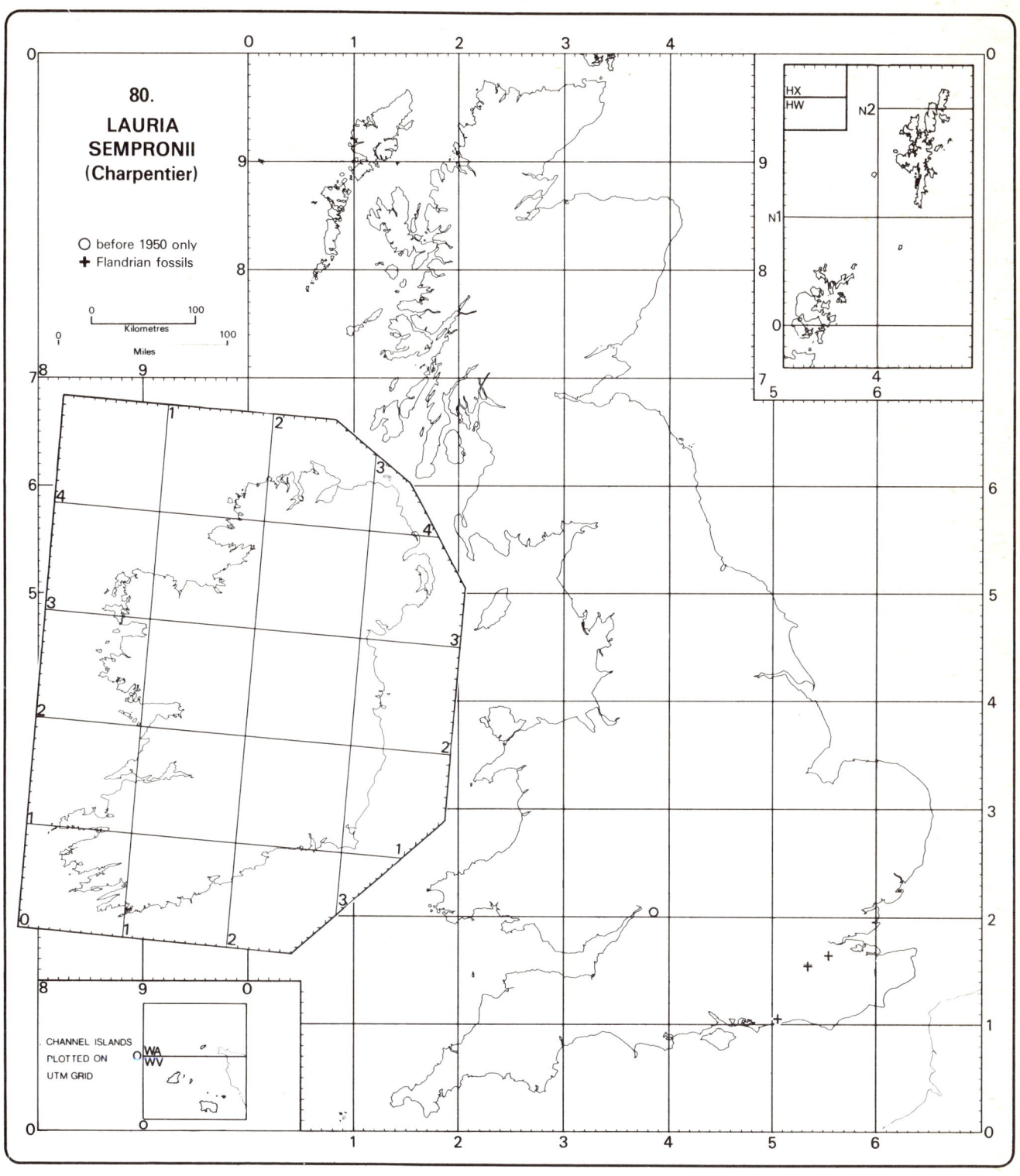

80.
**LAURIA
SEMPRONII**
(Charpentier)

○ before 1950 only
+ Flandrian fossils

Kilometres

Miles

CHANNEL ISLANDS
PLOTTED ON
UTM GRID

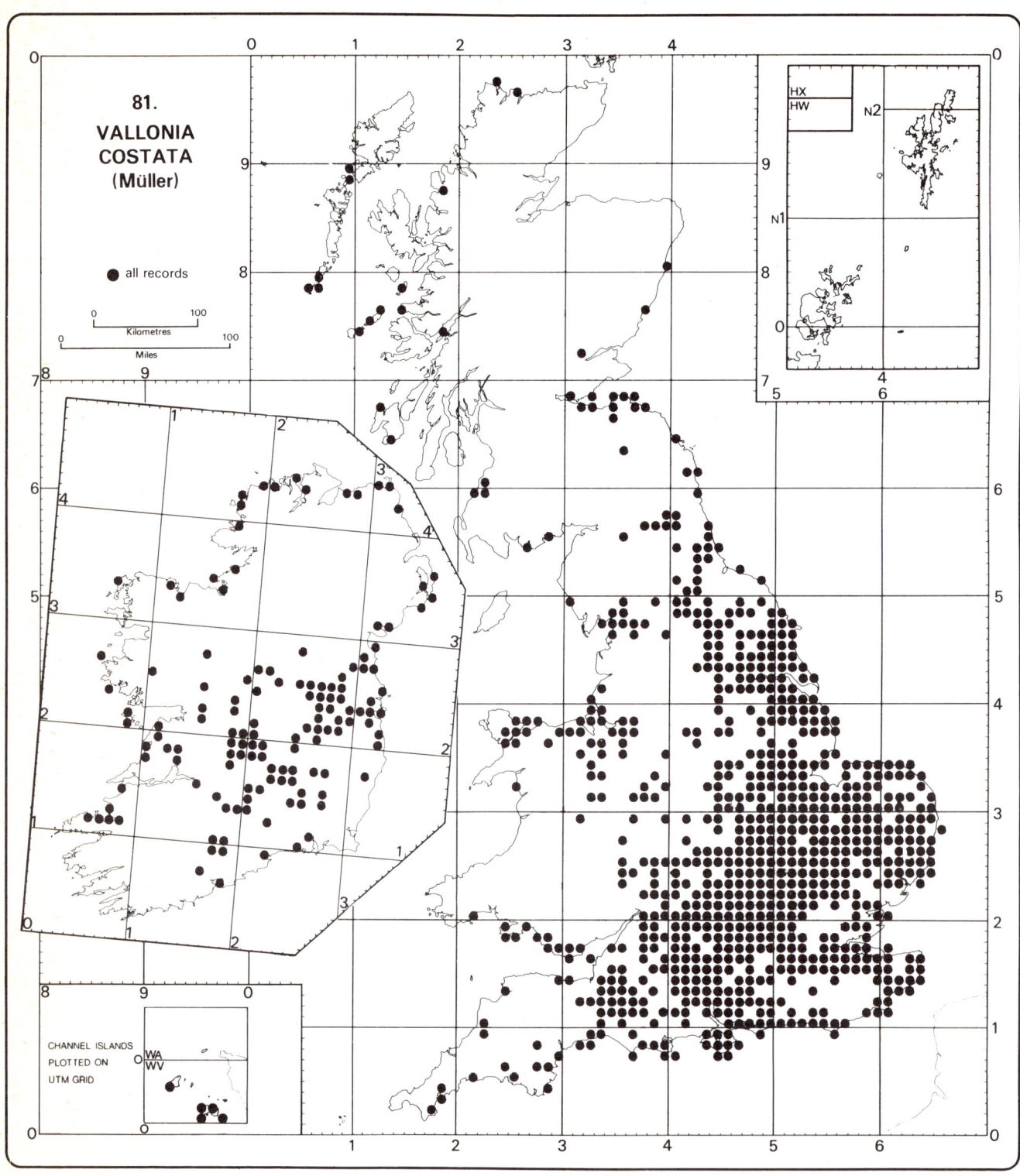

81.
VALLONIA
COSTATA
(Müller)

● all records

Kilometres

Miles

HX
HW

CHANNEL ISLANDS
PLOTTED ON
UTM GRID

WA
WV

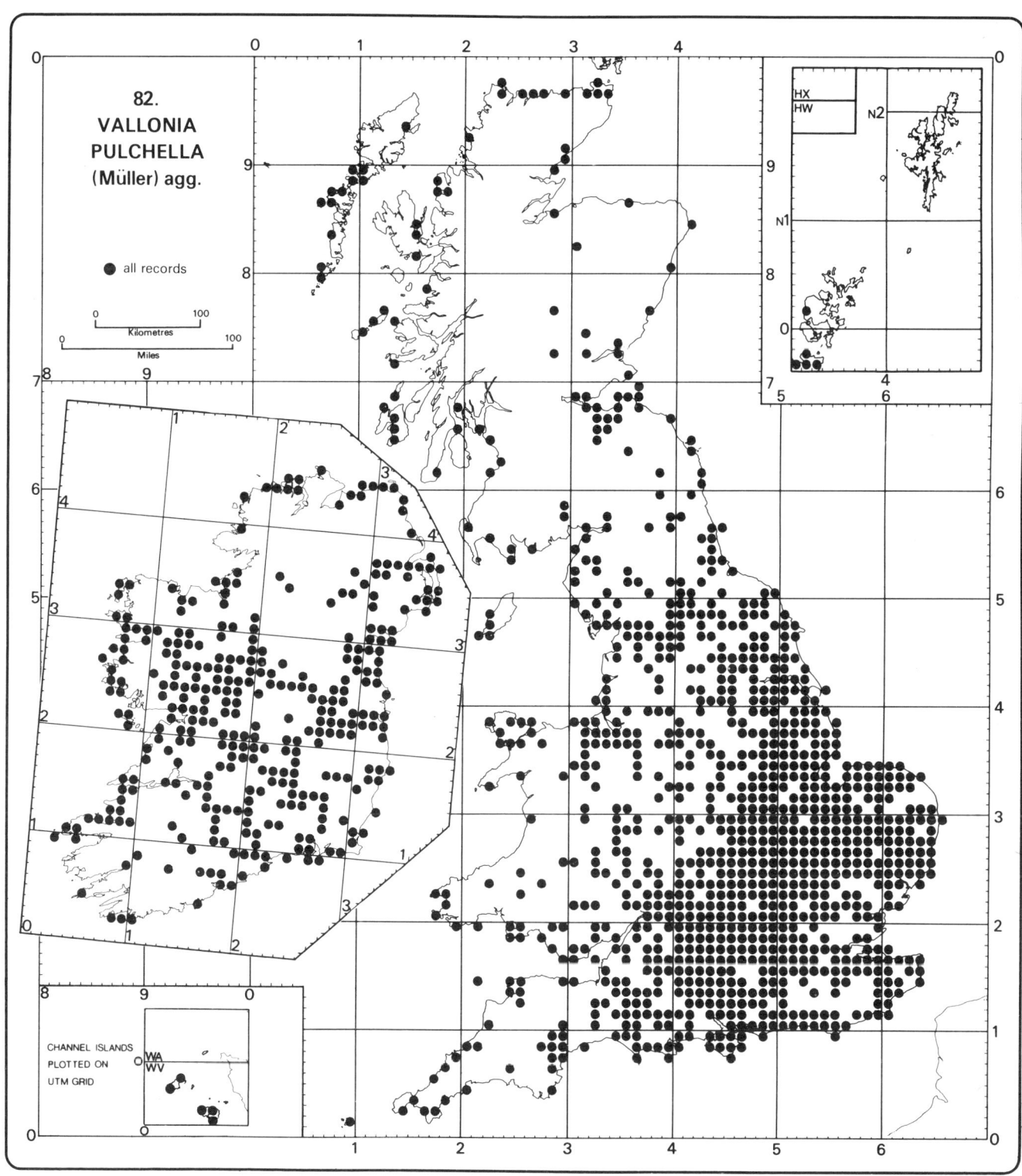

82.
VALLONIA
PULCHELLA
(Müller) agg.

● all records

Kilometres

Miles

CHANNEL ISLANDS
PLOTTED ON
UTM GRID

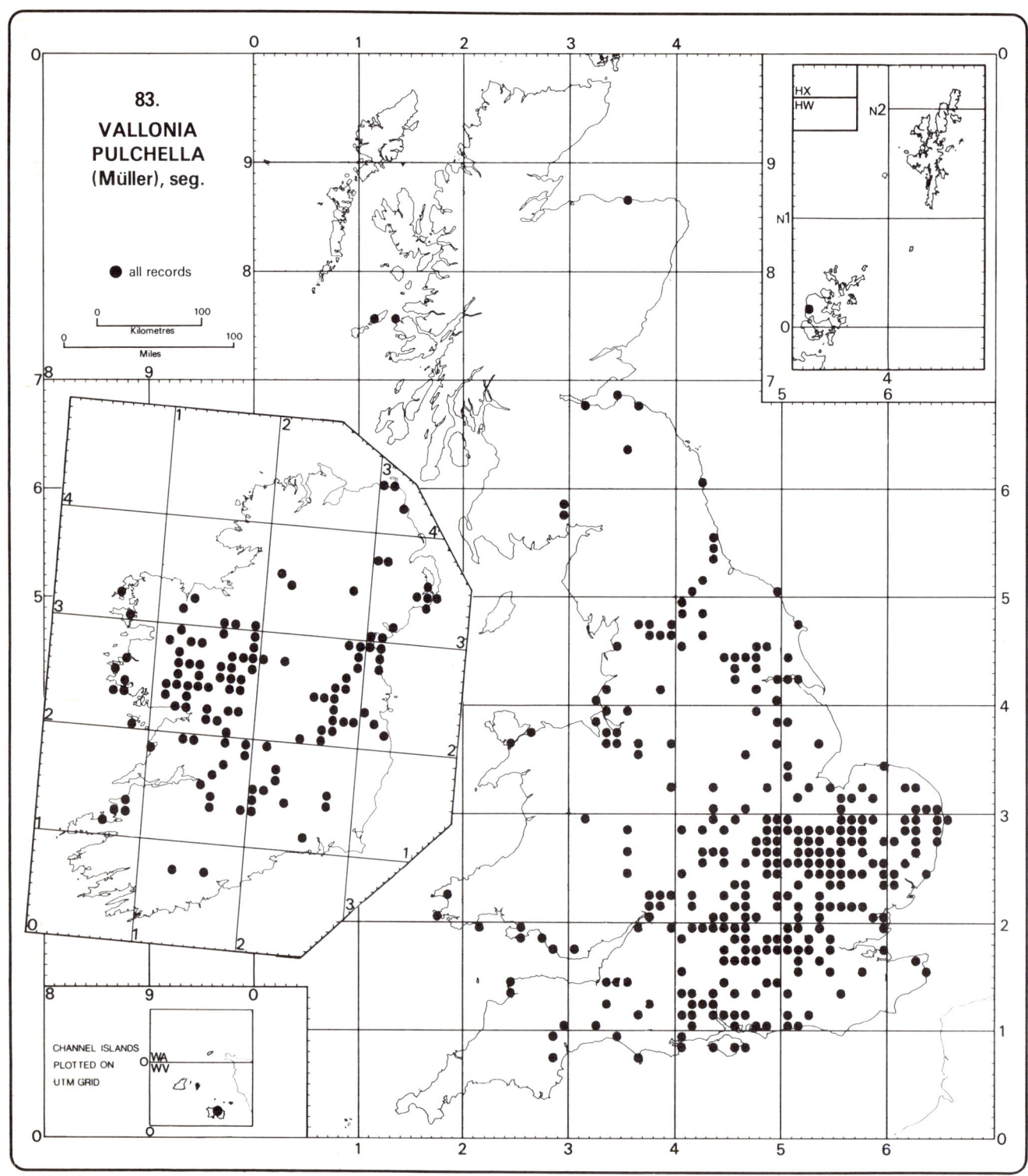

83.
VALLONIA
PULCHELLA
(Müller), seg.

● all records

Kilometres

Miles

HX
HW

CHANNEL ISLANDS
PLOTTED ON
UTM GRID

WA
WV

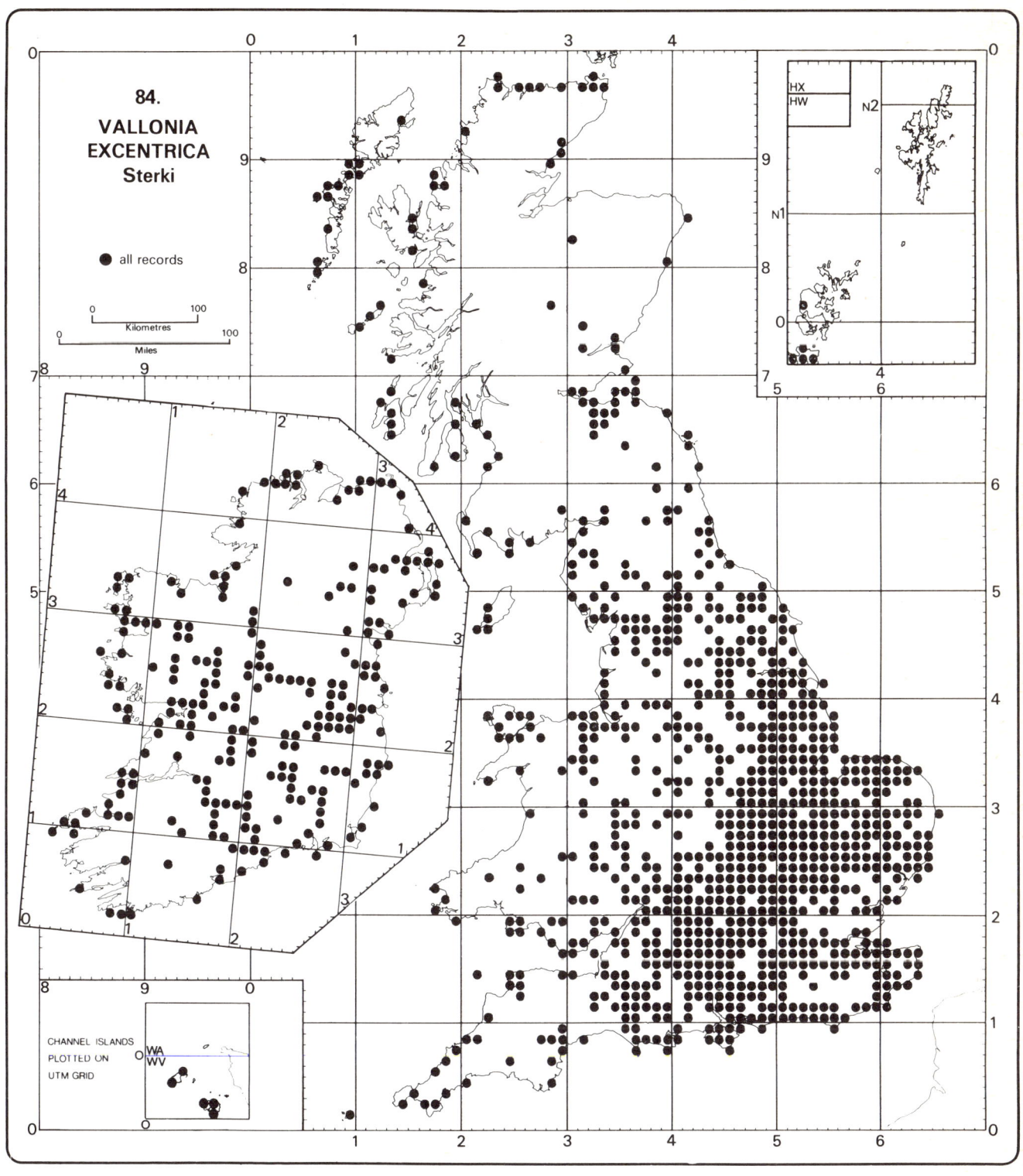

84.
VALLONIA
EXCENTRICA
Sterki

● all records

Kilometres
Miles

CHANNEL ISLANDS
PLOTTED ON
UTM GRID

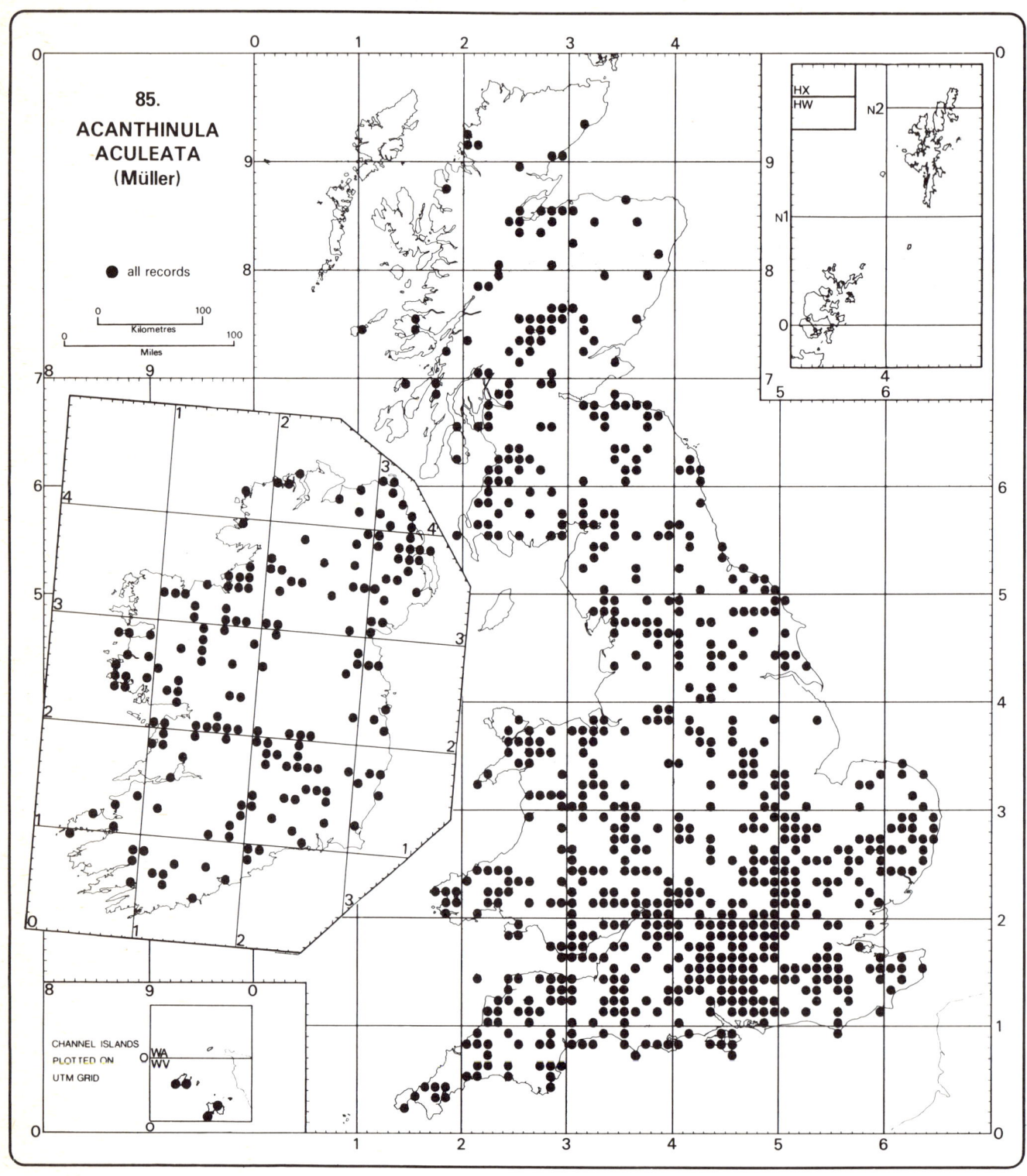

85.
ACANTHINULA
ACULEATA
(Müller)

● all records

Kilometres

Miles

CHANNEL ISLANDS
PLOTTED ON
UTM GRID

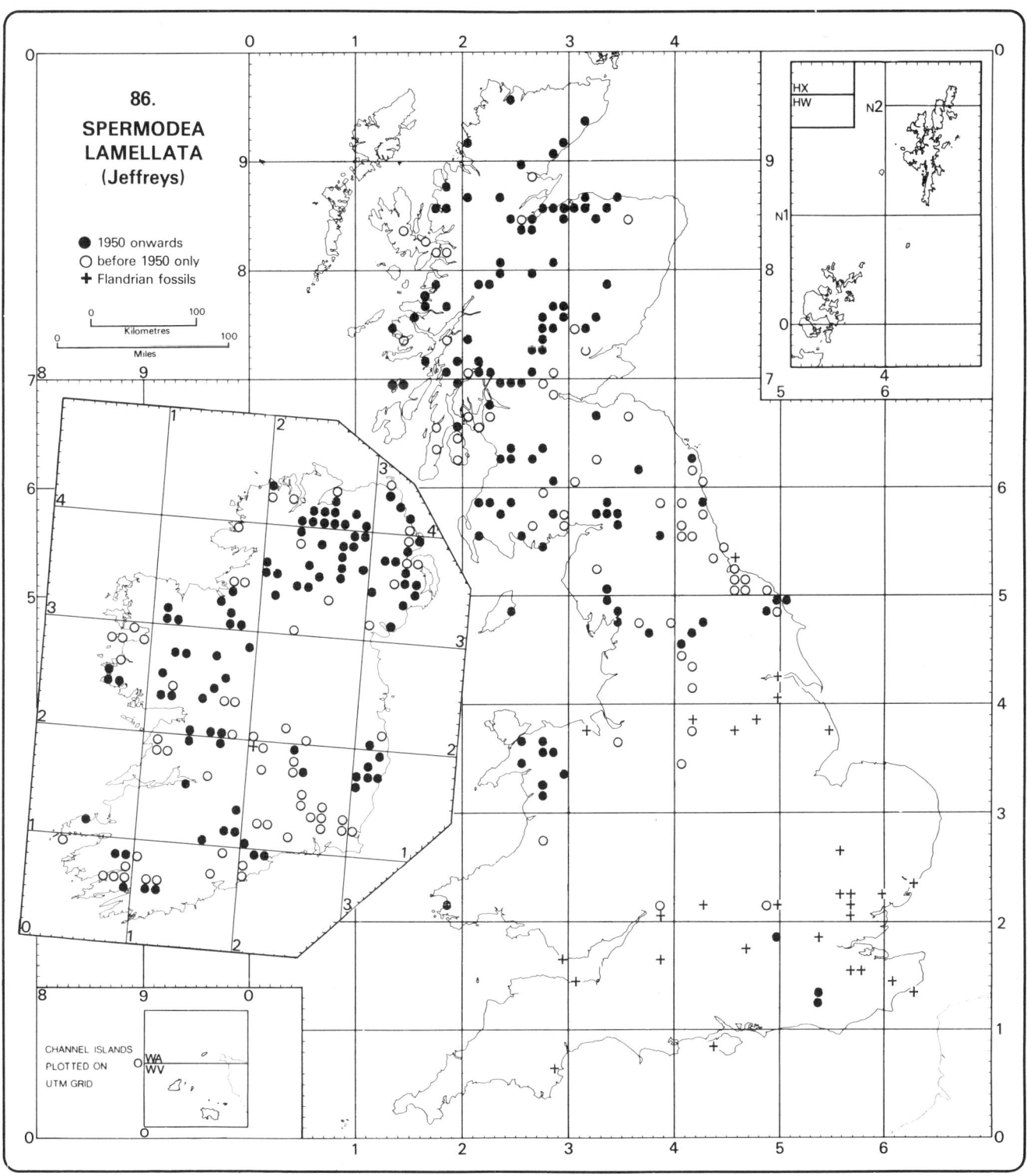

86.
SPERMODEA
LAMELLATA
(Jeffreys)

● 1950 onwards
○ before 1950 only
+ Flandrian fossils

0 _____ 100
Kilometres
0 _____ 100
Miles

CHANNEL ISLANDS
PLOTTED ON
UTM GRID

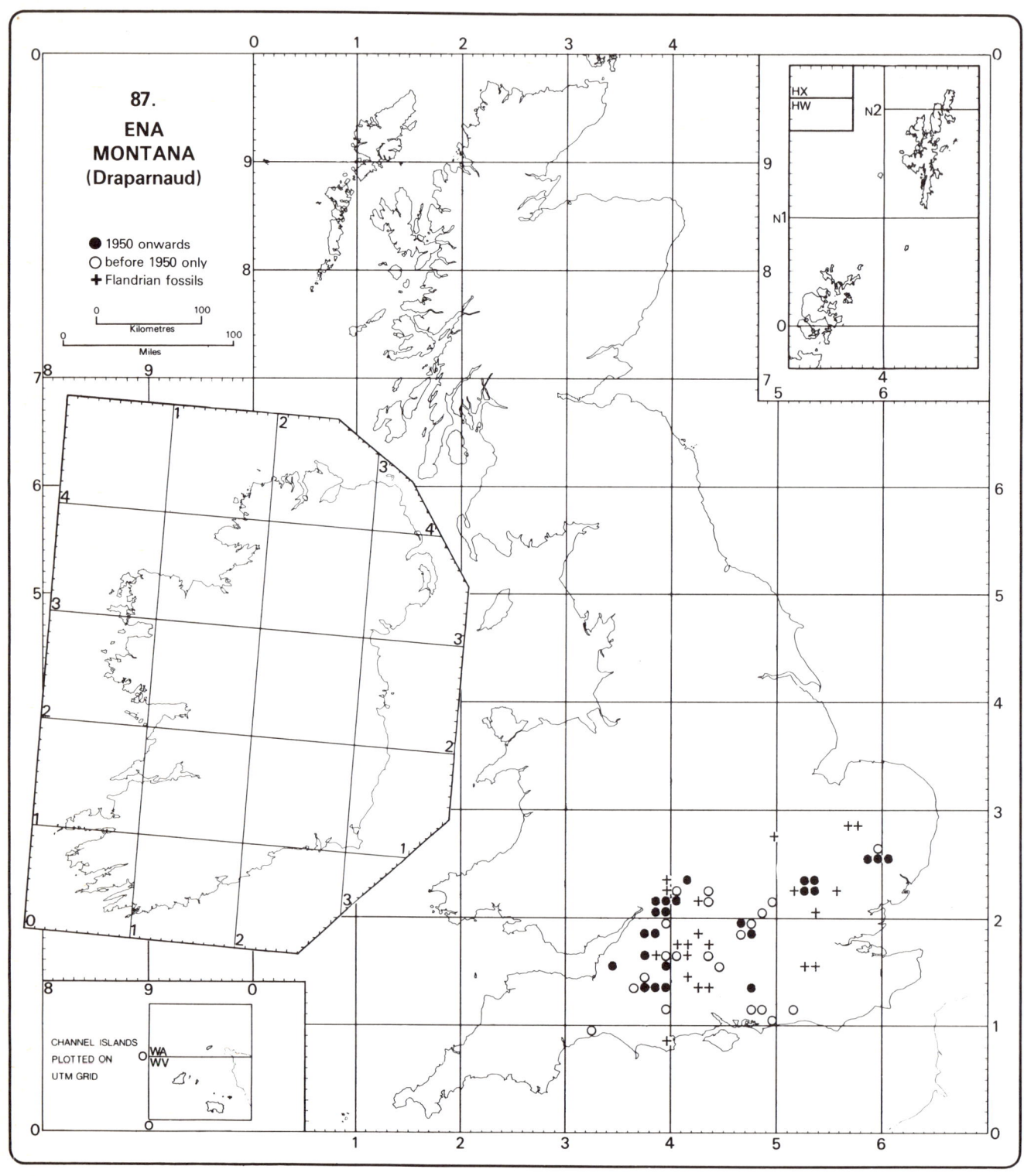

87.

**ENA
MONTANA
(Draparnaud)**

● 1950 onwards
○ before 1950 only
+ Flandrian fossils

Kilometres

Miles

CHANNEL ISLANDS
PLOTTED ON
UTM GRID

WA
WV

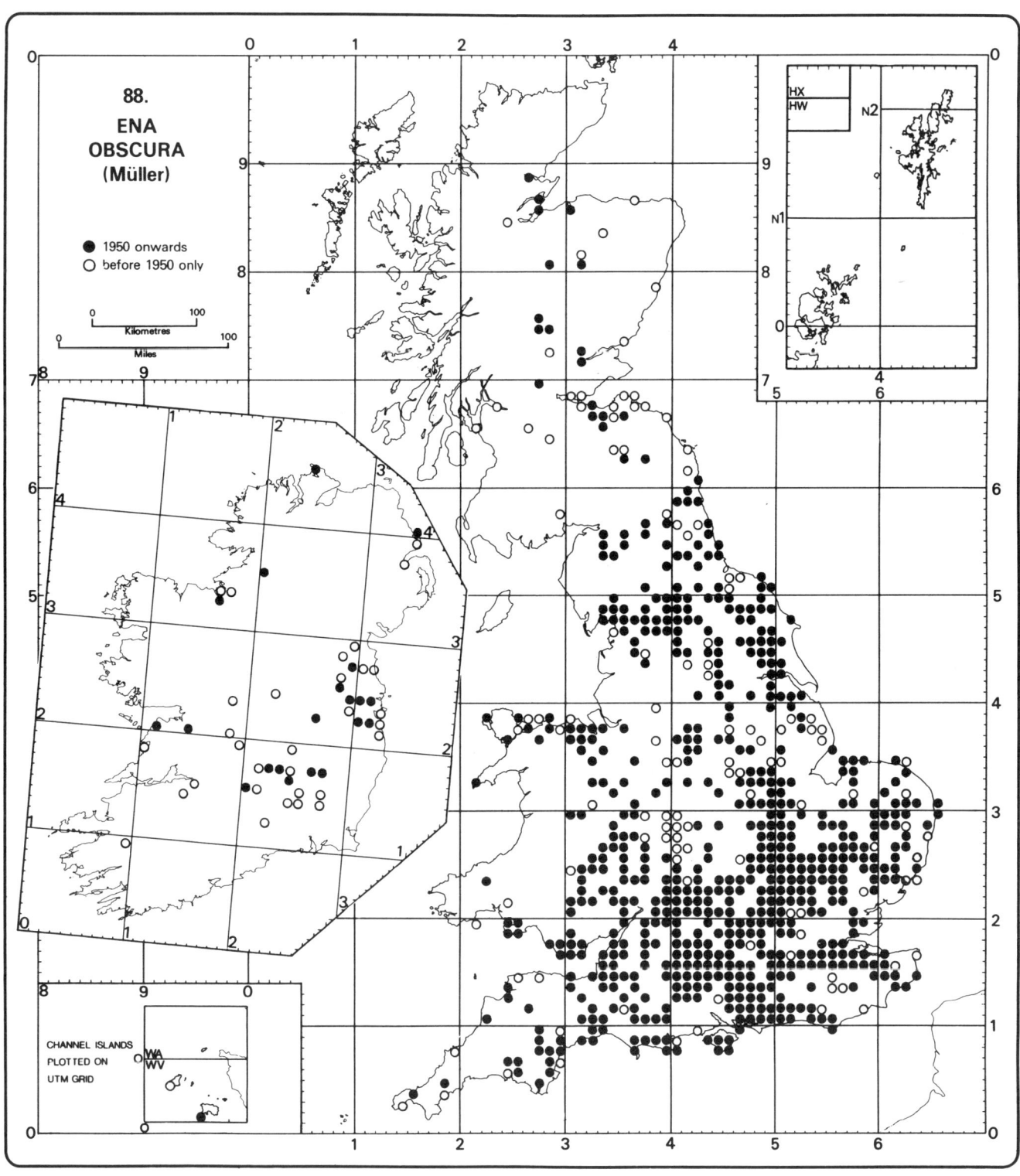

88.
ENA
OBSCURA
(Müller)

● 1950 onwards
○ before 1950 only

0 ___ 100
Kilometres
0 ___ 100
Miles

CHANNEL ISLANDS
PLOTTED ON
UTM GRID

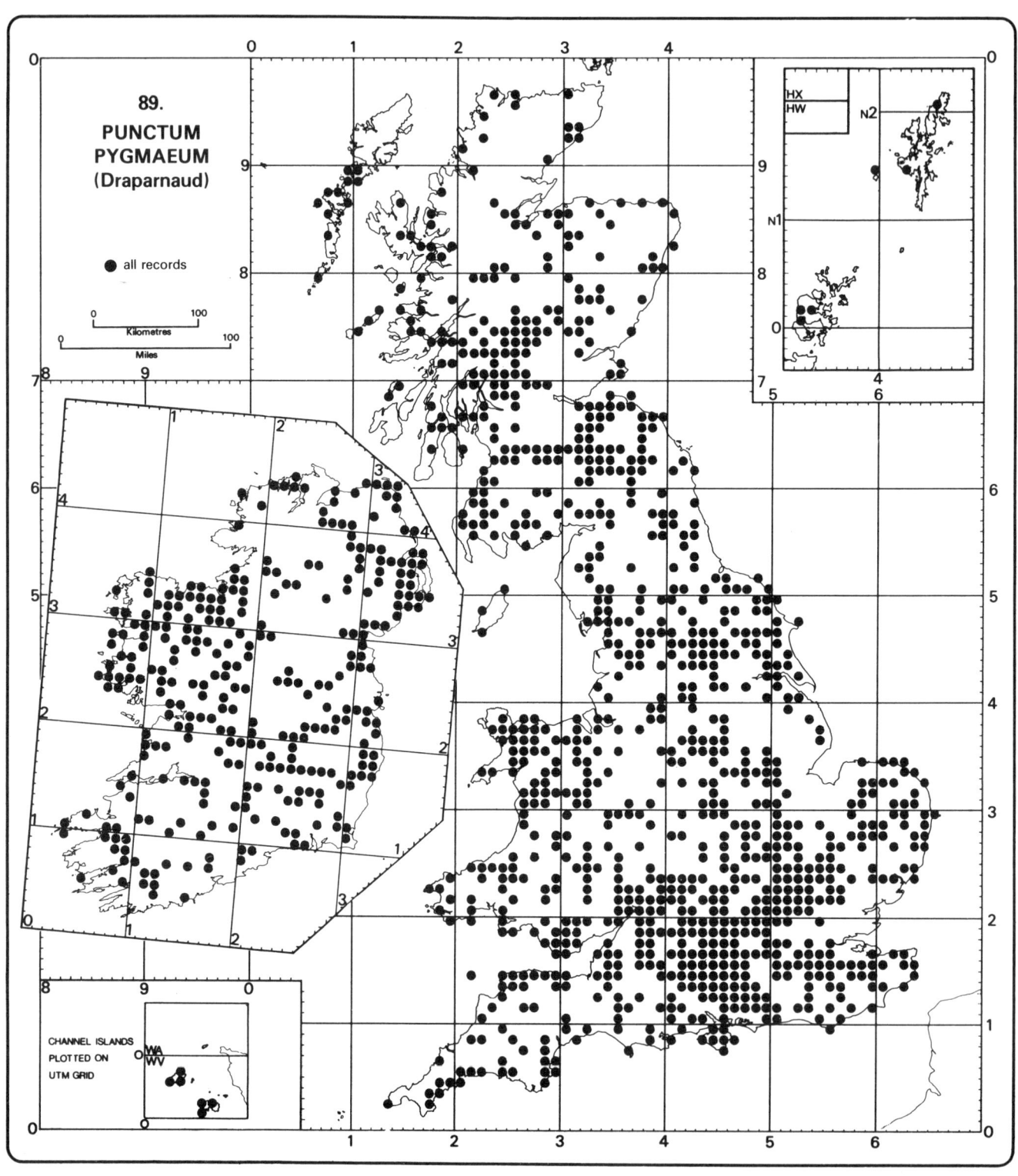

89.
PUNCTUM
PYGMAEUM
(Draparnaud)

● all records

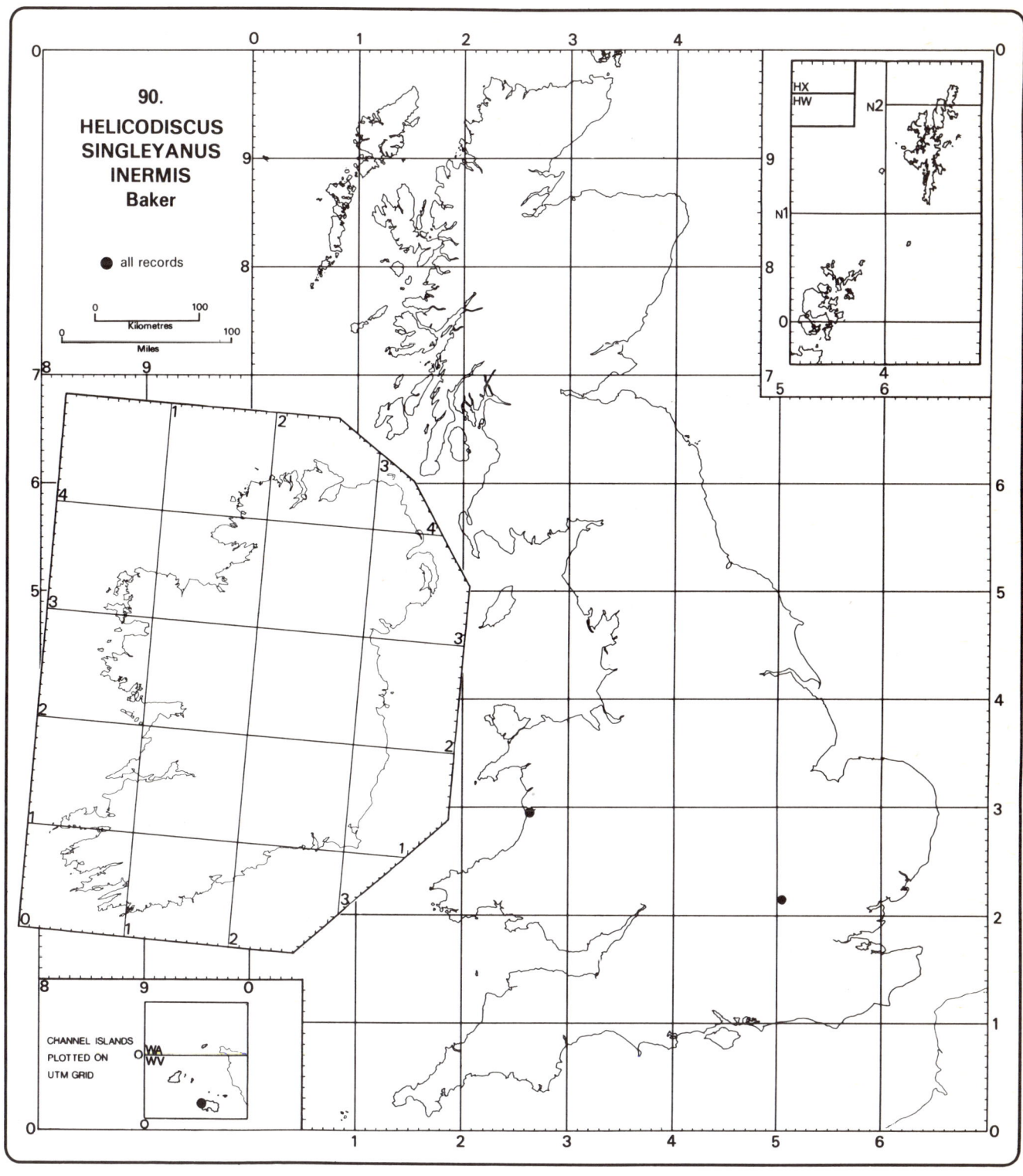

90.
HELICODISCUS
SINGLEYANUS
INERMIS
Baker

● all records

0   100
Kilometres
0   100
Miles

CHANNEL ISLANDS
PLOTTED ON
UTM GRID

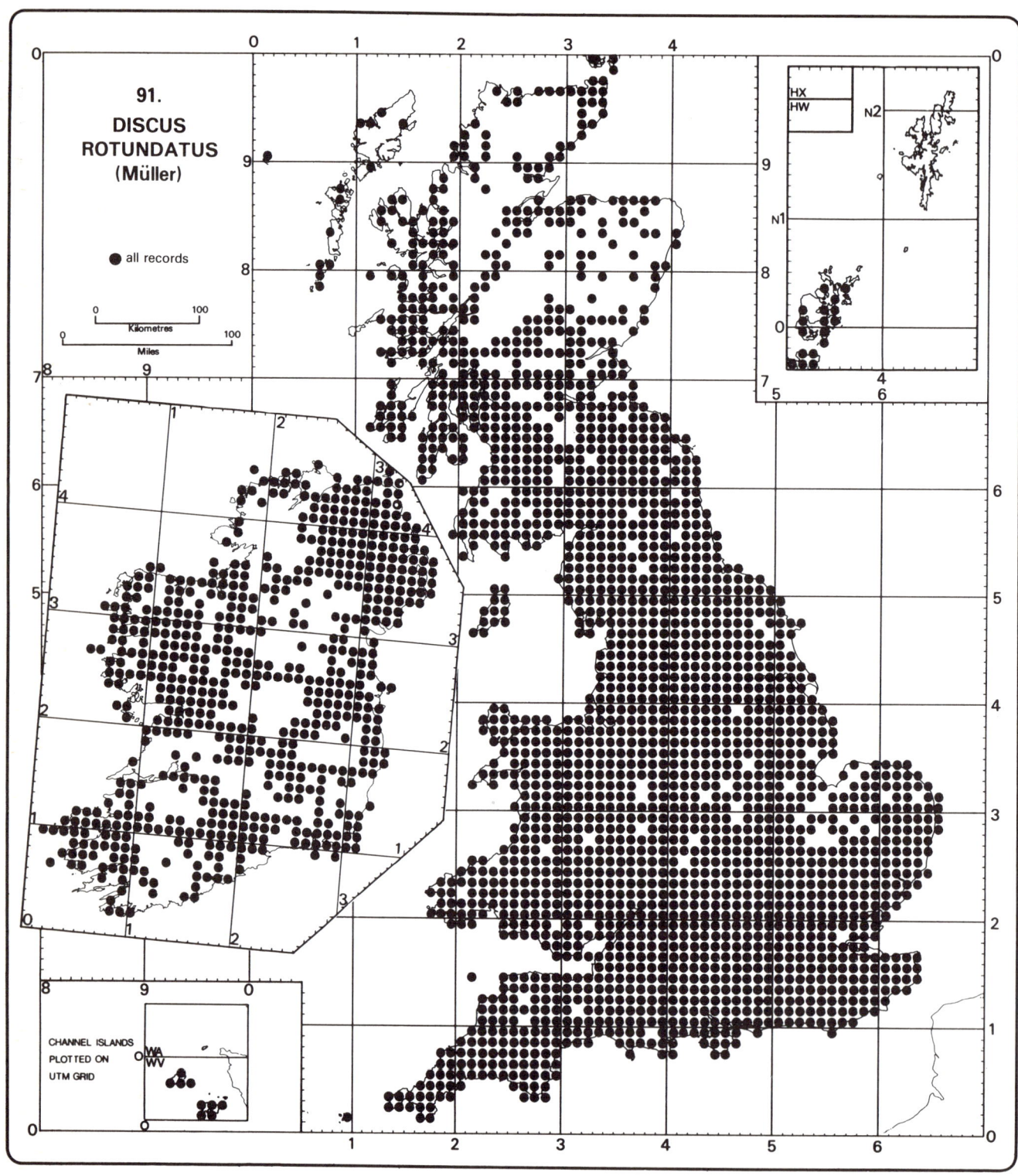

91.
DISCUS
ROTUNDATUS
(Müller)

● all records

Kilometres
Miles

CHANNEL ISLANDS
PLOTTED ON
UTM GRID

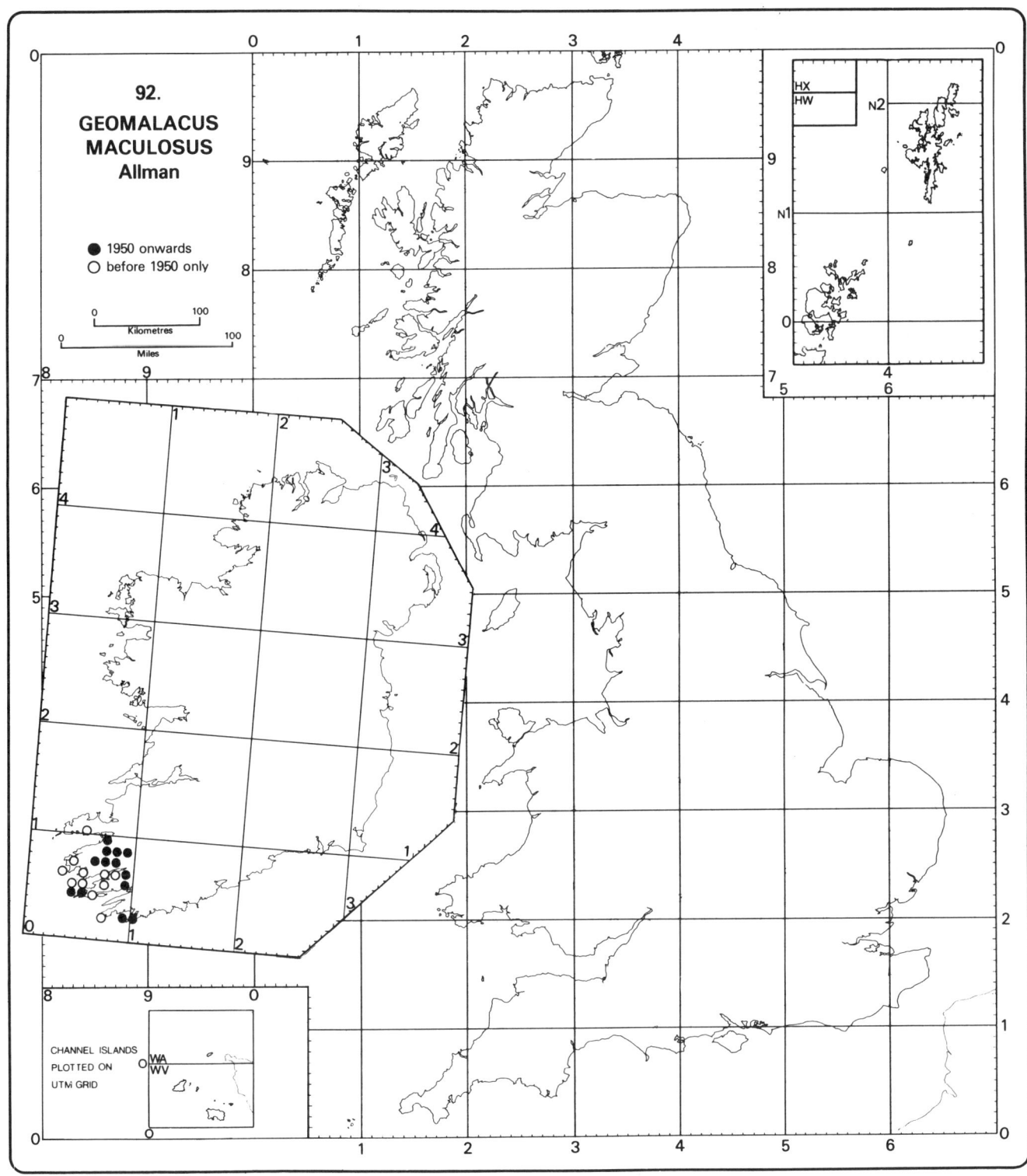

92.
GEOMALACUS
MACULOSUS
Allman

● 1950 onwards
○ before 1950 only

Kilometres
Miles

CHANNEL ISLANDS
PLOTTED ON
UTM GRID

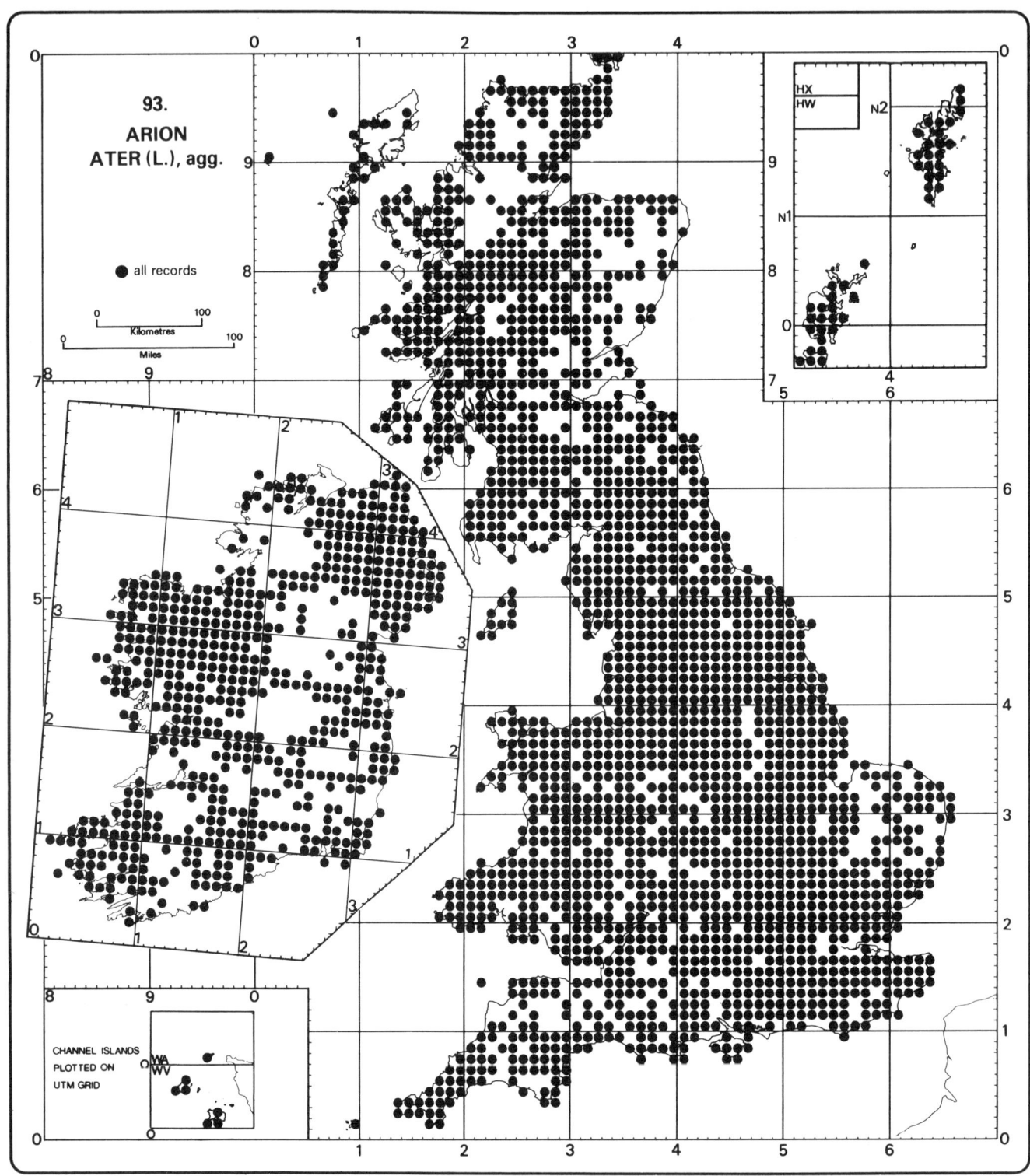

93.
ARION
ATER (L.), agg.

● all records

Kilometres
Miles

CHANNEL ISLANDS
PLOTTED ON
UTM GRID

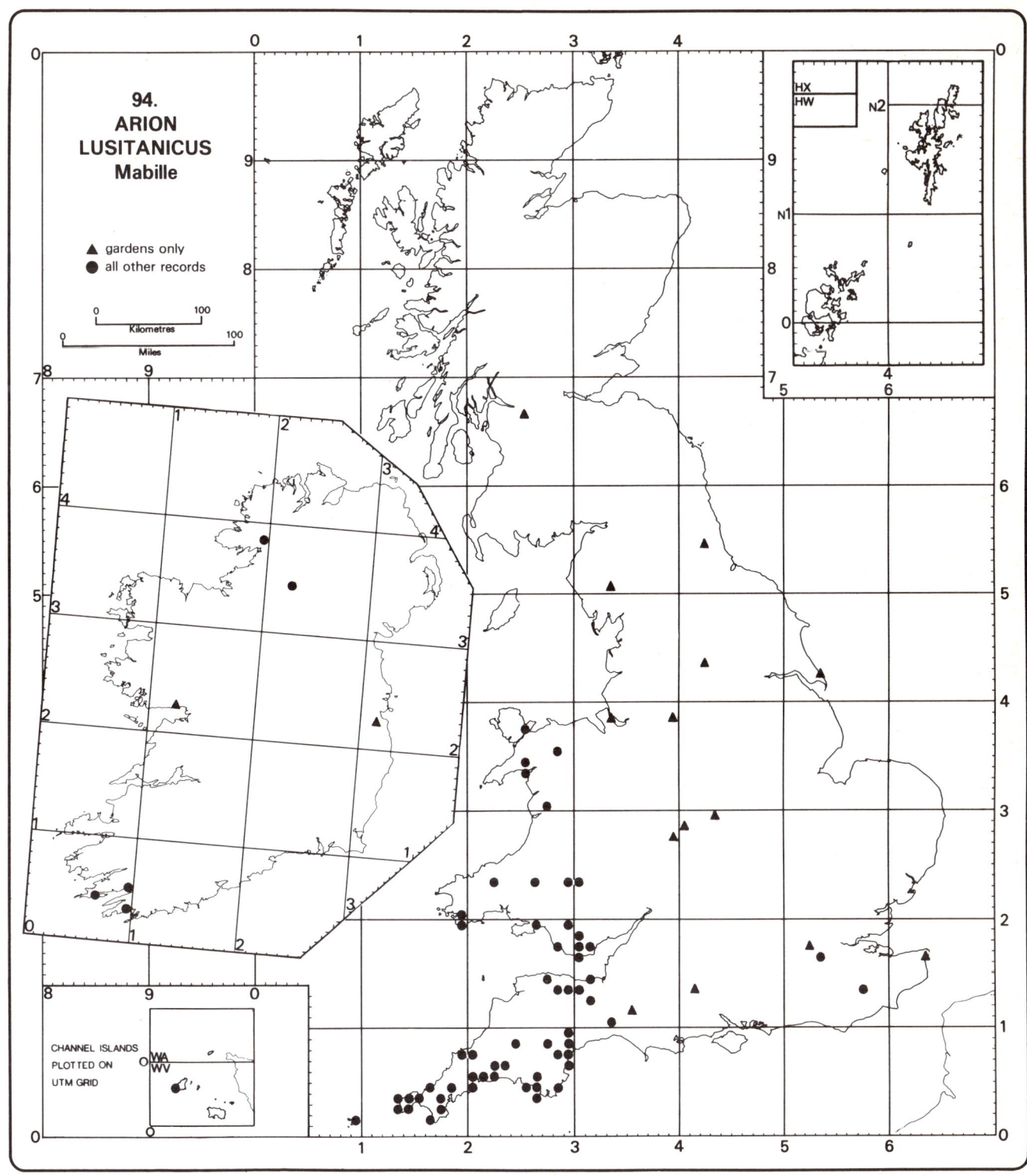

94.
ARION
LUSITANICUS
Mabille

▲ gardens only
● all other records

0 _____ 100
Kilometres
0 _____ 100
Miles

CHANNEL ISLANDS
PLOTTED ON
UTM GRID

WA
WV

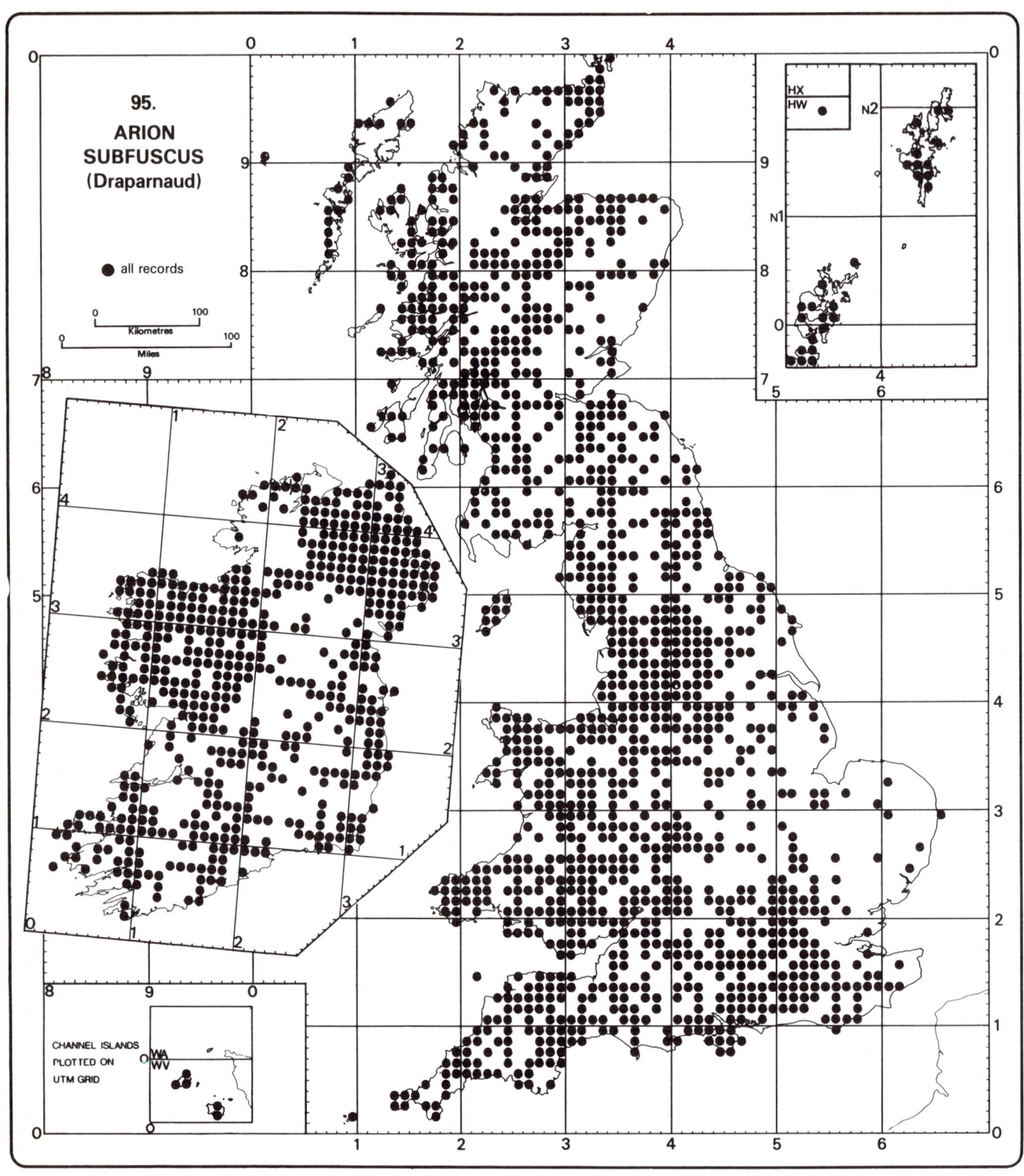

95.
ARION
SUBFUSCUS
(Draparnaud)

● all records

Kilometres
Miles

CHANNEL ISLANDS
PLOTTED ON
UTM GRID

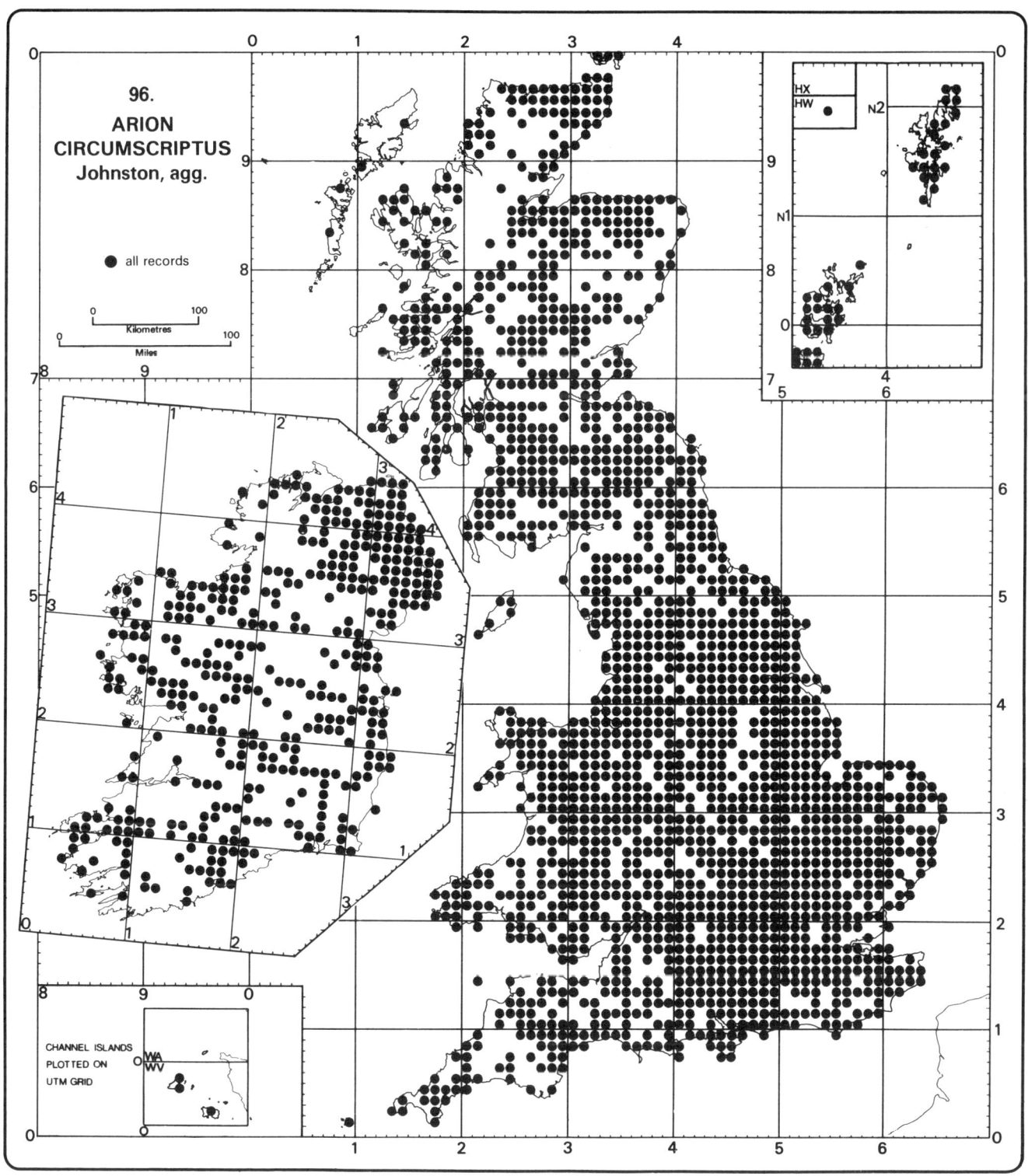

96.
ARION
CIRCUMSCRIPTUS
Johnston, agg.

● all records

Kilometres

Miles

CHANNEL ISLANDS
PLOTTED ON
UTM GRID

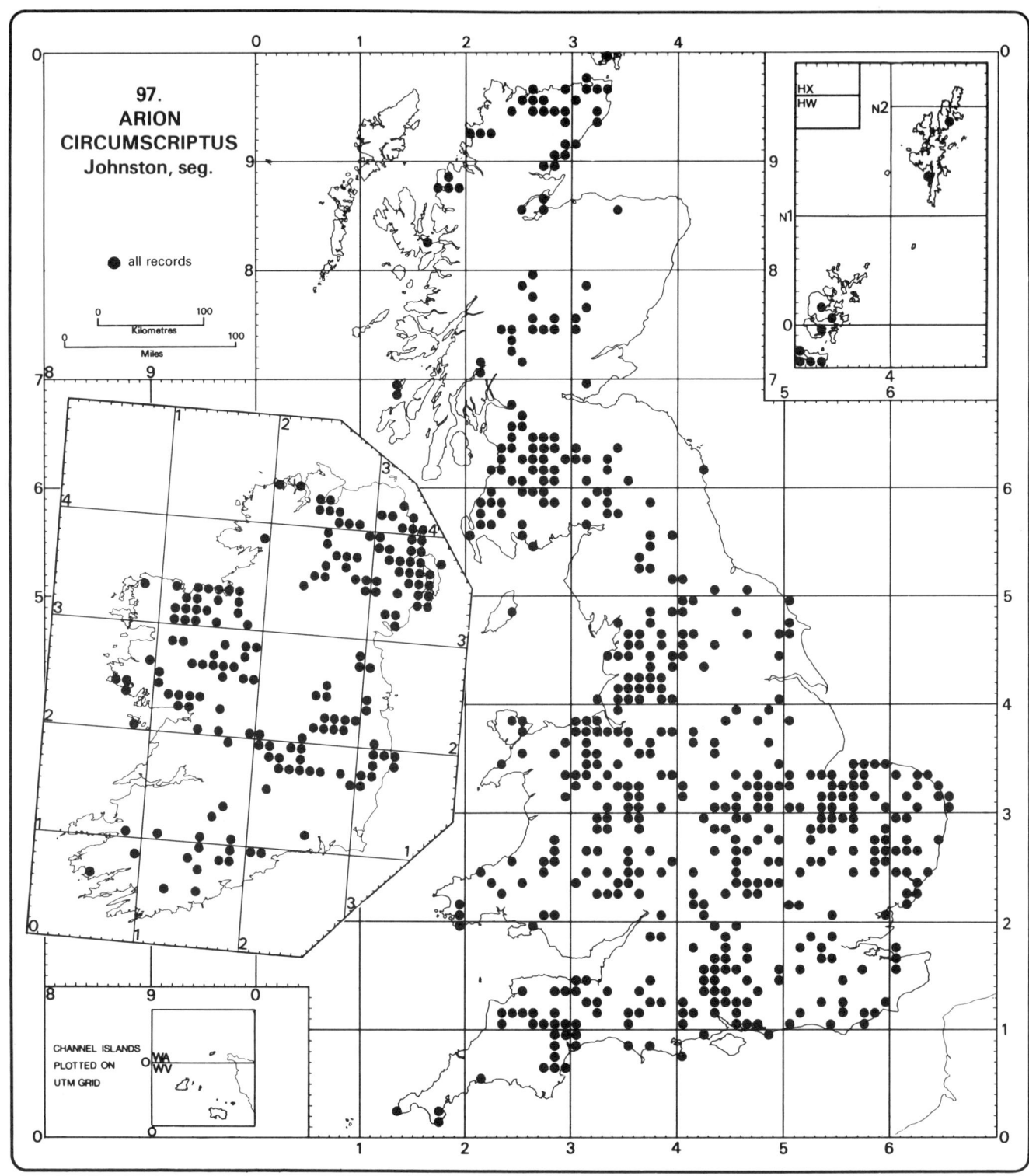

97.
ARION
CIRCUMSCRIPTUS
Johnston, seg.

● all records

Kilometres
Miles

CHANNEL ISLANDS
PLOTTED ON
UTM GRID

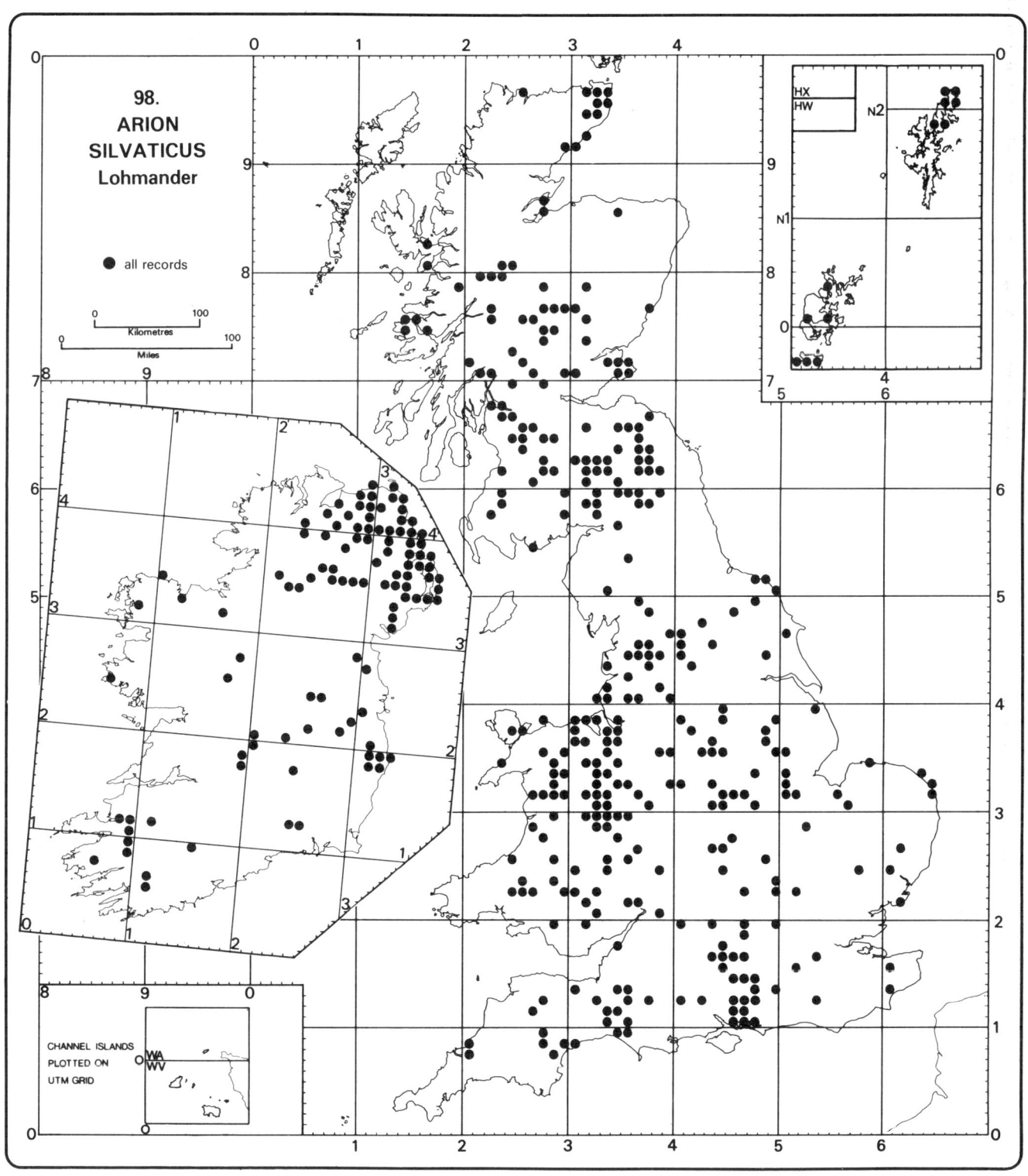

98.
ARION
SILVATICUS
Lohmander

● all records

Kilometres

Miles

CHANNEL ISLANDS
PLOTTED ON
UTM GRID

WA
WV

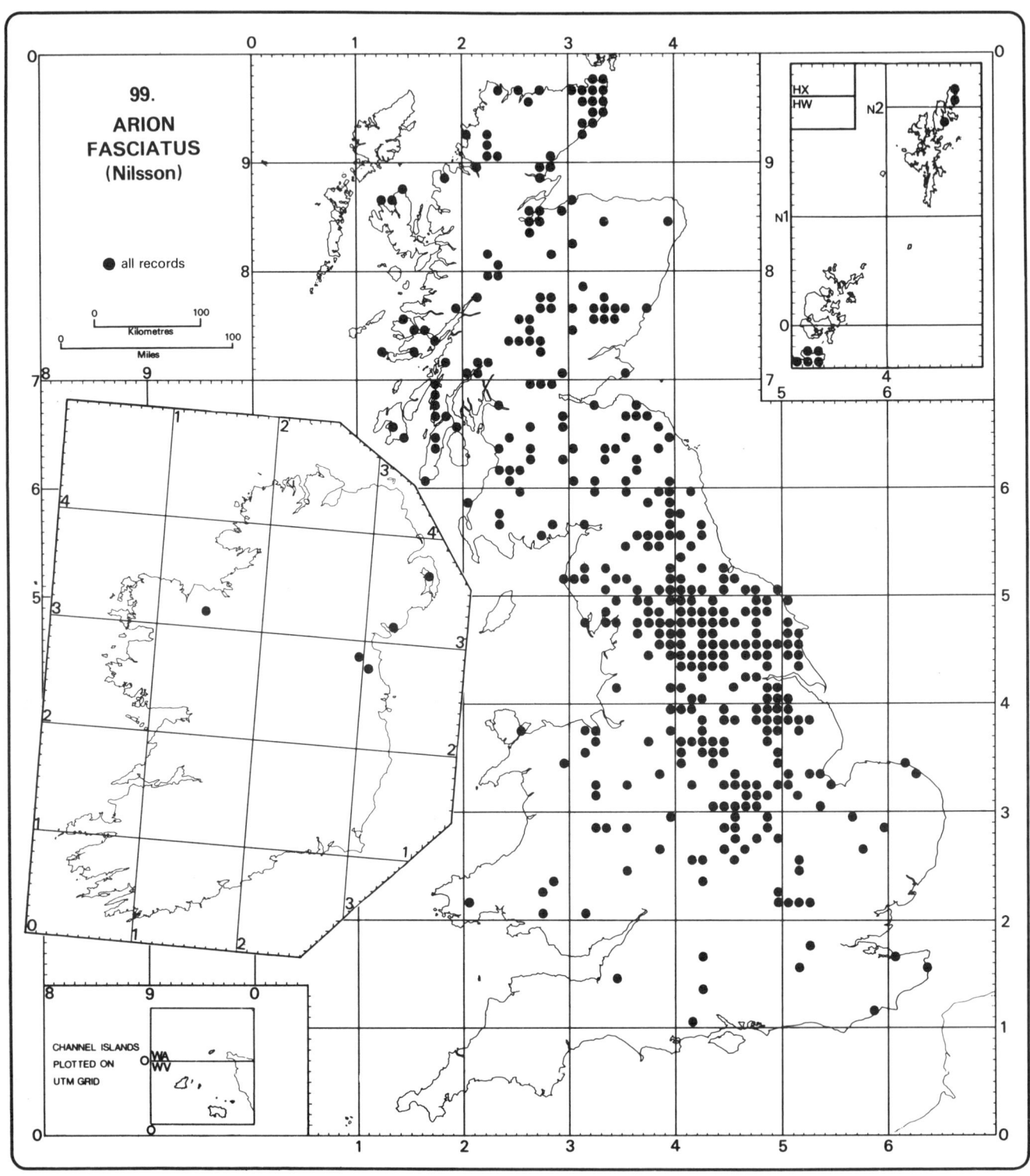

99.
ARION
FASCIATUS
(Nilsson)

● all records

0 — 100
Kilometres
0 — 100
Miles

CHANNEL ISLANDS
PLOTTED ON
UTM GRID

WA
WV

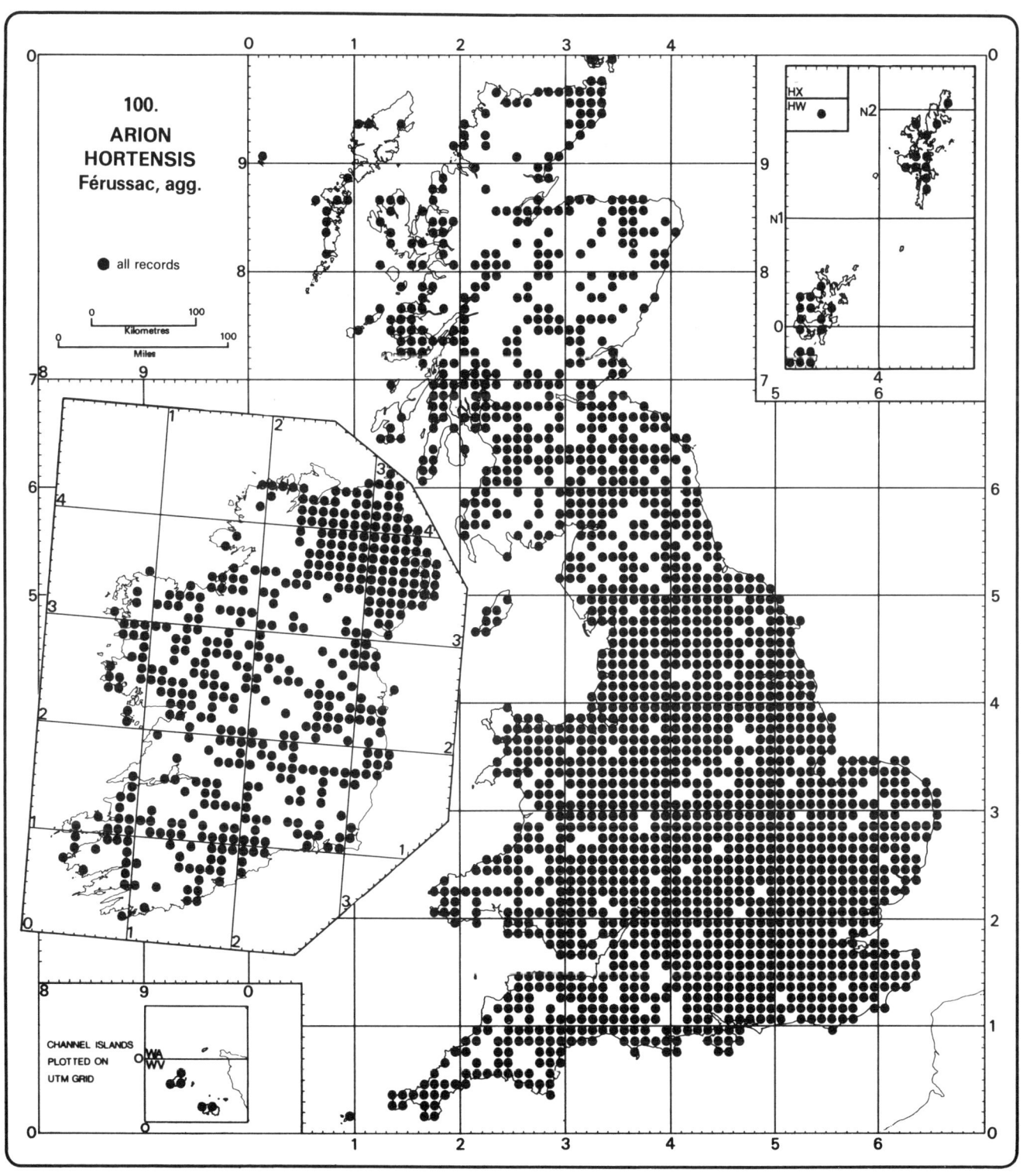

100.
ARION
HORTENSIS
Férussac, agg.

● all records

0 Kilometres 100
0 Miles 100

CHANNEL ISLANDS
PLOTTED ON
UTM GRID

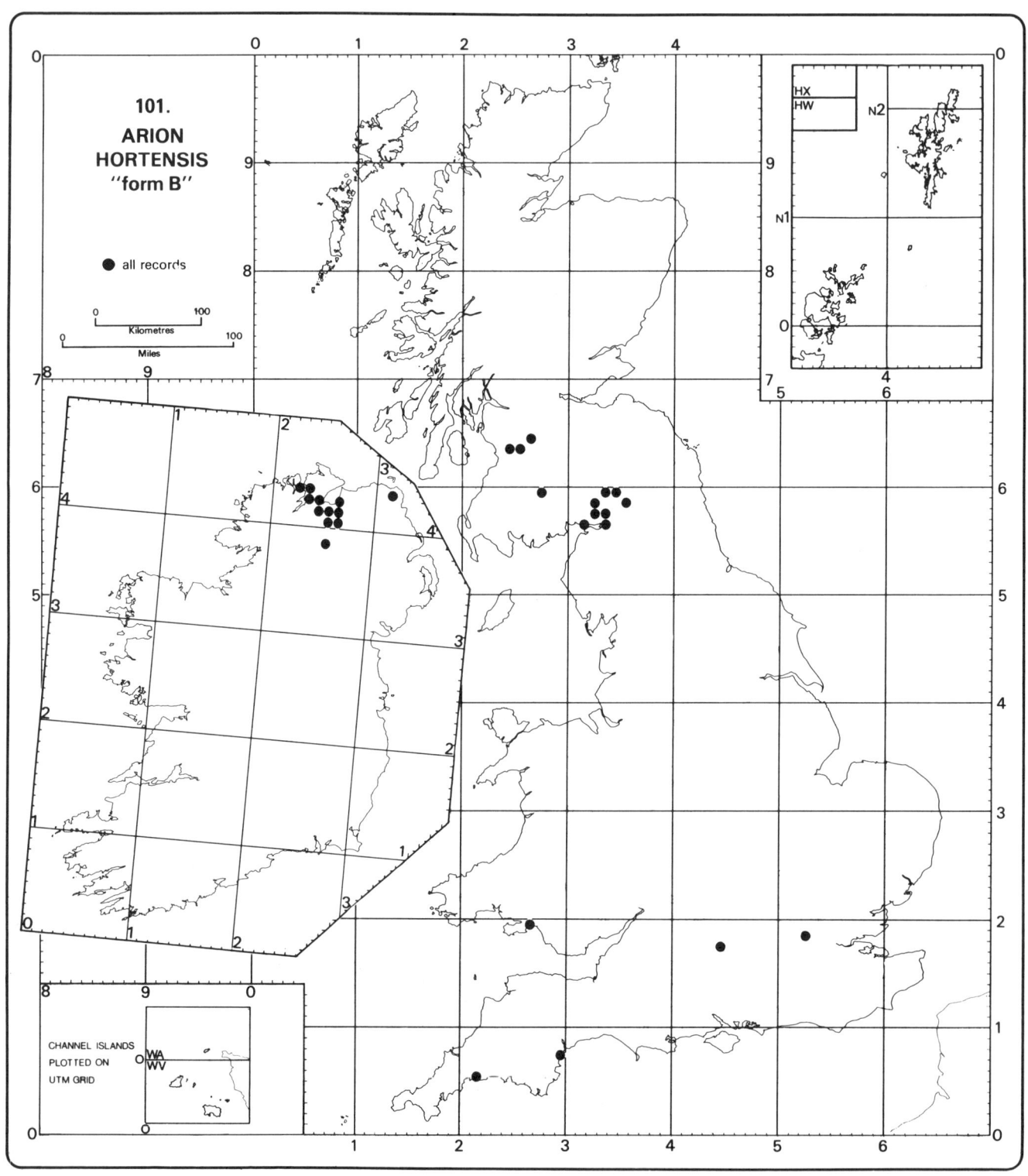

101.
ARION
HORTENSIS
"form B"

● all records

Kilometres
Miles

CHANNEL ISLANDS
PLOTTED ON
UTM GRID

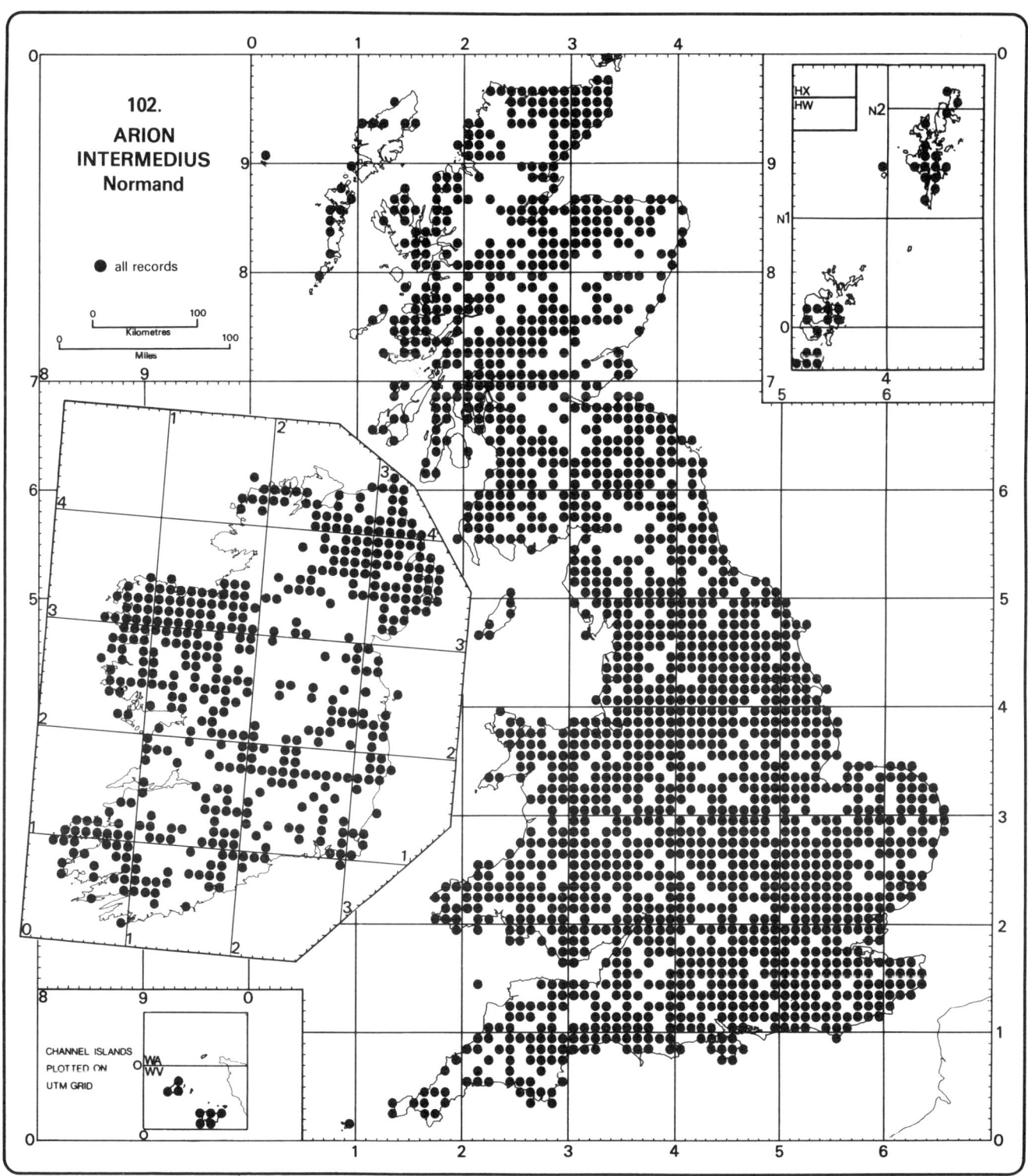

102.
ARION
INTERMEDIUS
Normand

● all records

Kilometres

Miles

CHANNEL ISLANDS
PLOTTED ON
UTM GRID

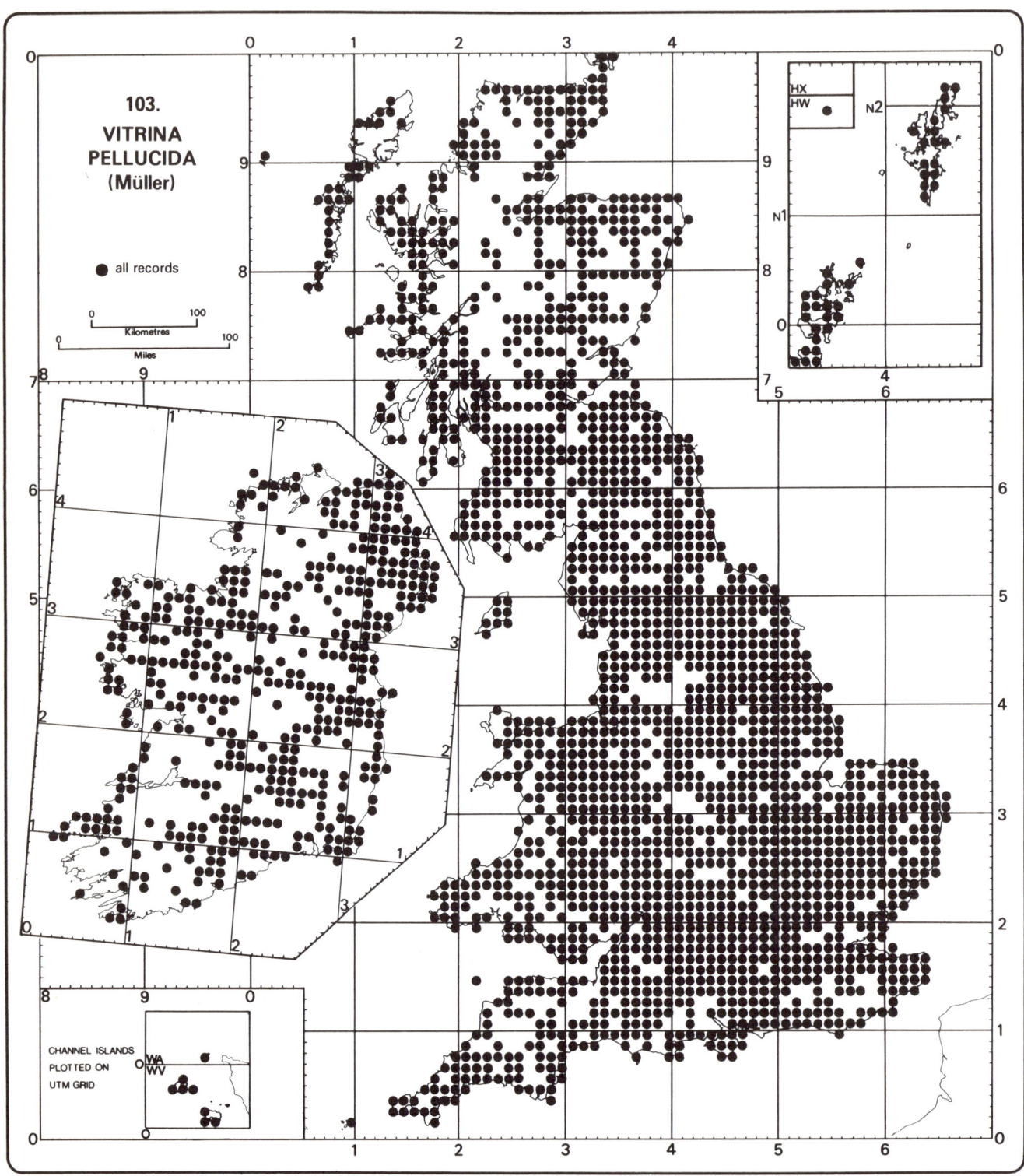

103.
VITRINA
PELLUCIDA
(Müller)

● all records

Kilometres

Miles

HX
HW ●

CHANNEL ISLANDS
PLOTTED ON
UTM GRID

WA
WV

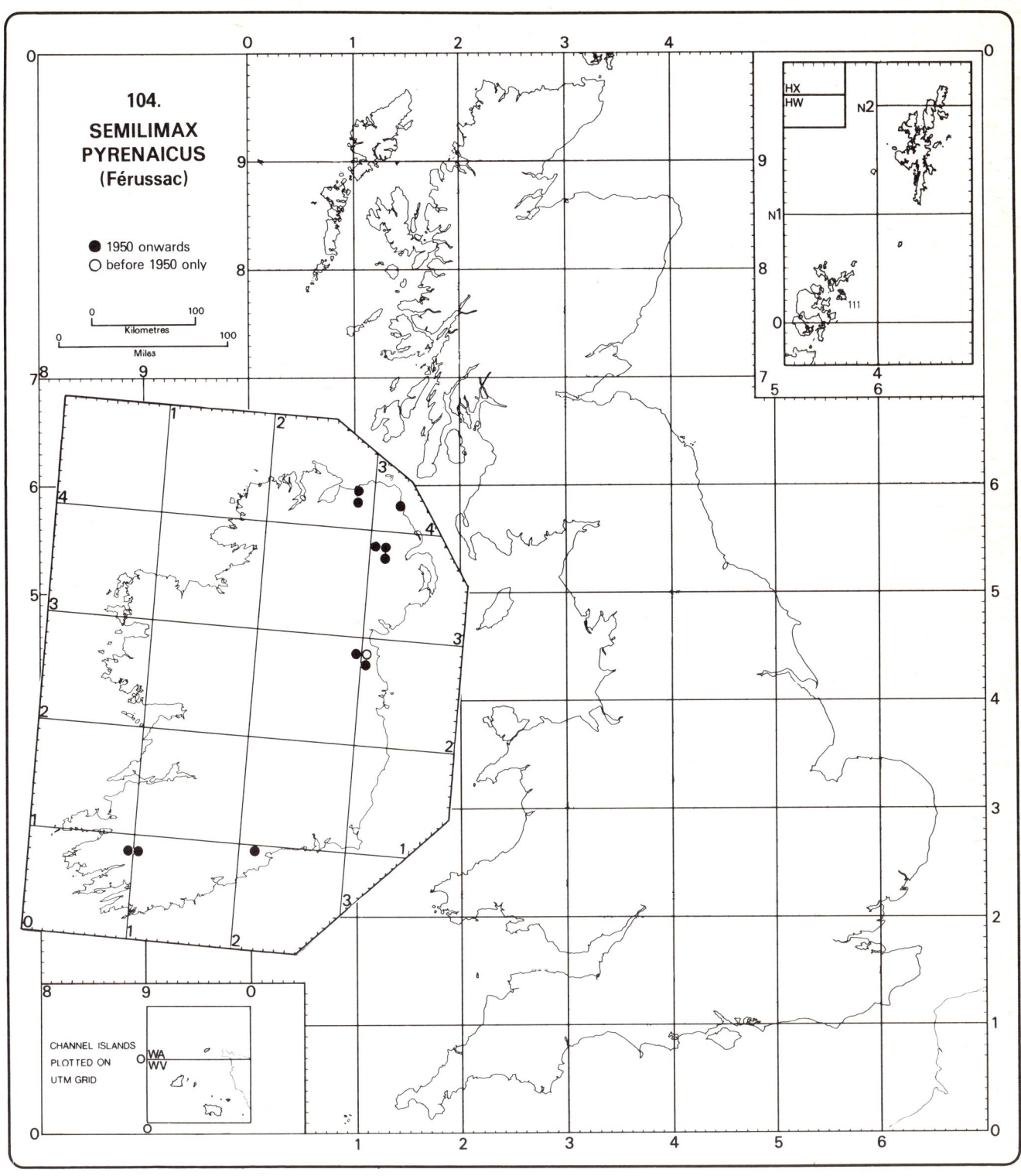

104.
SEMILIMAX
PYRENAICUS
(Férussac)

● 1950 onwards
○ before 1950 only

0 — 100
Kilometres
0 — 100
Miles

CHANNEL ISLANDS
PLOTTED ON
UTM GRID

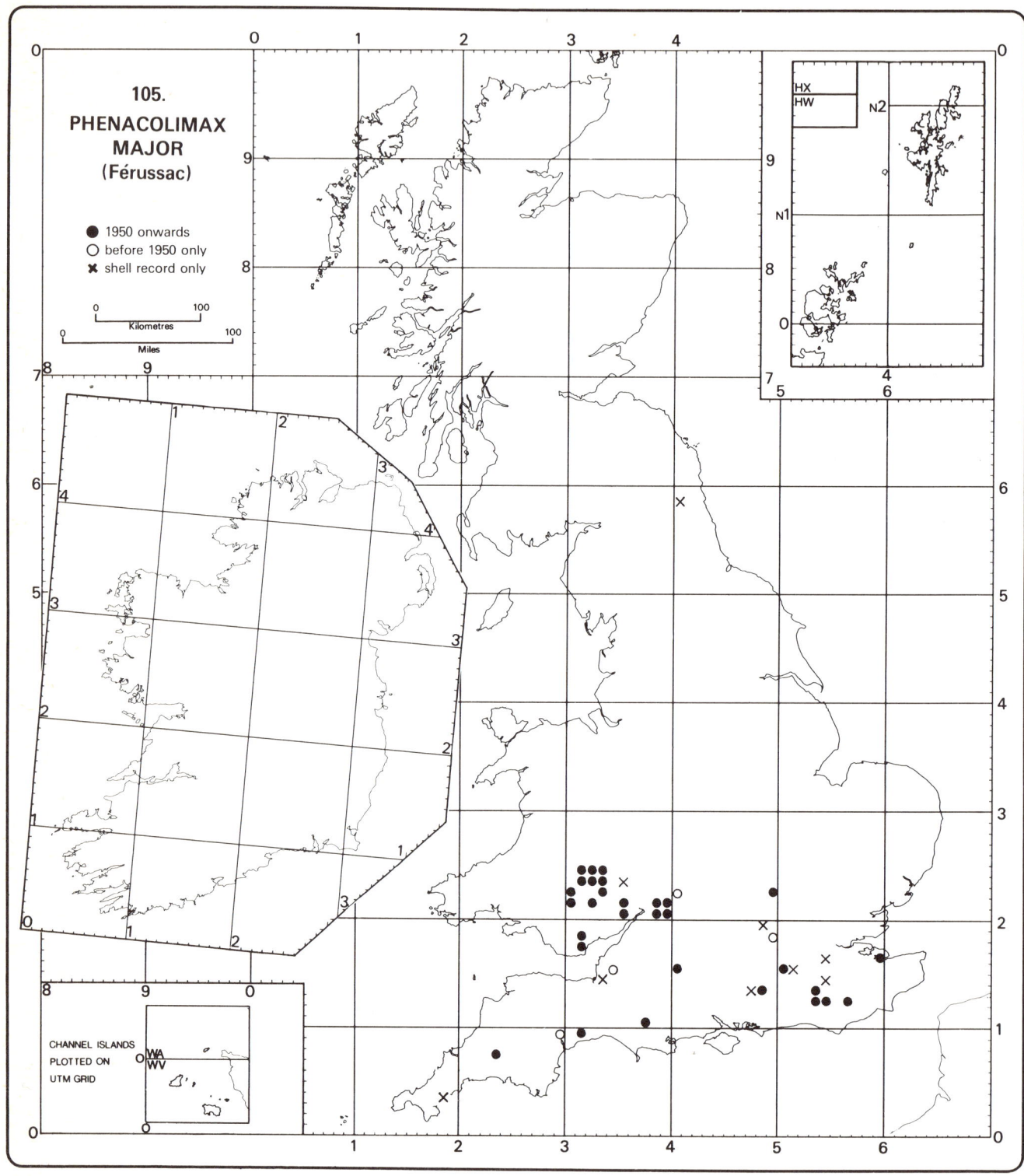

105.
PHENACOLIMAX
MAJOR
(Férussac)

● 1950 onwards
○ before 1950 only
✘ shell record only

0   100
Kilometres
0   100
Miles

CHANNEL ISLANDS
PLOTTED ON
UTM GRID

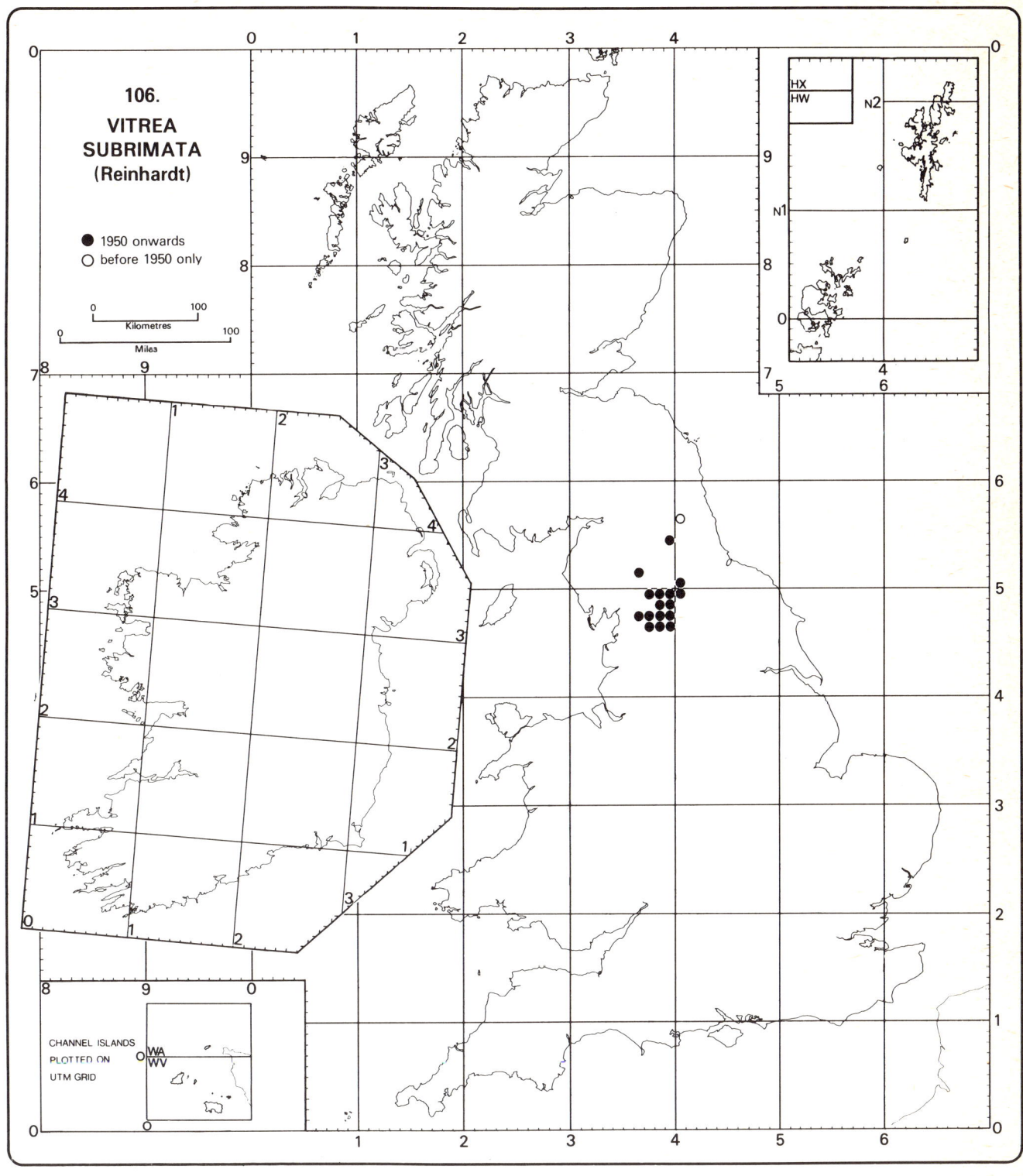

106.
VITREA
SUBRIMATA
(Reinhardt)

● 1950 onwards
○ before 1950 only

0   100
Kilometres
0   100
Miles

CHANNEL ISLANDS
PLOTTED ON
UTM GRID

WA
WV

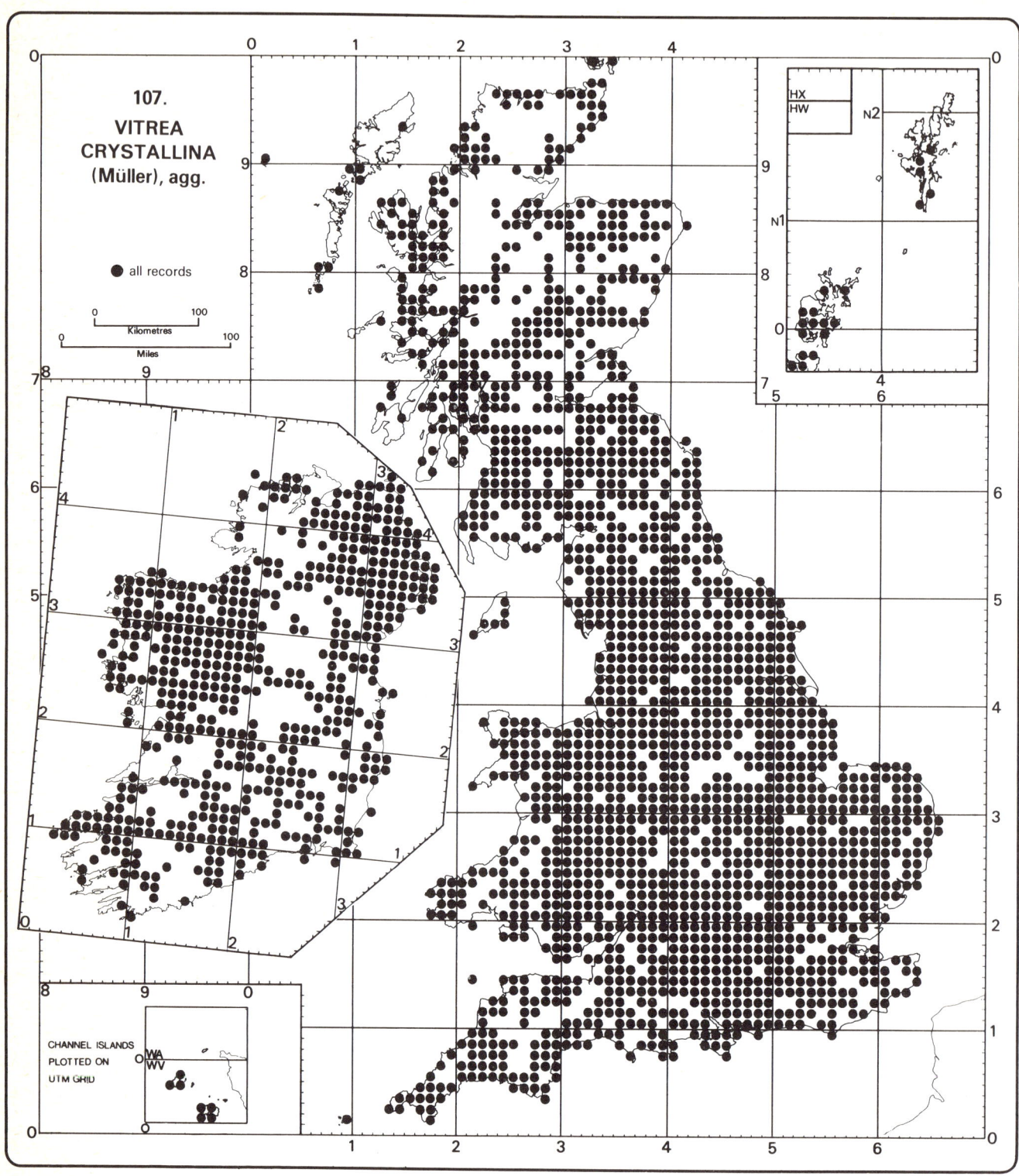

107.
VITREA
CRYSTALLINA
(Müller), agg.

● all records

0   100
Kilometres
0   100
Miles

CHANNEL ISLANDS
PLOTTED ON
UTM GRID

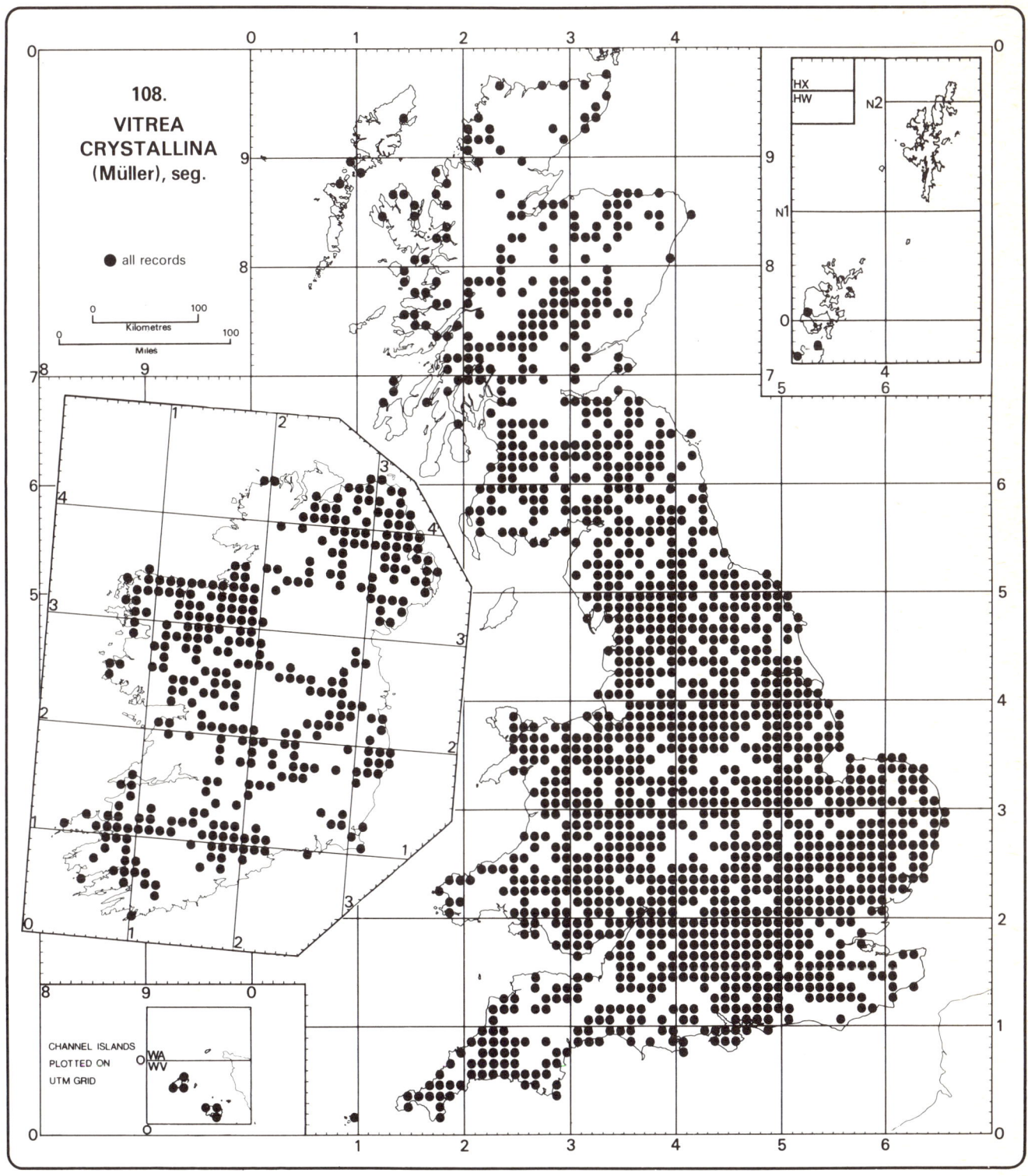

108.
VITREA
CRYSTALLINA
(Müller), seg.

● all records

Kilometres

Miles

CHANNEL ISLANDS
PLOTTED ON
UTM GRID

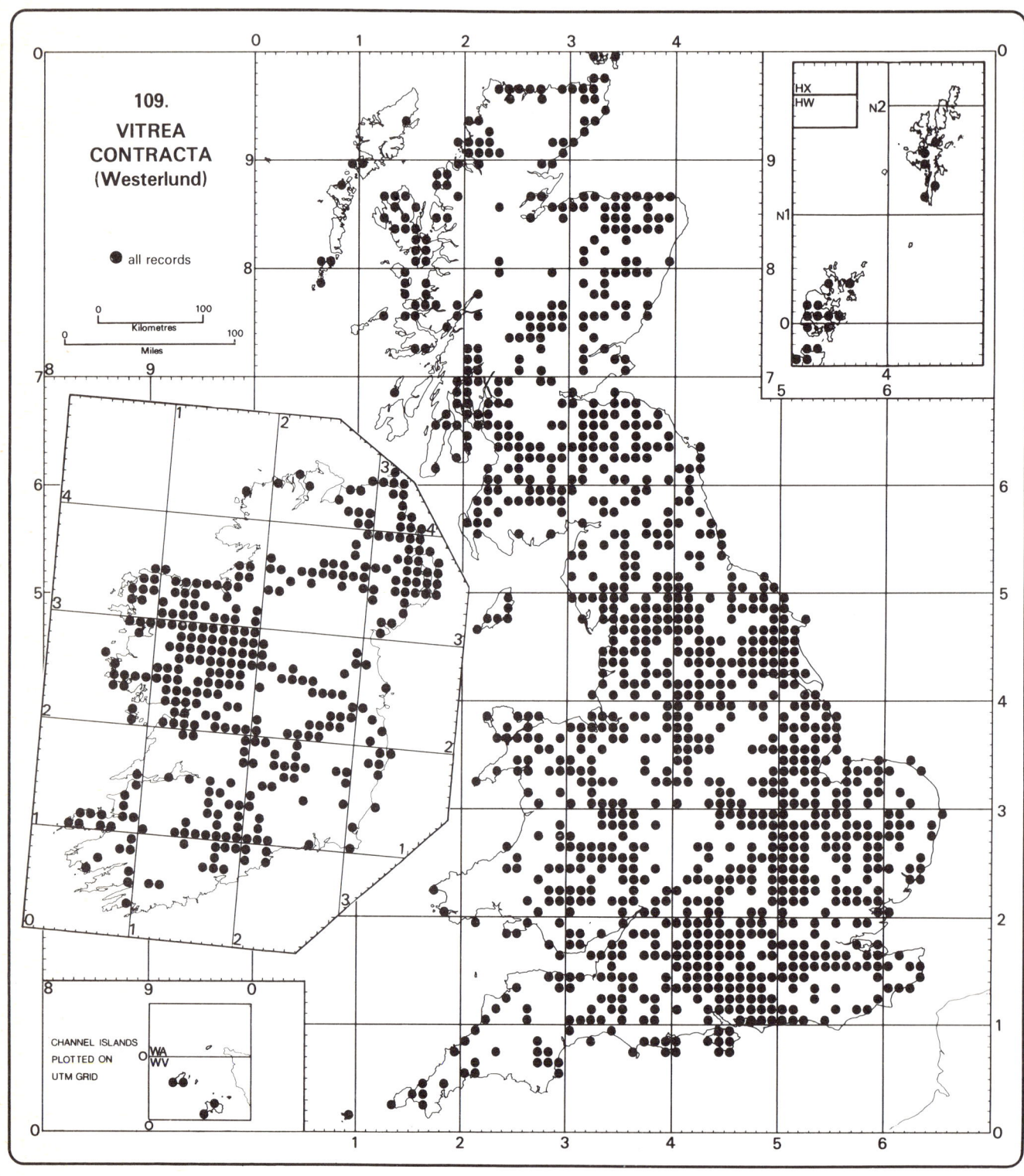

109.
VITREA
CONTRACTA
(Westerlund)

● all records

0 _____ 100
Kilometres
0 _____ 100
Miles

CHANNEL ISLANDS
PLOTTED ON
UTM GRID

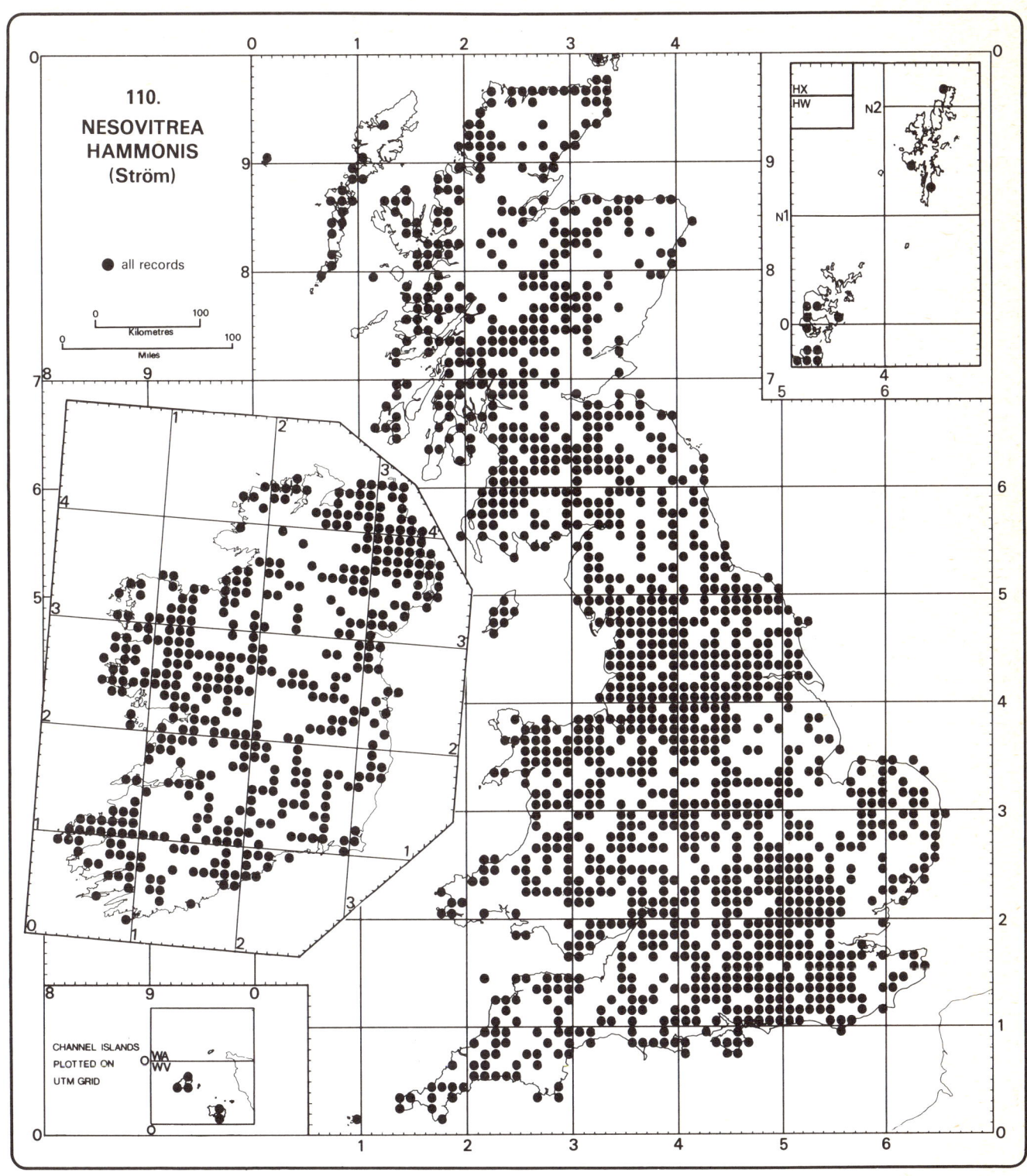

110.
NESOVITREA
HAMMONIS
(Ström)

● all records

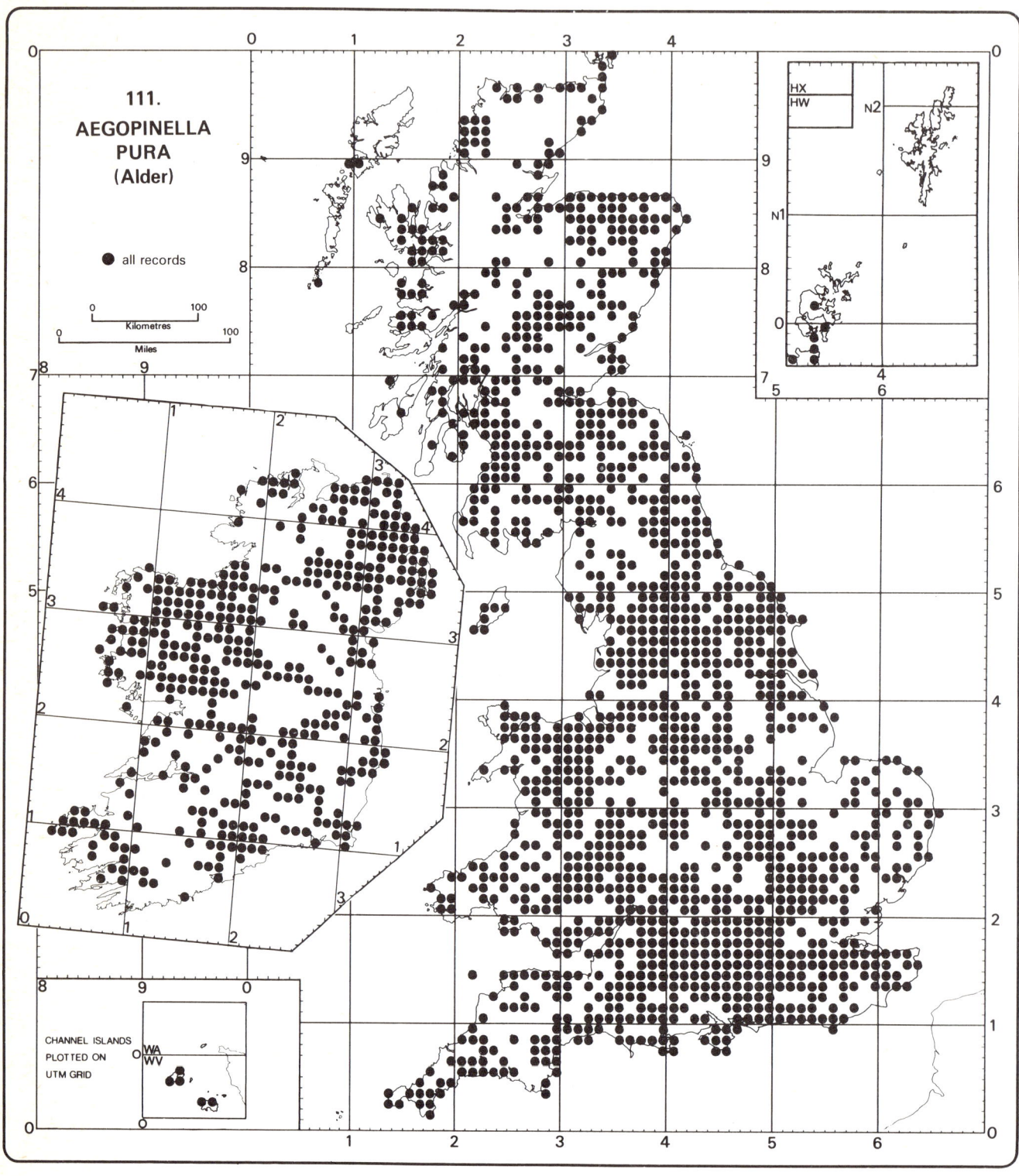

111.
AEGOPINELLA
PURA
(Alder)

● all records

Kilometres
0   100

Miles
0   100

HX
HW
N2
N1

CHANNEL ISLANDS
PLOTTED ON
UTM GRID

WA
WV

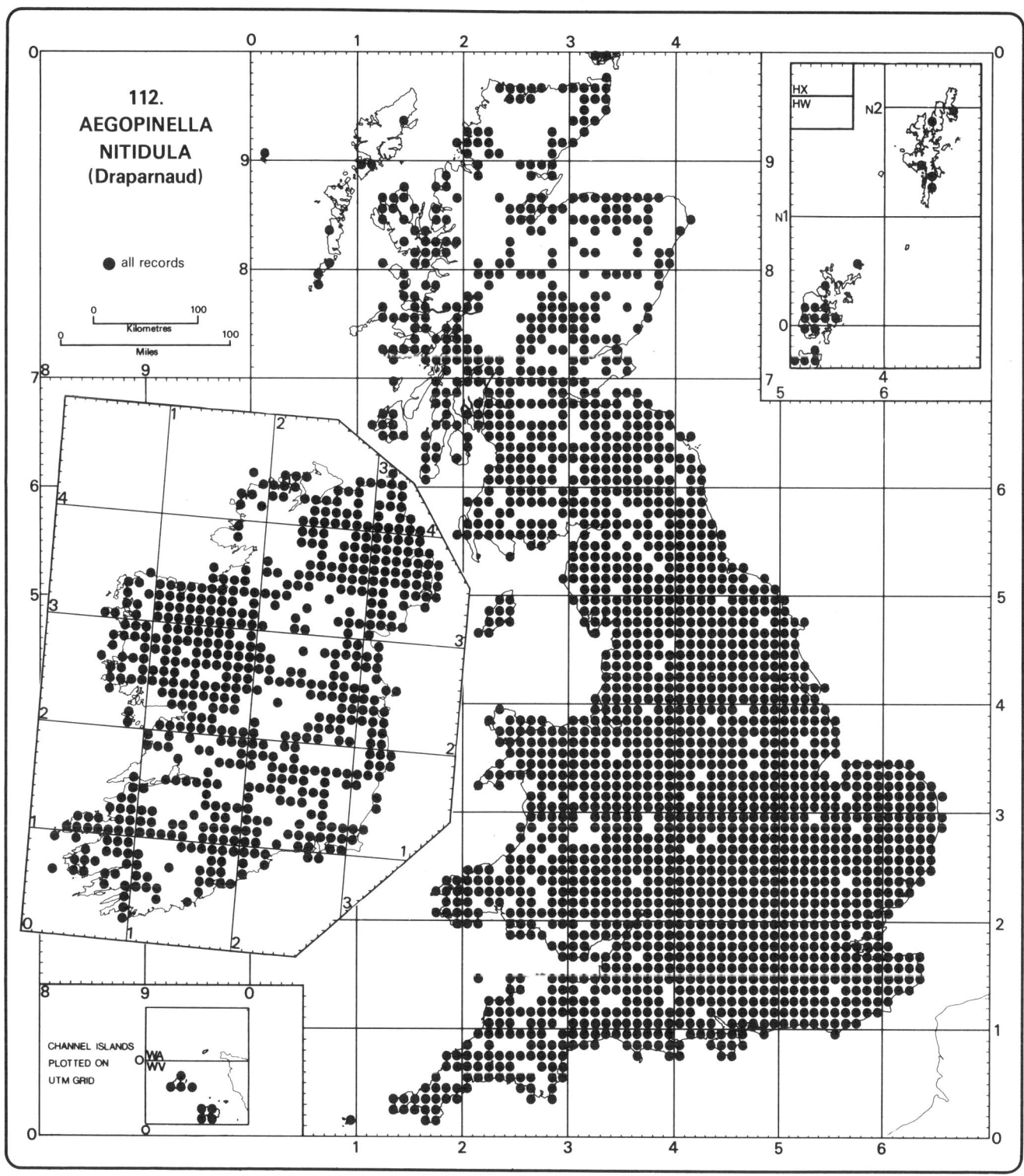

112.
AEGOPINELLA
NITIDULA
(Draparnaud)

● all records

Kilometres

Miles

CHANNEL ISLANDS
PLOTTED ON
UTM GRID

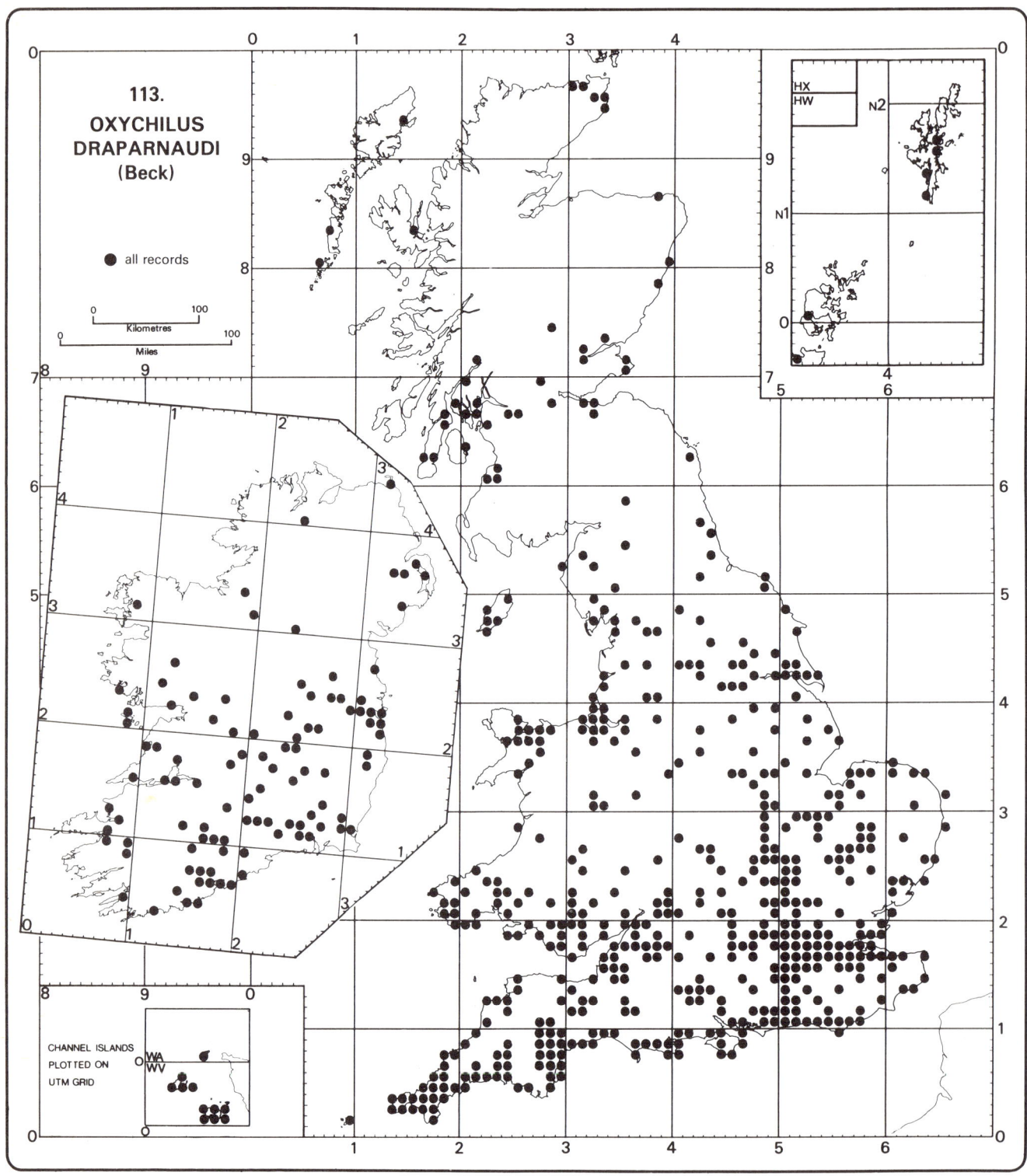

113.
OXYCHILUS
DRAPARNAUDI
(Beck)

● all records

Kilometres

Miles

HX
HW

N2

N1

0

CHANNEL ISLANDS
PLOTTED ON
UTM GRID

WA
WV

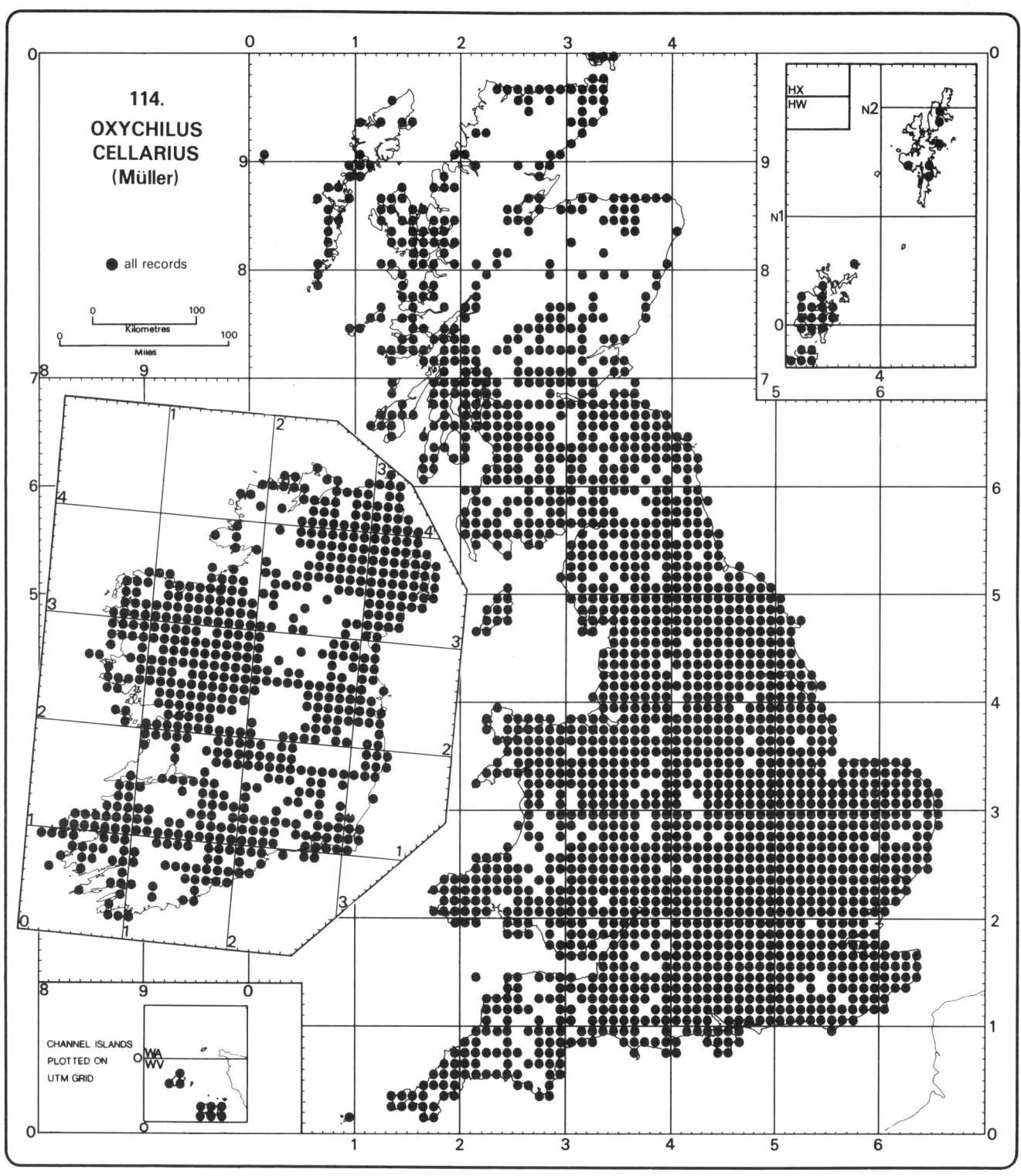

114.
OXYCHILUS
CELLARIUS
(Müller)

● all records

Kilometres

Miles

CHANNEL ISLANDS
PLOTTED ON
UTM GRID

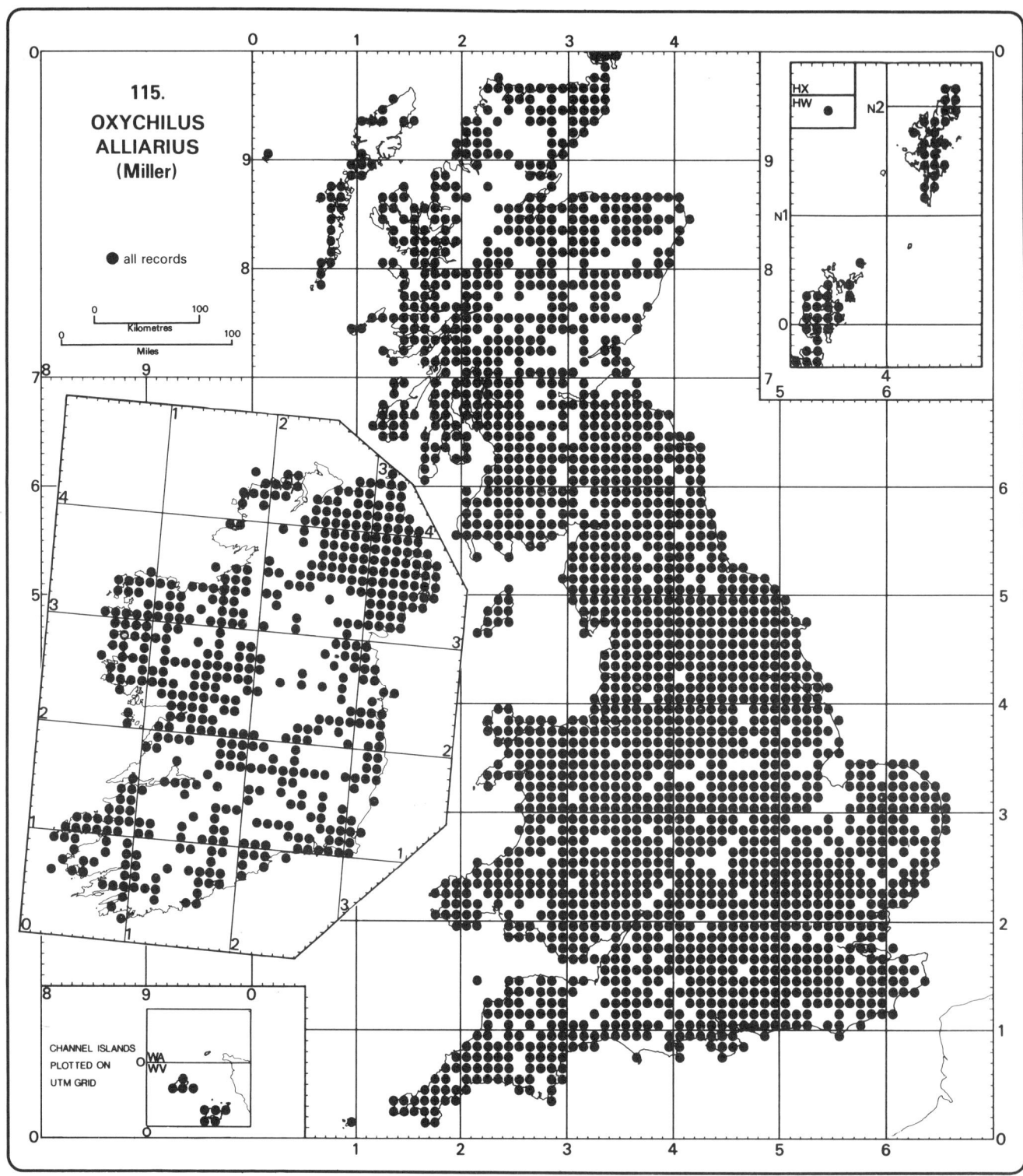

115.
OXYCHILUS
ALLIARIUS
(Miller)

● all records

Kilometres

Miles

HX
HW ●

CHANNEL ISLANDS
PLOTTED ON
UTM GRID

WA
WV

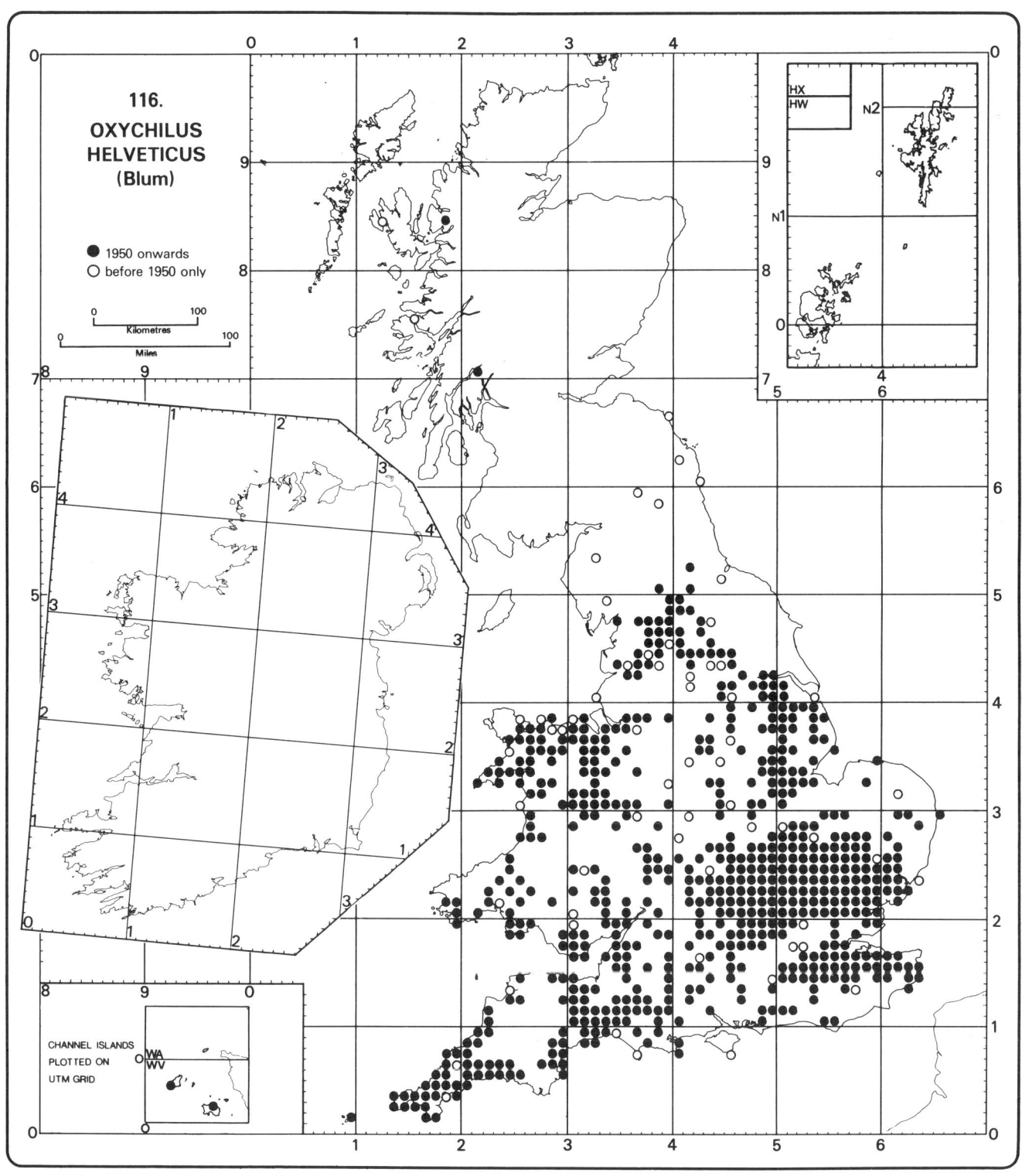

116.
OXYCHILUS
HELVETICUS
(Blum)

● 1950 onwards
○ before 1950 only

0   100
Kilometres
0   100
Miles

HX
HW

N2

N1

CHANNEL ISLANDS
PLOTTED ON
UTM GRID

WA
WV

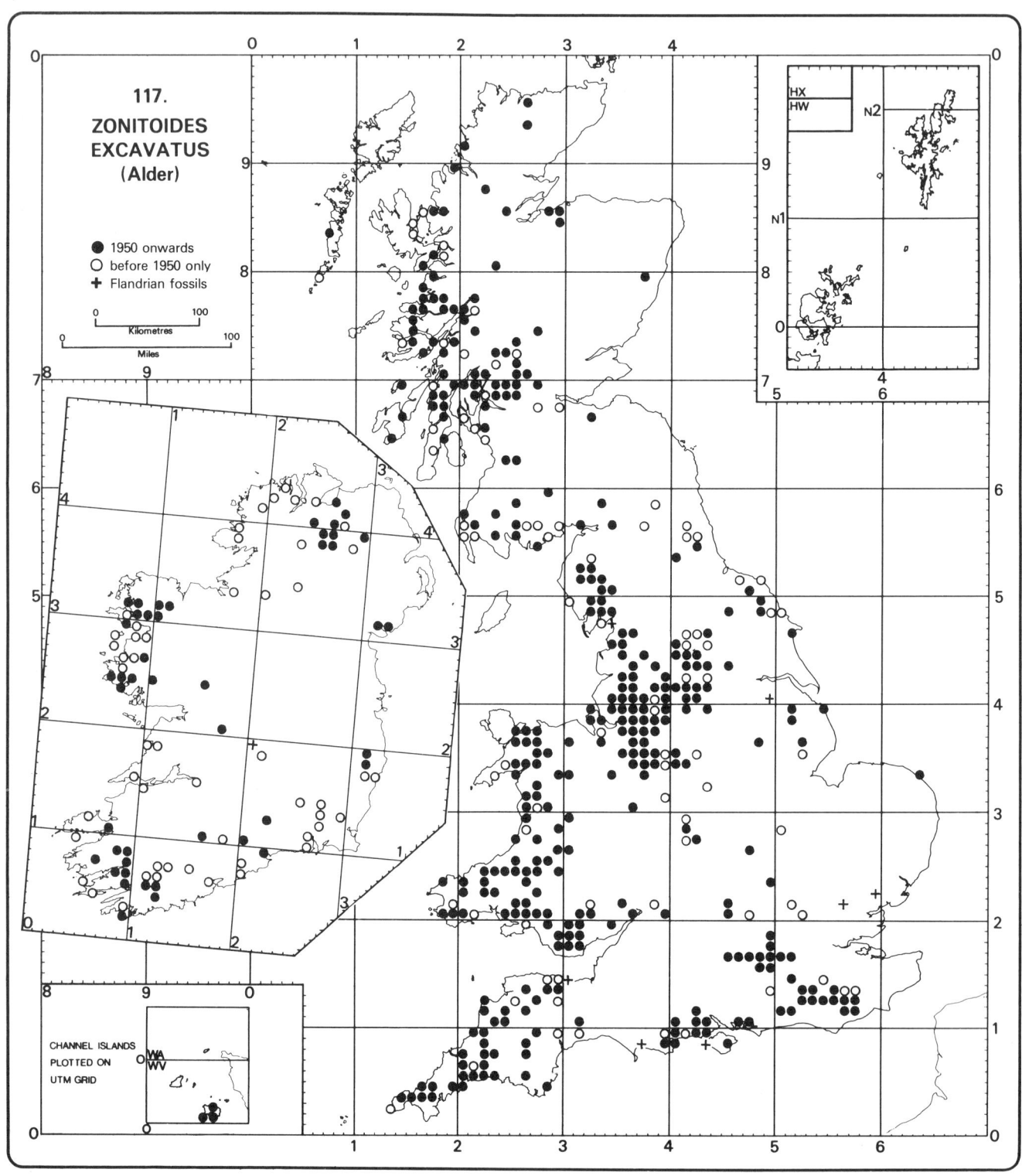

117.
ZONITOIDES
EXCAVATUS
(Alder)

● 1950 onwards
○ before 1950 only
+ Flandrian fossils

0   100
Kilometres
0   100
Miles

CHANNEL ISLANDS
PLOTTED ON
UTM GRID

WA
WV

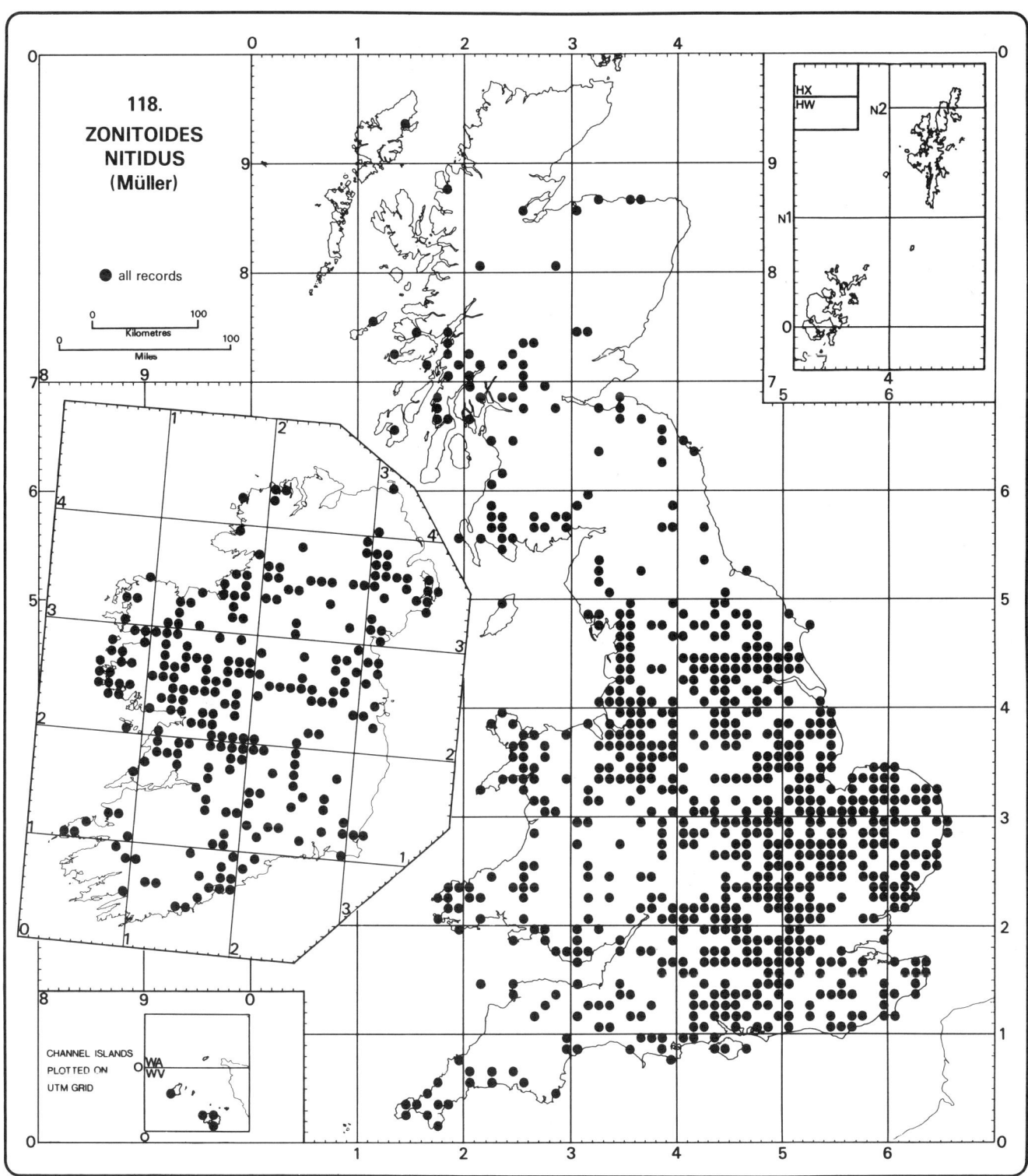

118.
ZONITOIDES
NITIDUS
(Müller)

● all records

Kilometres

Miles

CHANNEL ISLANDS
PLOTTED ON
UTM GRID

WA
WV

HX
HW

N2

N1

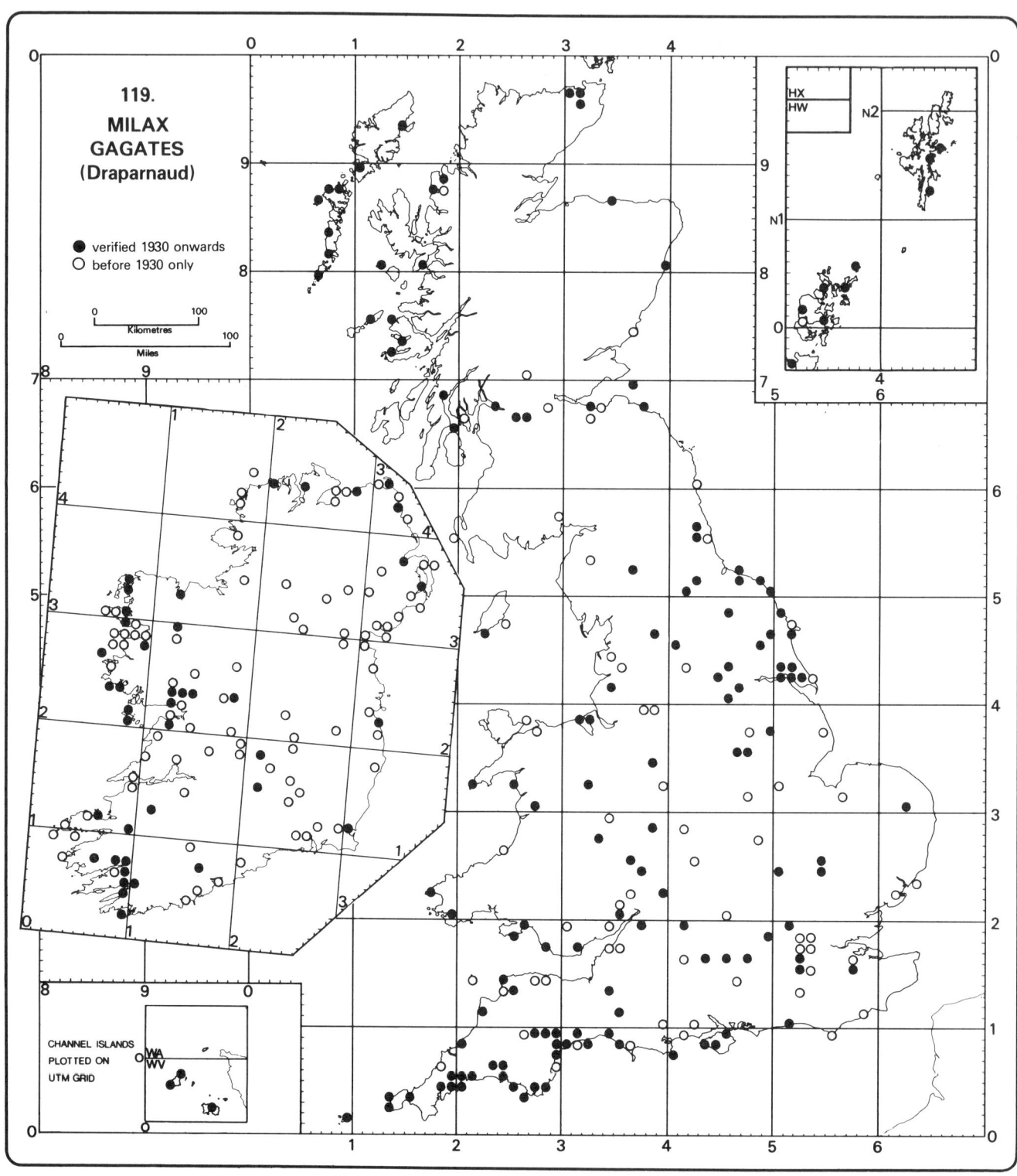

119.

MILAX
GAGATES
(Draparnaud)

● verified 1930 onwards
○ before 1930 only

Kilometres

Miles

CHANNEL ISLANDS
PLOTTED ON
UTM GRID

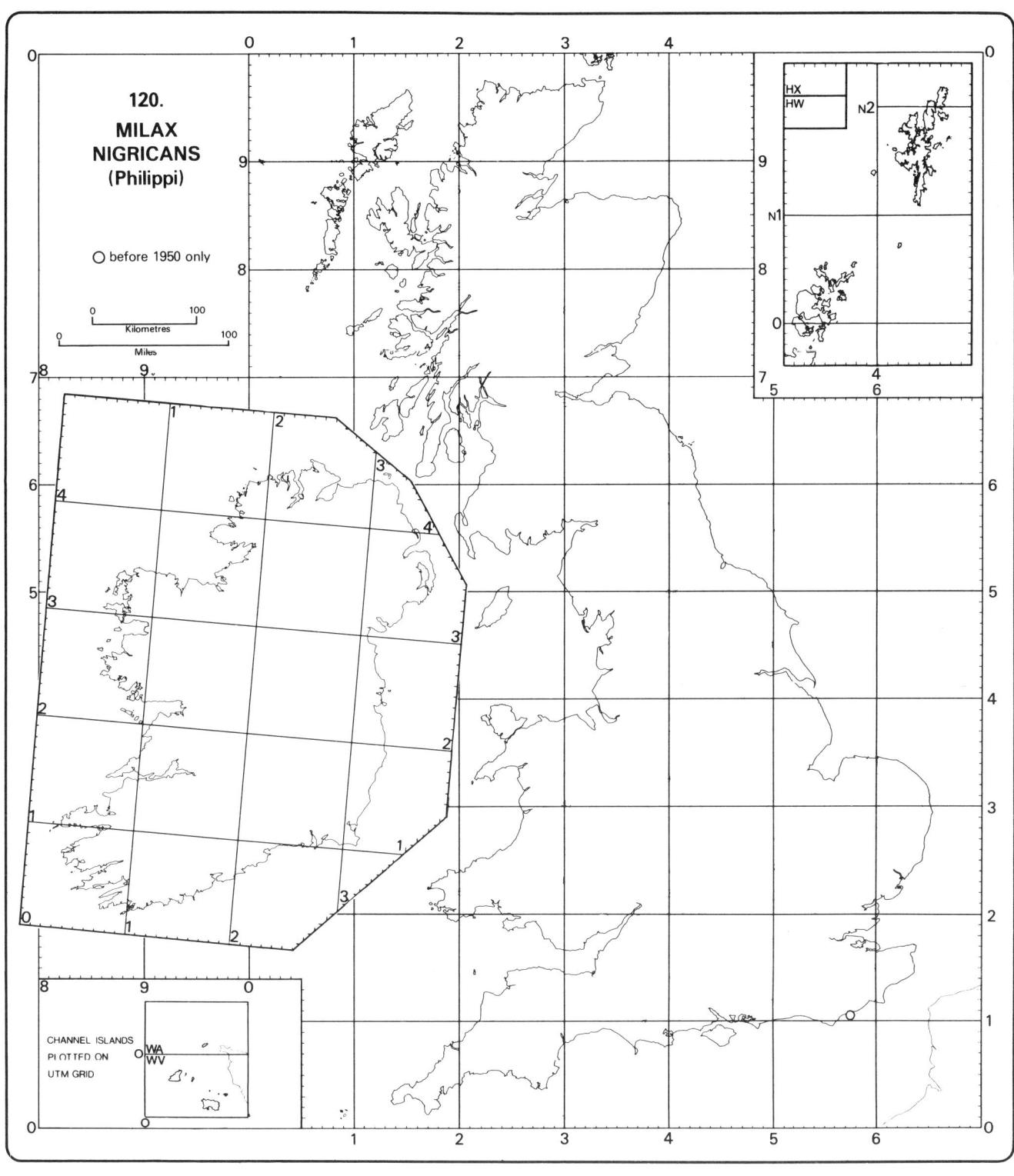

120.

MILAX
NIGRICANS
(Philippi)

○ before 1950 only

Kilometres

Miles

CHANNEL ISLANDS
PLOTTED ON
UTM GRID

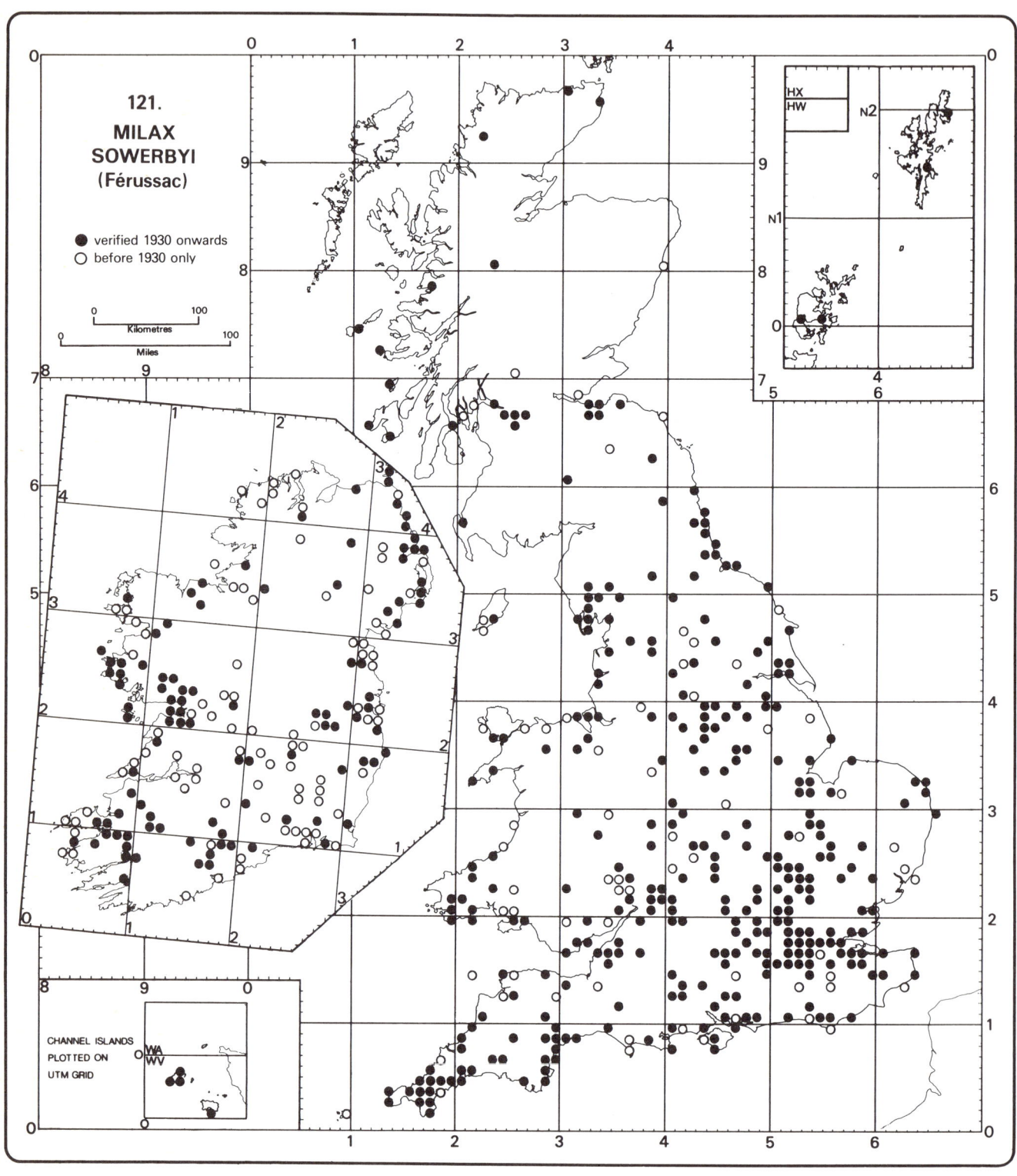

121.
MILAX
SOWERBYI
(Férussac)

● verified 1930 onwards
○ before 1930 only

0   100
Kilometres
0   100
Miles

CHANNEL ISLANDS
PLOTTED ON
UTM GRID

WA
WV

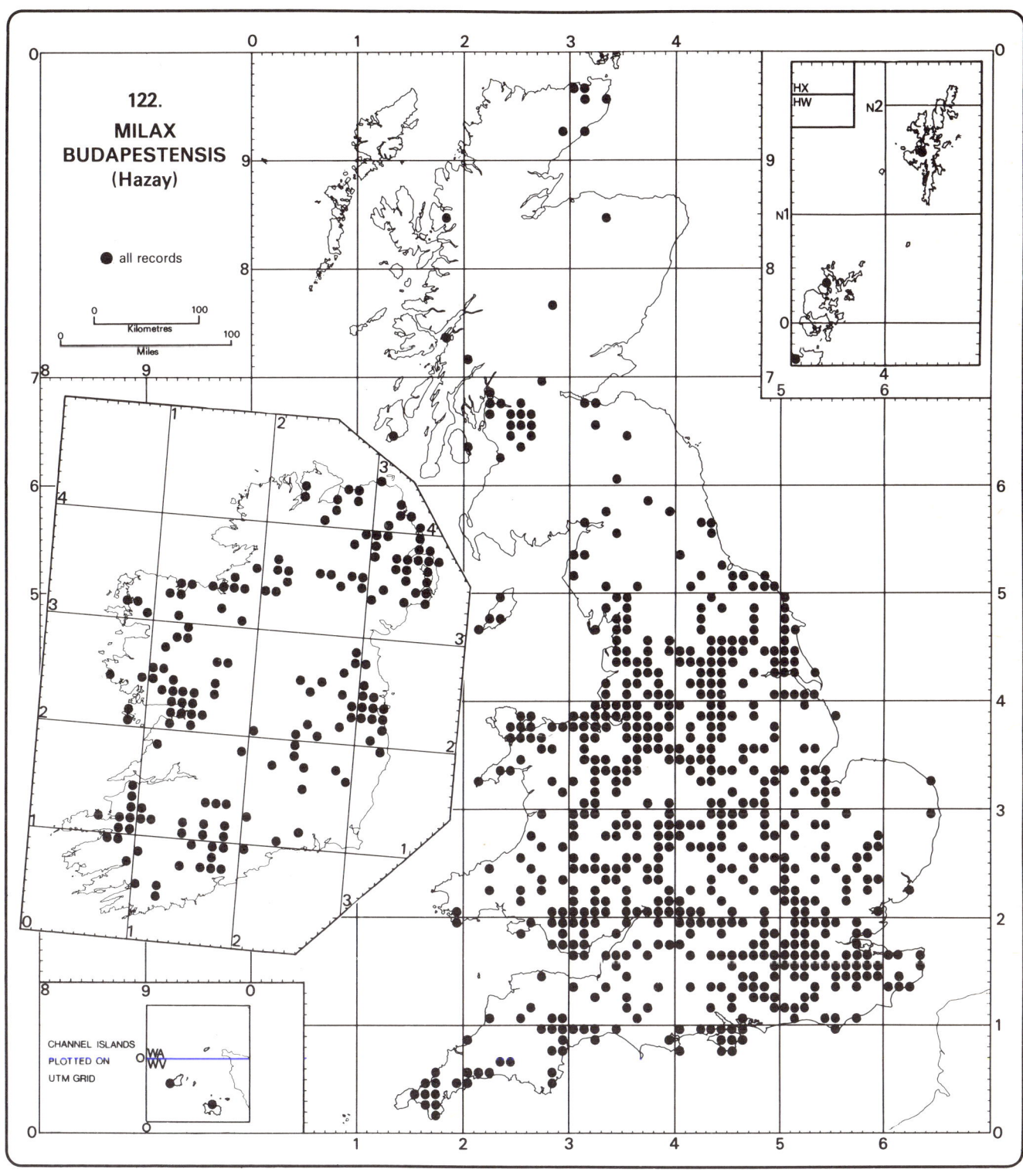

122.

MILAX
BUDAPESTENSIS
(Hazay)

● all records

Kilometres
Miles

HX
HW

CHANNEL ISLANDS
PLOTTED ON
UTM GRID

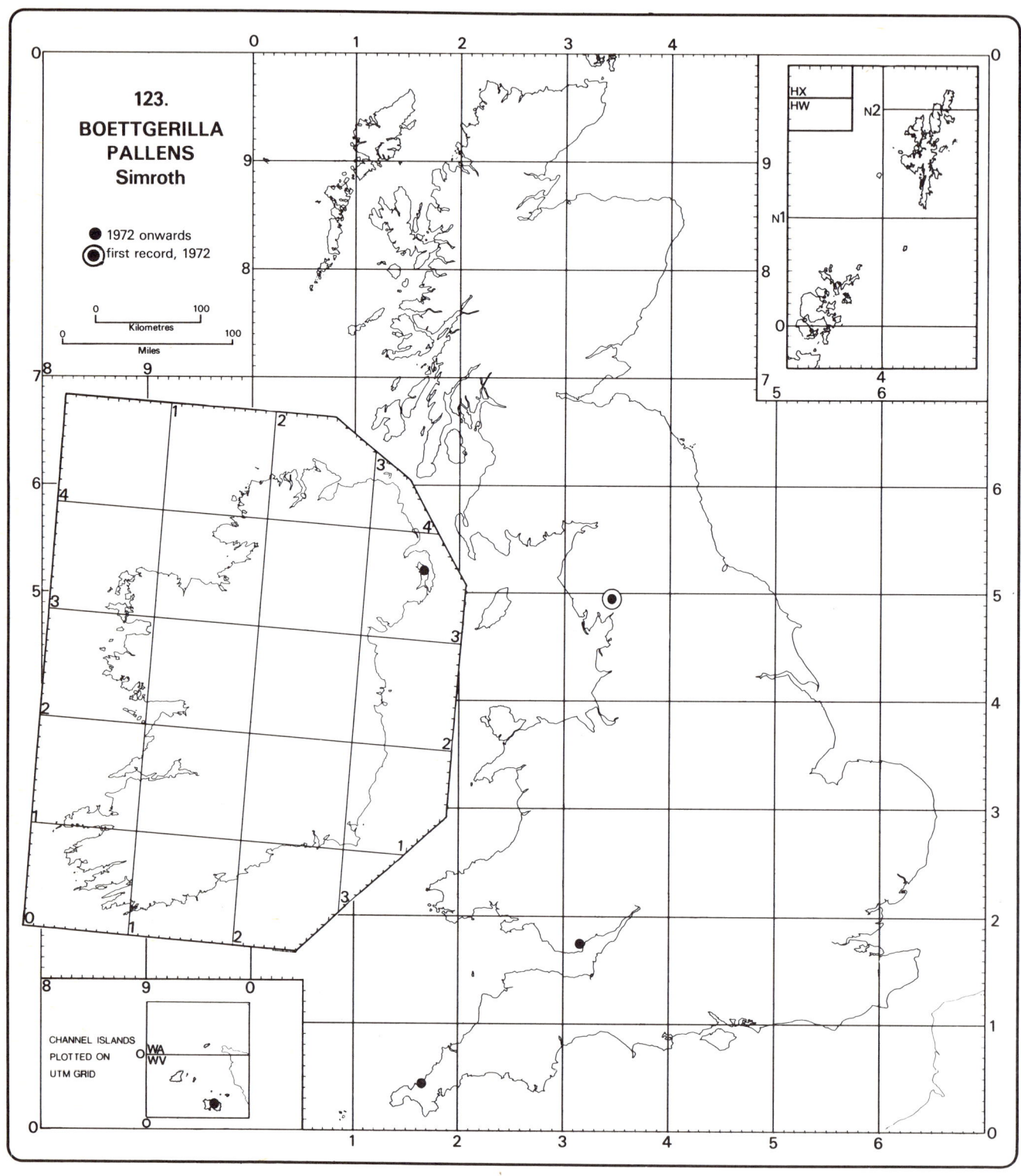

123.
**BOETTGERILLA
PALLENS**
Simroth

● 1972 onwards
◉ first record, 1972

Kilometres

Miles

CHANNEL ISLANDS
PLOTTED ON
UTM GRID

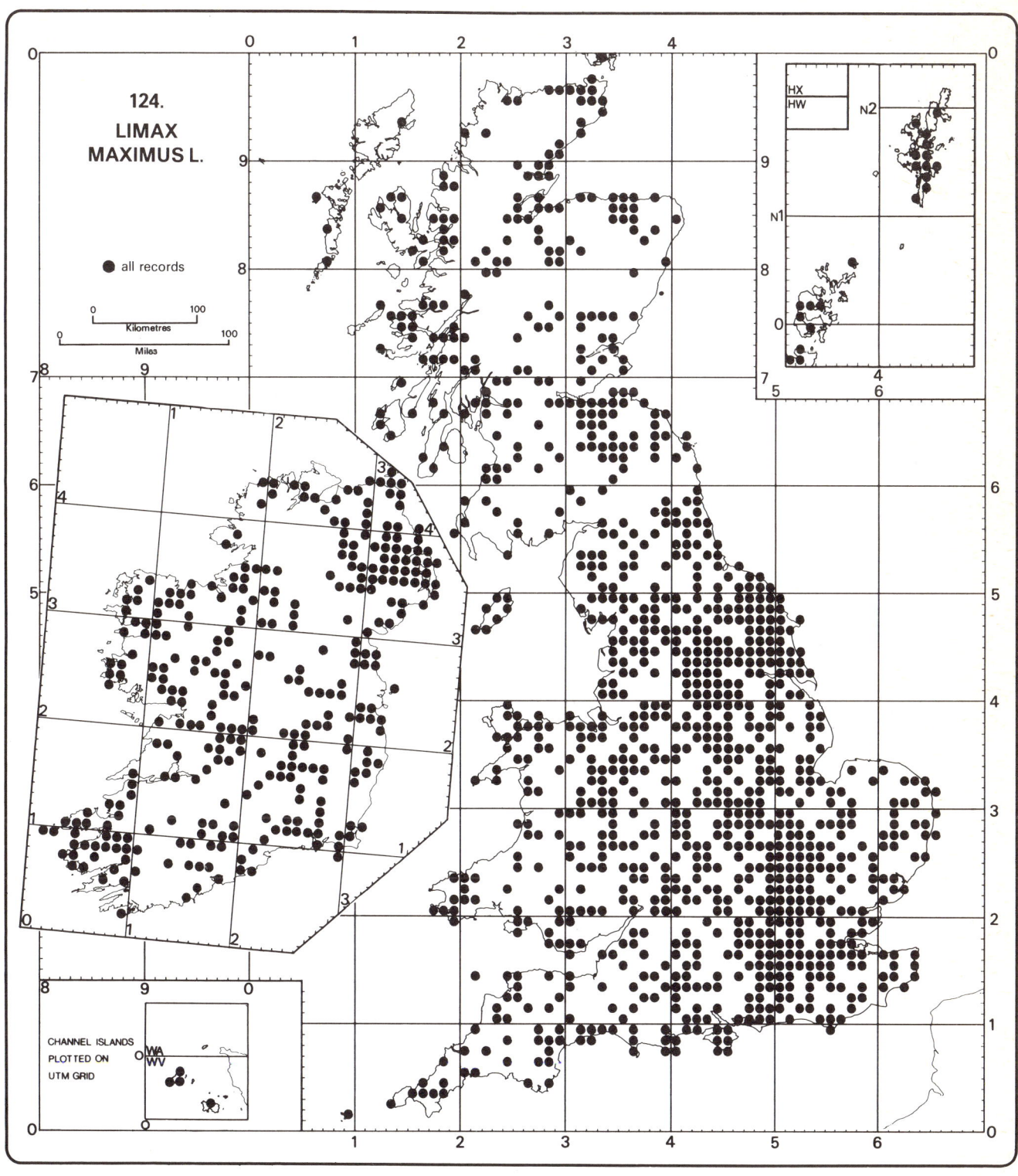

124.
LIMAX
MAXIMUS L.

● all records

Kilometres

Miles

CHANNEL ISLANDS
PLOTTED ON
UTM GRID

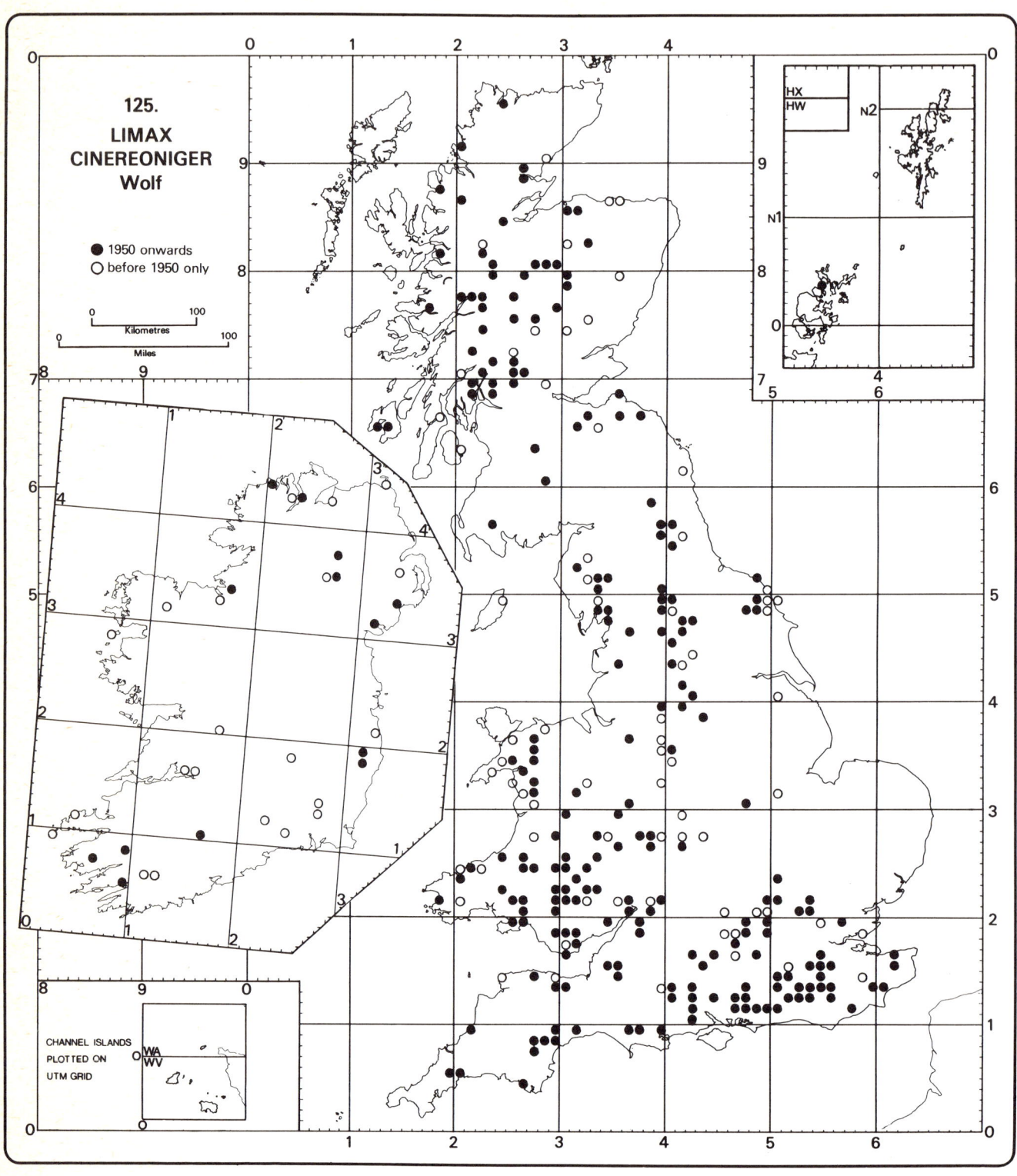

125.
LIMAX
CINEREONIGER
Wolf

● 1950 onwards
○ before 1950 only

Kilometres

Miles

CHANNEL ISLANDS
PLOTTED ON
UTM GRID

WA
WV

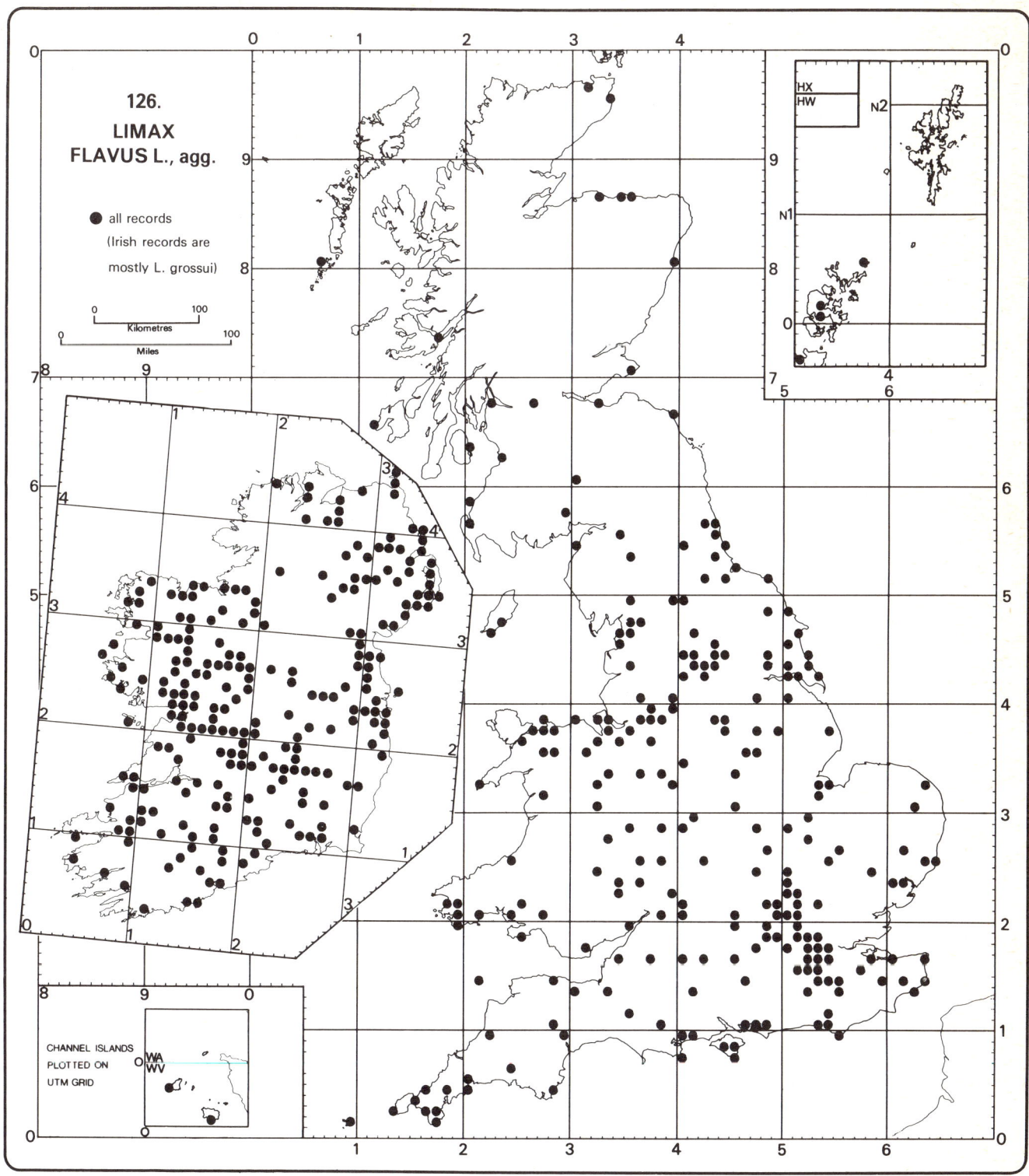

126.

LIMAX
FLAVUS L., agg.

● all records
(Irish records are
mostly L. grossui)

Kilometres

Miles

CHANNEL ISLANDS
PLOTTED ON
UTM GRID

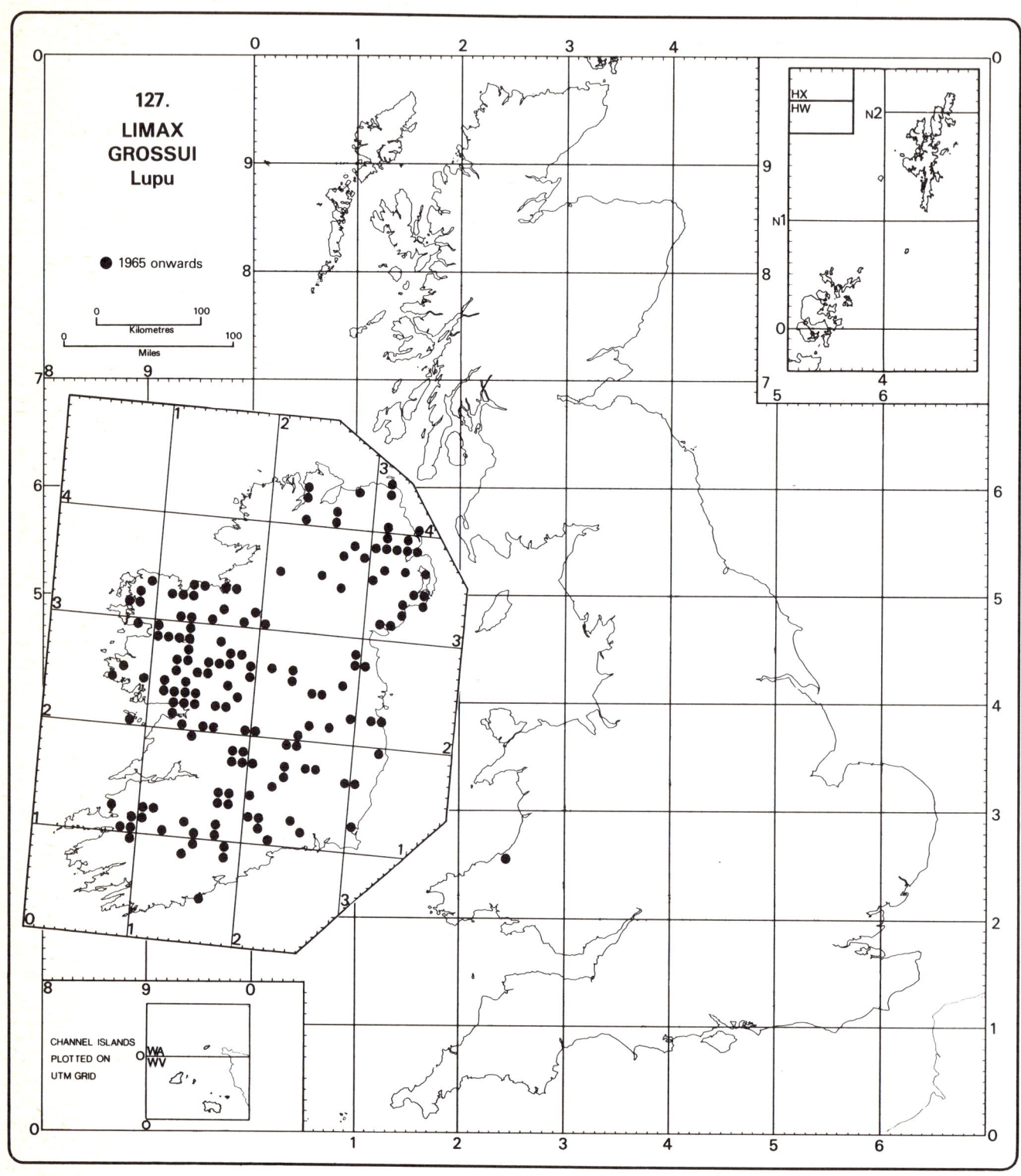

127.
LIMAX
GROSSUI
Lupu

● 1965 onwards

Kilometres

Miles

CHANNEL ISLANDS
PLOTTED ON
UTM GRID

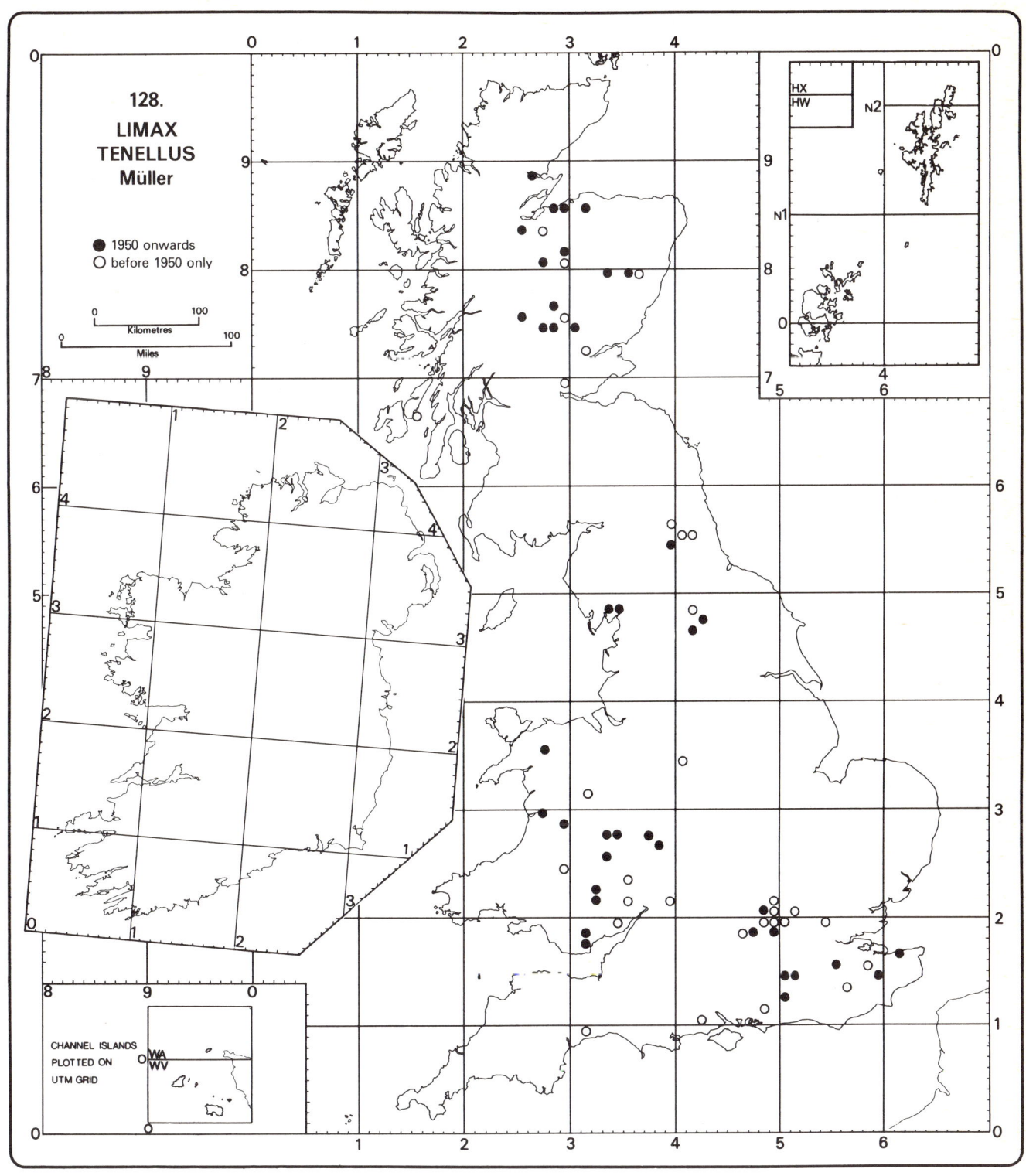

128.
LIMAX
TENELLUS
Müller

● 1950 onwards
○ before 1950 only

Kilometres
Miles

CHANNEL ISLANDS
PLOTTED ON
UTM GRID

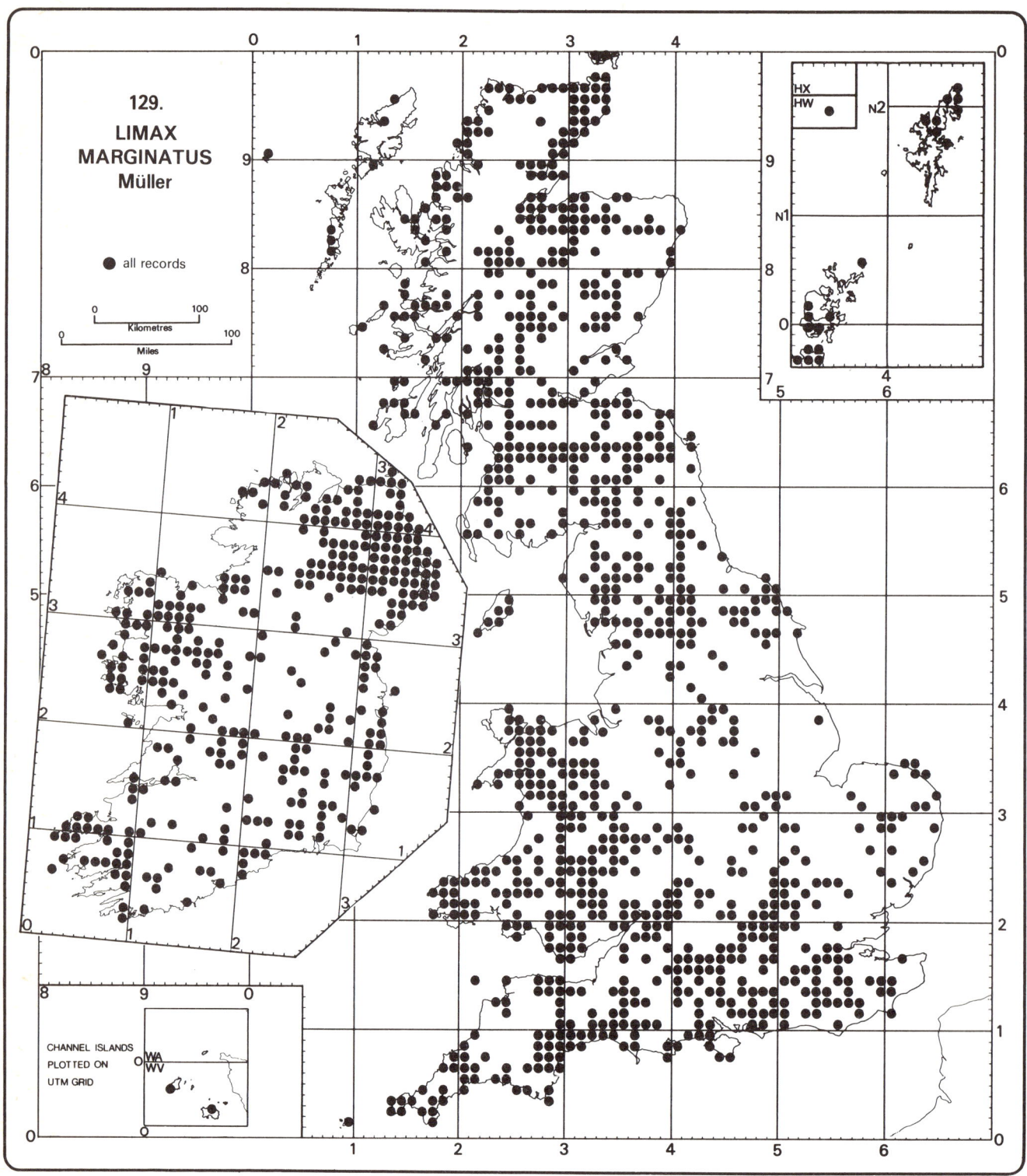

129.
LIMAX
MARGINATUS
Müller

● all records

Kilometres
Miles

CHANNEL ISLANDS
PLOTTED ON
UTM GRID

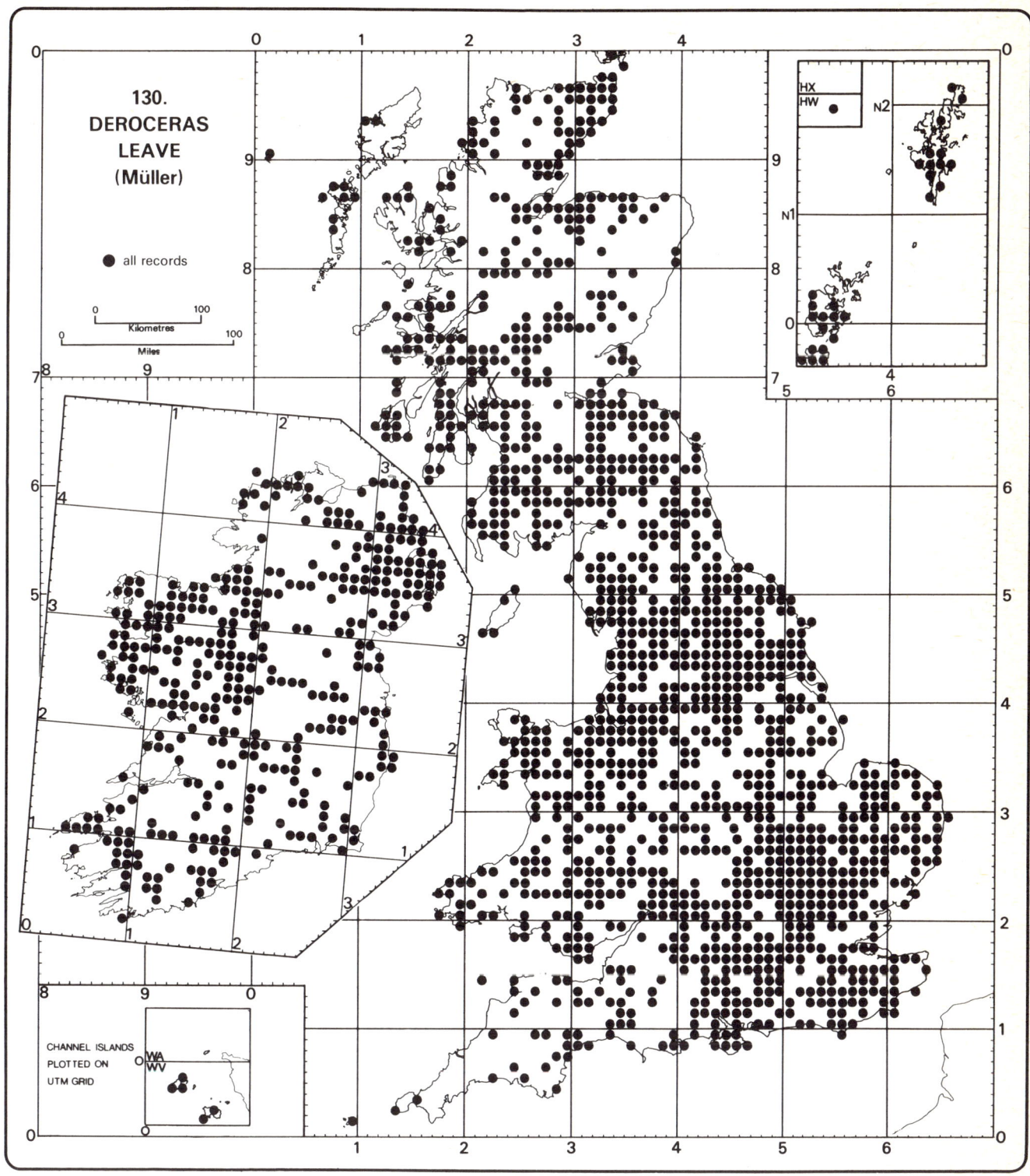

130.
DEROCERAS
LEAVE
(Müller)

● all records

Kilometres

Miles

CHANNEL ISLANDS
PLOTTED ON
UTM GRID

WA
WV

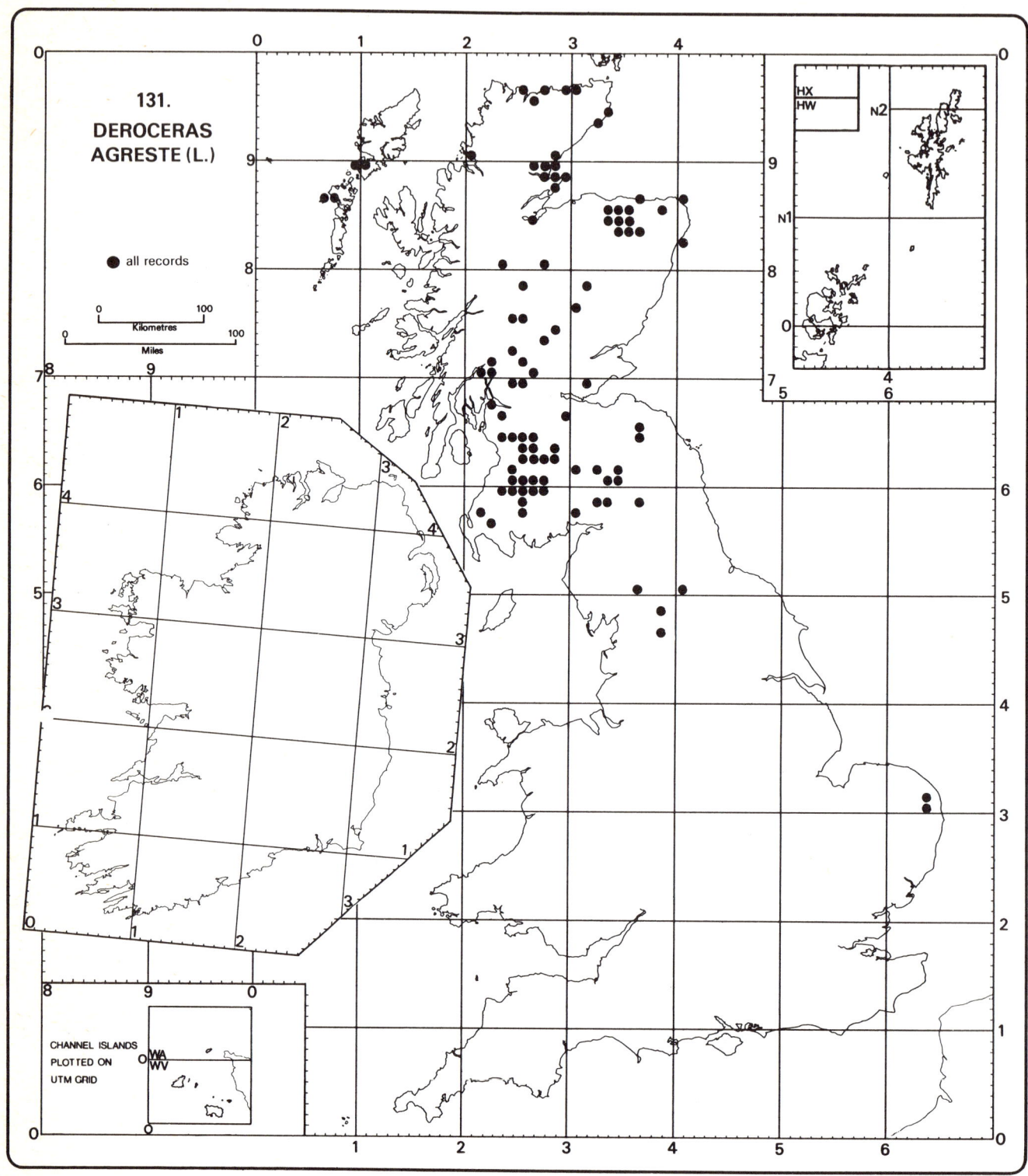

131.
DEROCERAS
AGRESTE (L.)

● all records

0   100
Kilometres
0   100
Miles

CHANNEL ISLANDS
PLOTTED ON
UTM GRID

WA
WV

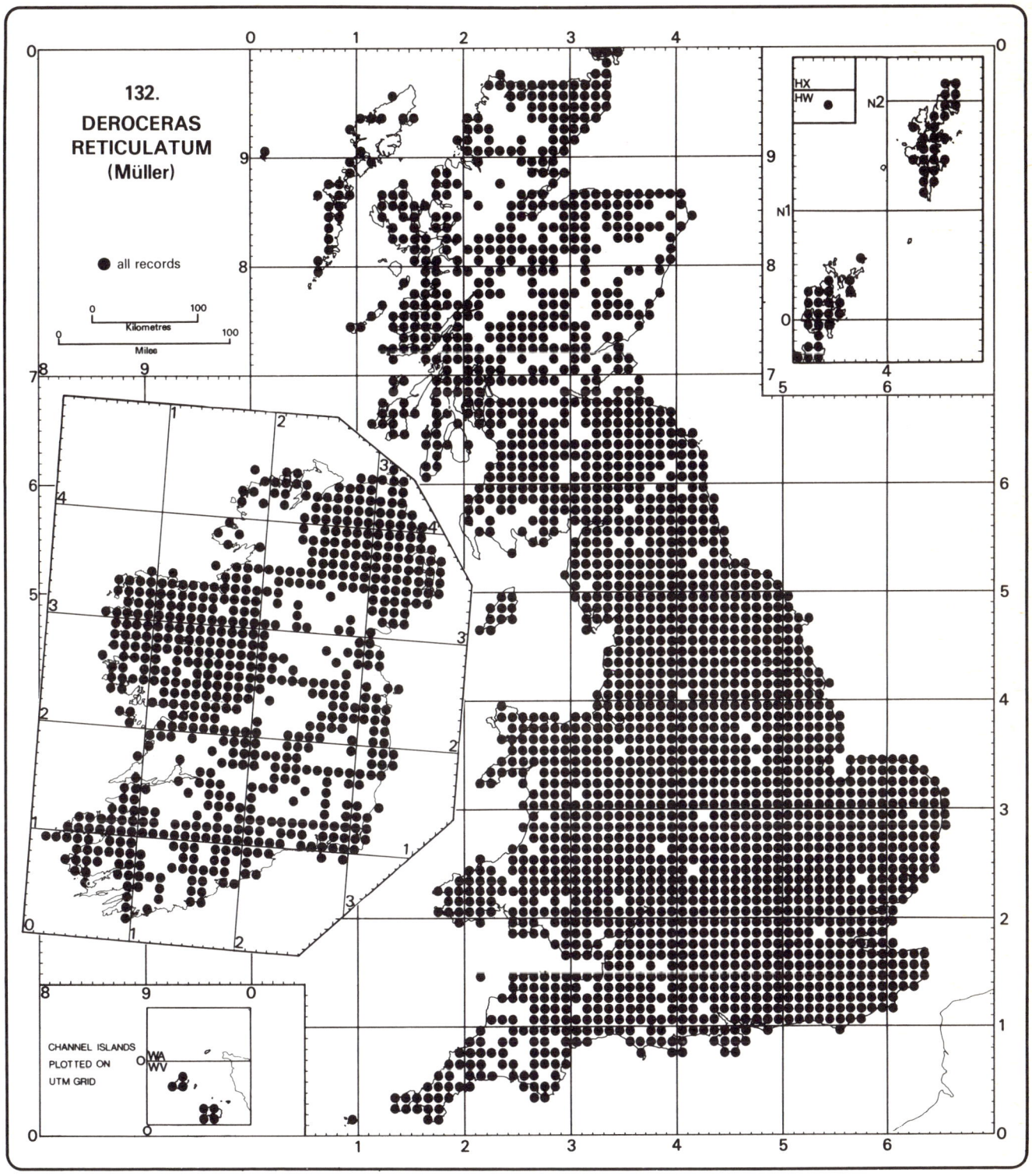

132.
DEROCERAS
RETICULATUM
(Müller)

● all records

Kilometres

Miles

CHANNEL ISLANDS
PLOTTED ON
UTM GRID

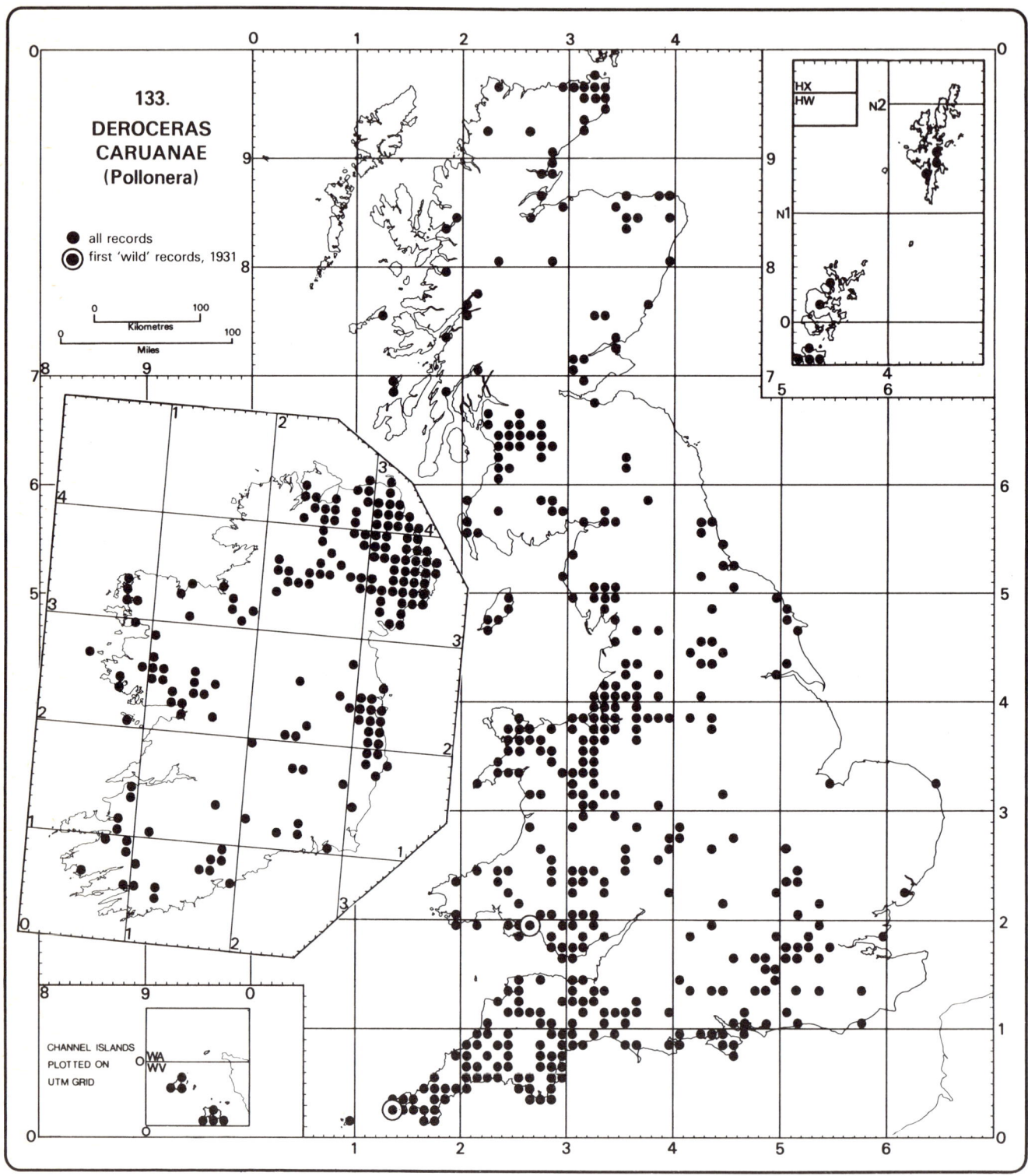

133.
DEROCERAS
CARUANAE
(Pollonera)

● all records
⦿ first 'wild' records, 1931

Kilometres
Miles

CHANNEL ISLANDS
PLOTTED ON
UTM GRID

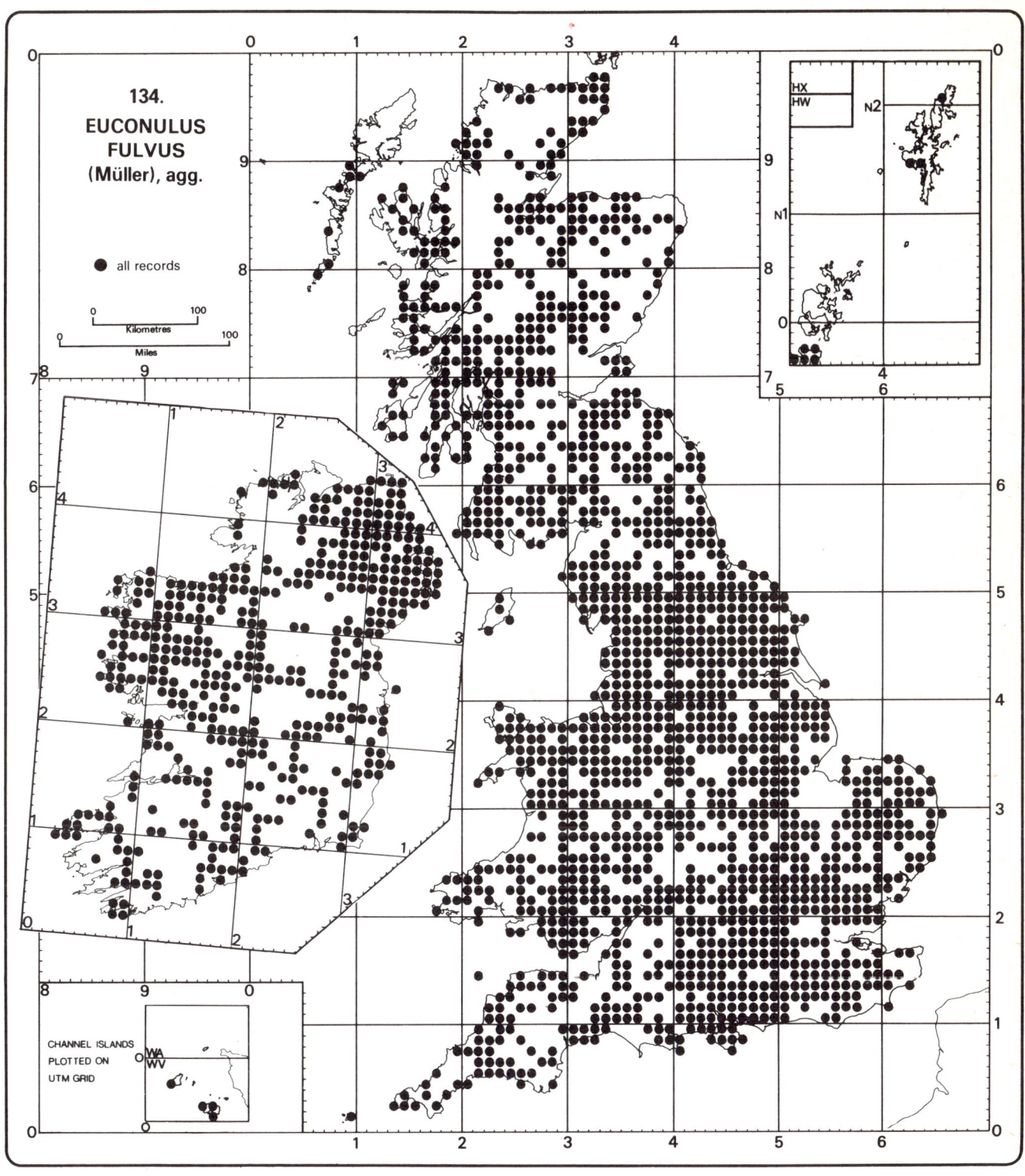

134.
EUCONULUS
FULVUS
(Müller), agg.

● all records

Kilometres

Miles

CHANNEL ISLANDS
PLOTTED ON
UTM GRID

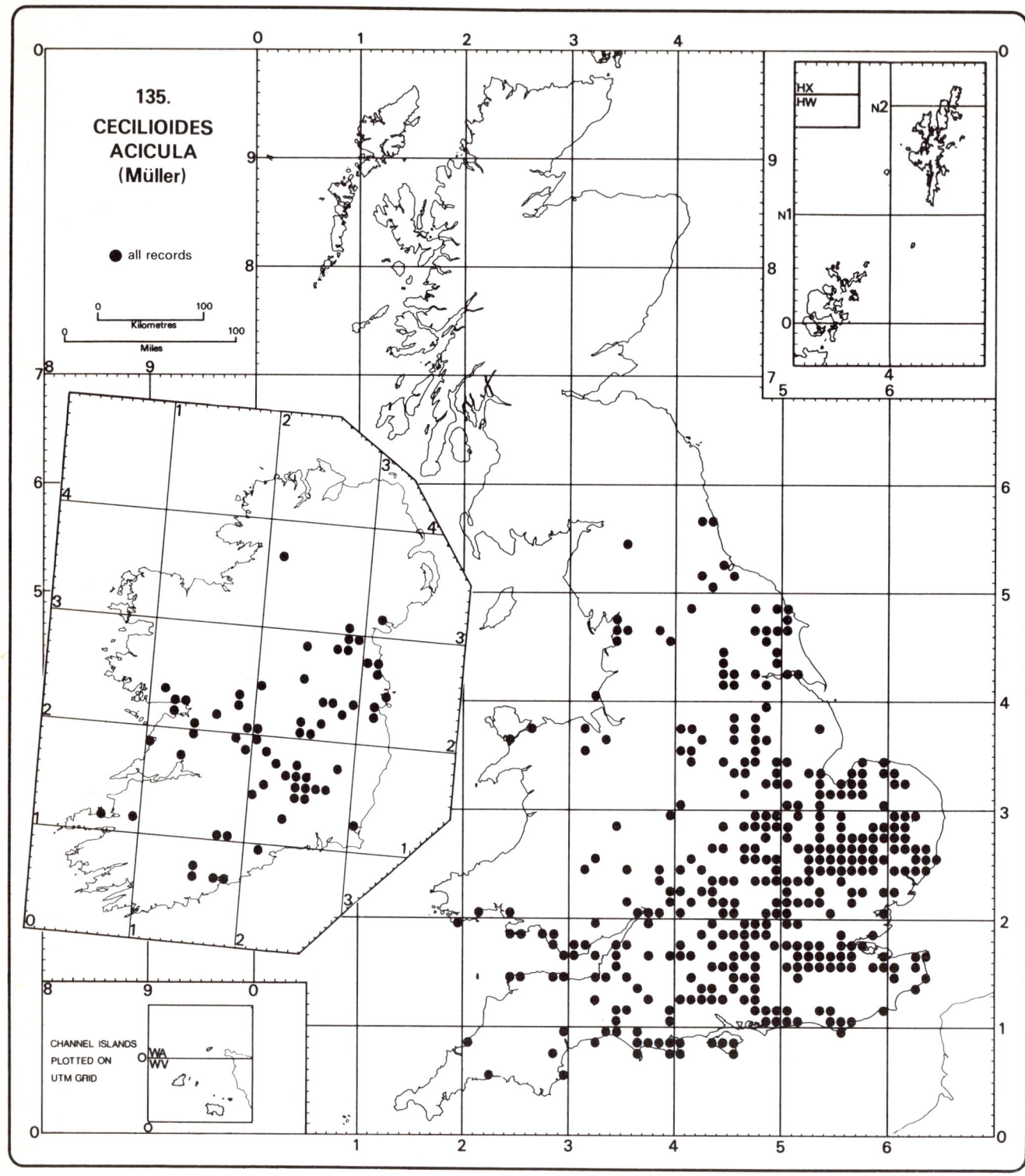

135.
CECILIOIDES
ACICULA
(Müller)

● all records

Kilometres
Miles

CHANNEL ISLANDS
PLOTTED ON
UTM GRID

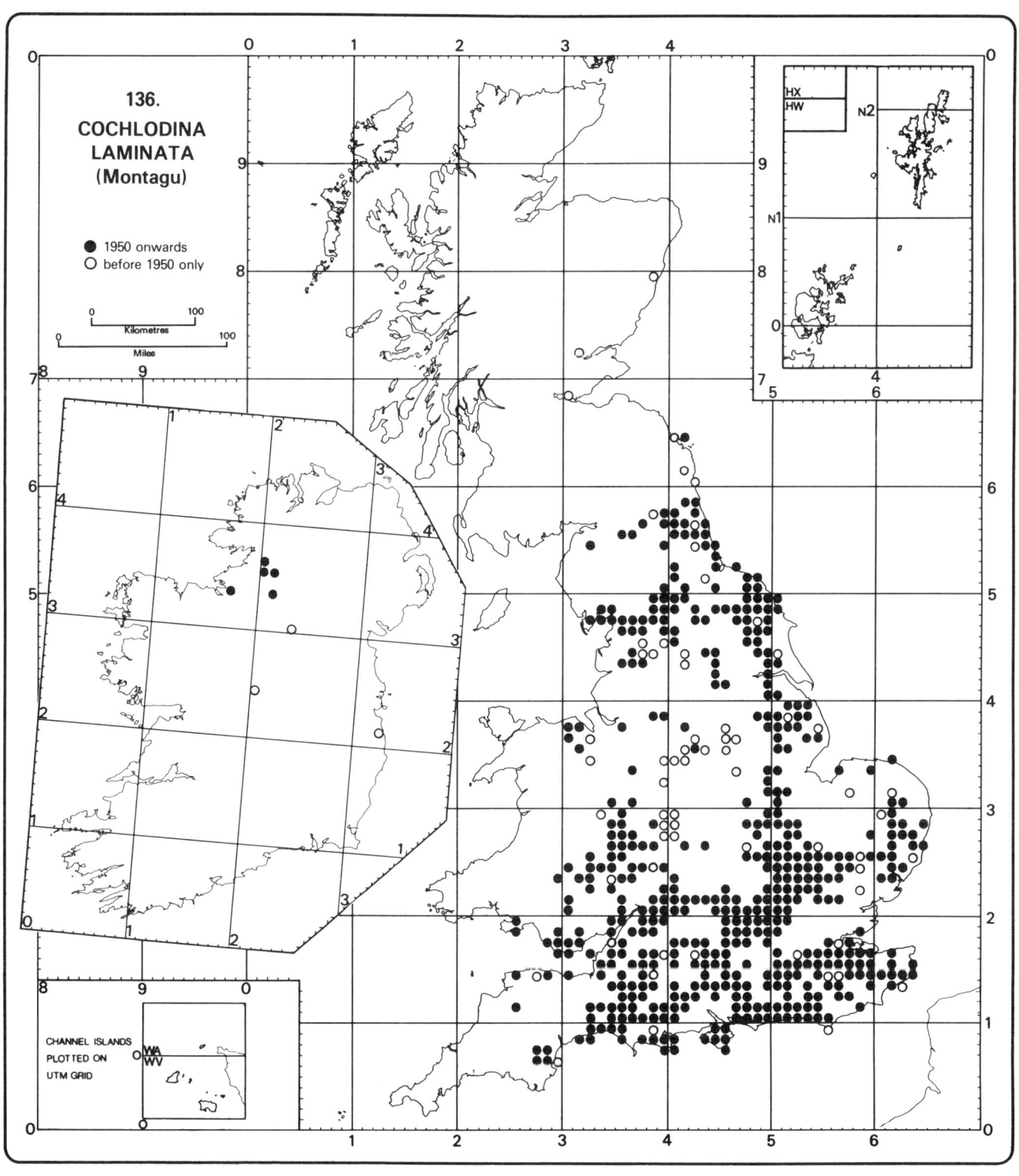

136.
COCHLODINA
LAMINATA
(Montagu)

● 1950 onwards
○ before 1950 only

0   100
Kilometres
0   100
Miles

CHANNEL ISLANDS
PLOTTED ON
UTM GRID

WA
WV

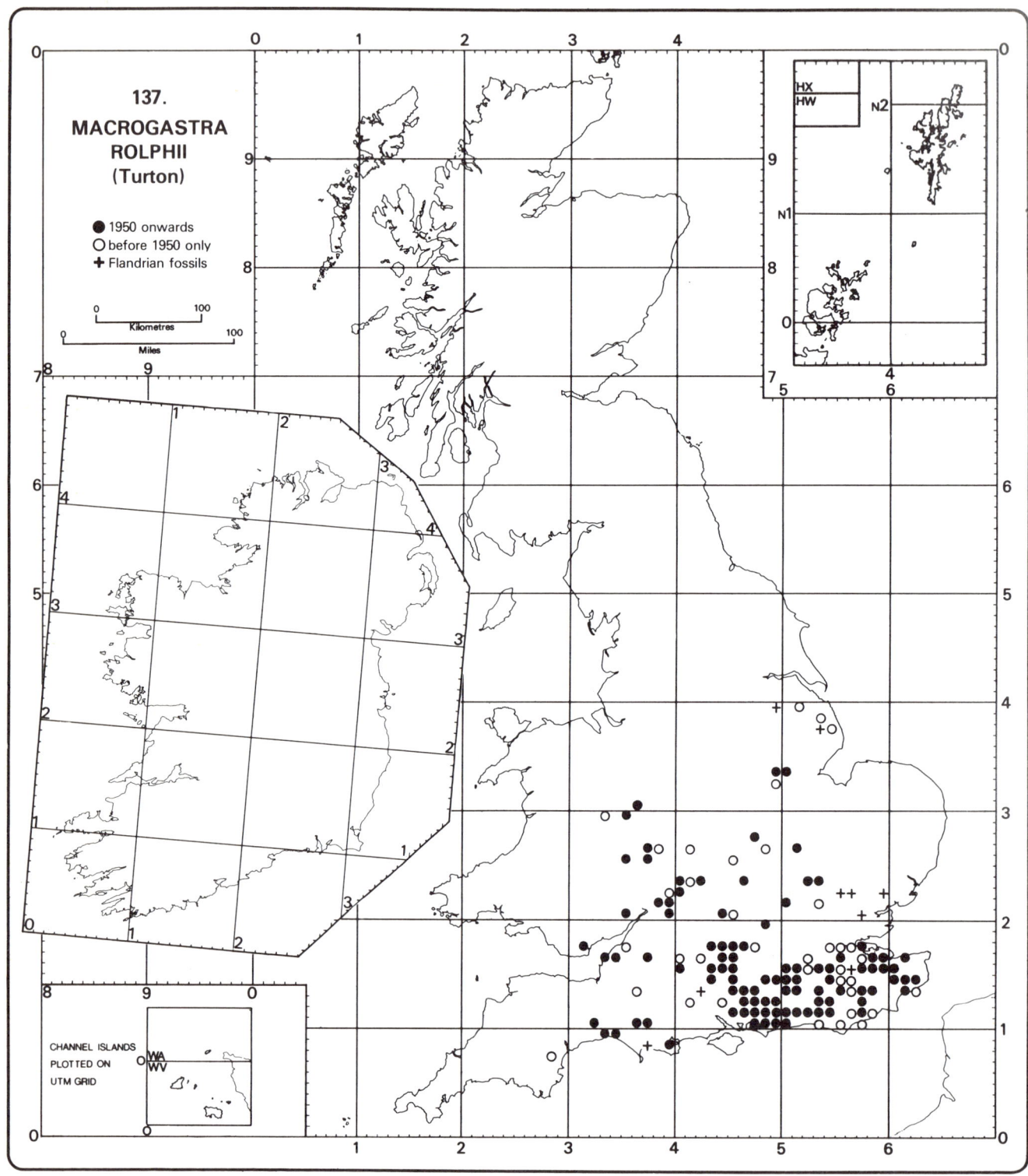

137.
MACROGASTRA
ROLPHII
(Turton)

● 1950 onwards
○ before 1950 only
+ Flandrian fossils

0   100
Kilometres
0   100
Miles

CHANNEL ISLANDS
PLOTTED ON
UTM GRID

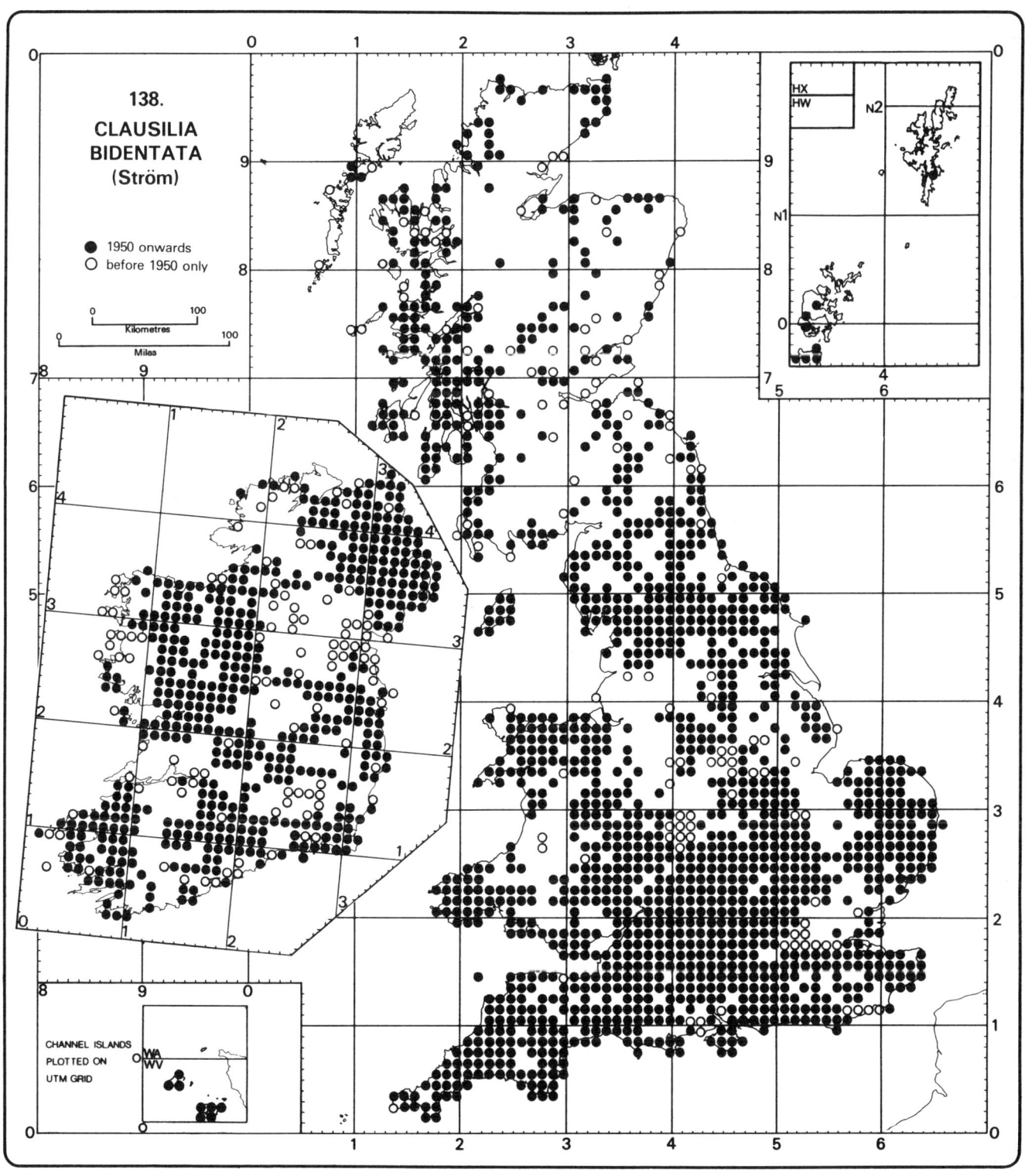

138.
CLAUSILIA
BIDENTATA
(Ström)

● 1950 onwards
○ before 1950 only

Kilometres
Miles

CHANNEL ISLANDS
PLOTTED ON
UTM GRID

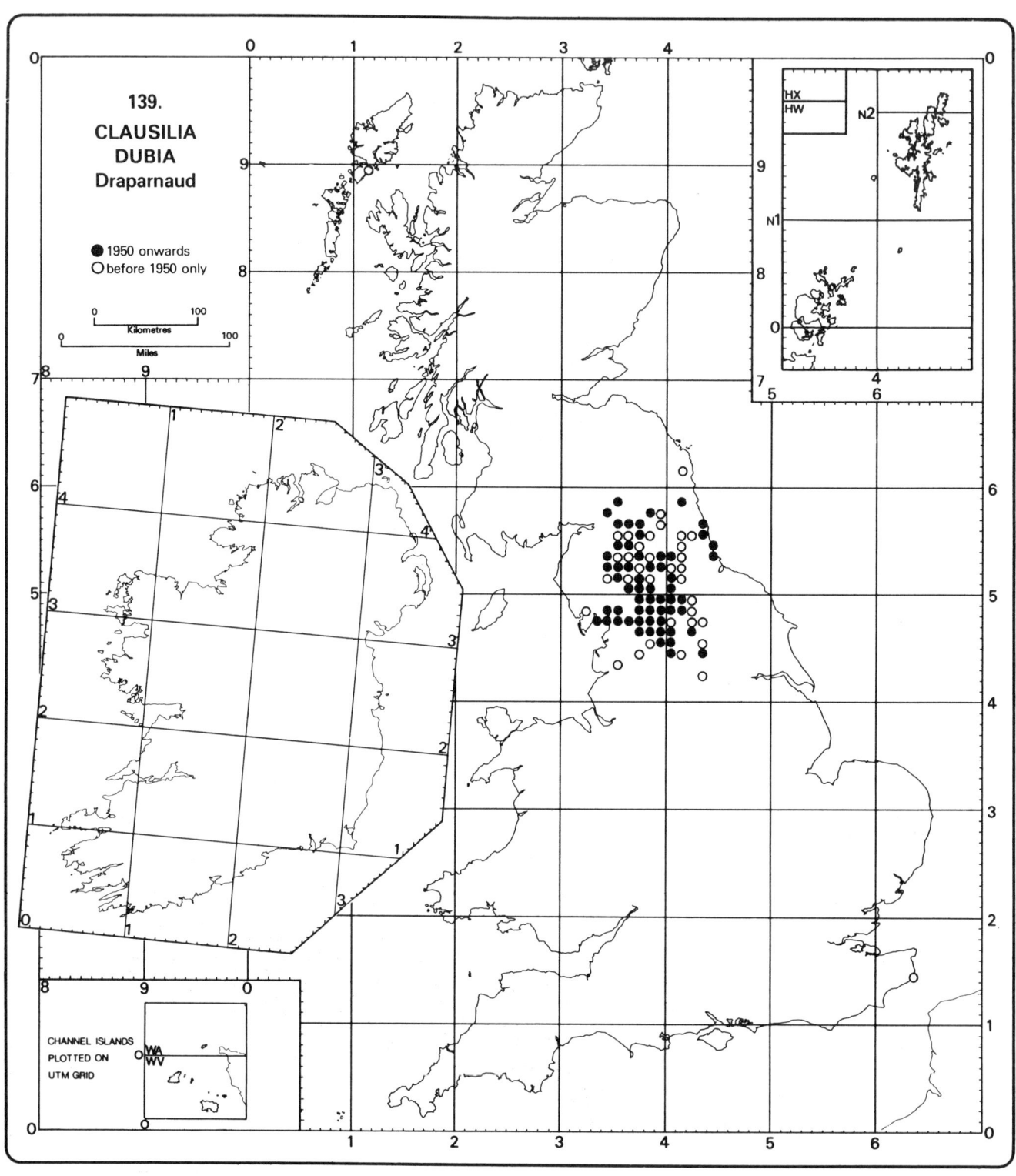

139.
CLAUSILIA
DUBIA
Draparnaud

● 1950 onwards
○ before 1950 only

Kilometres

Miles

HX
HW

N2

N1

CHANNEL ISLANDS
PLOTTED ON
UTM GRID

WA
WV

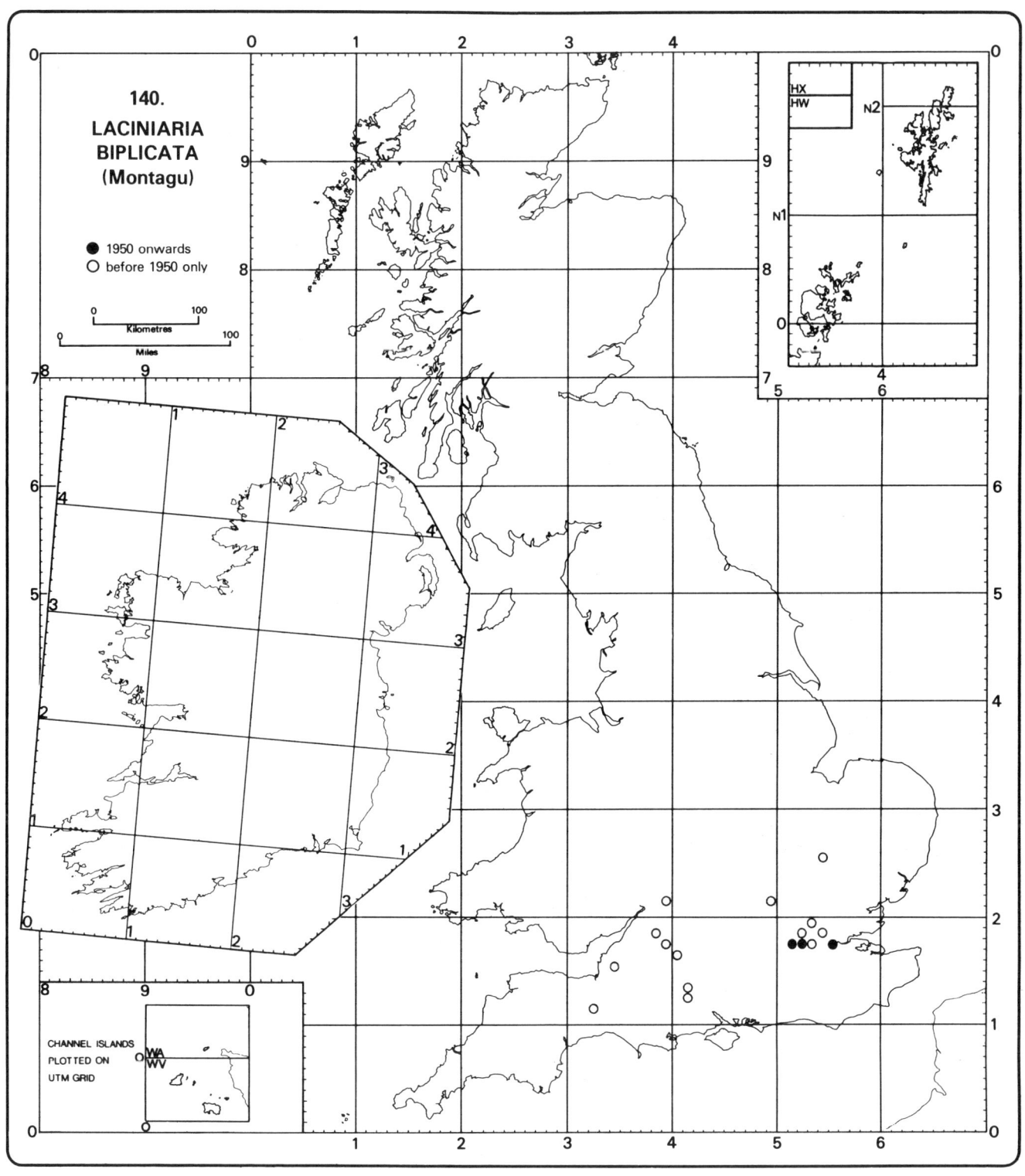

140.
LACINIARIA
BIPLICATA
(Montagu)

● 1950 onwards
○ before 1950 only

Kilometres
Miles

CHANNEL ISLANDS
PLOTTED ON
UTM GRID

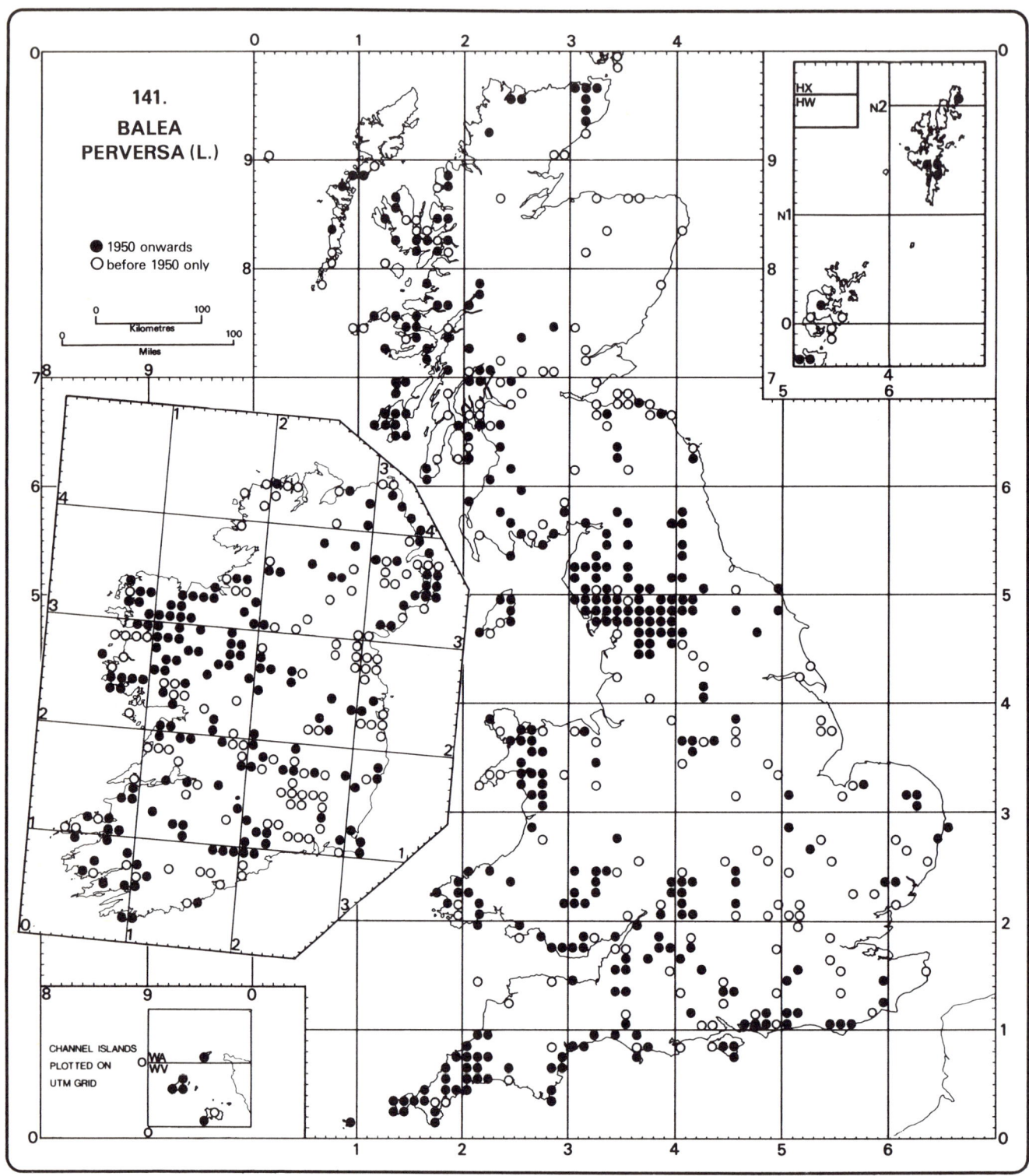

141.

BALEA
PERVERSA (L.)

● 1950 onwards
○ before 1950 only

0   100
Kilometres
0   100
Miles

CHANNEL ISLANDS
PLOTTED ON
UTM GRID

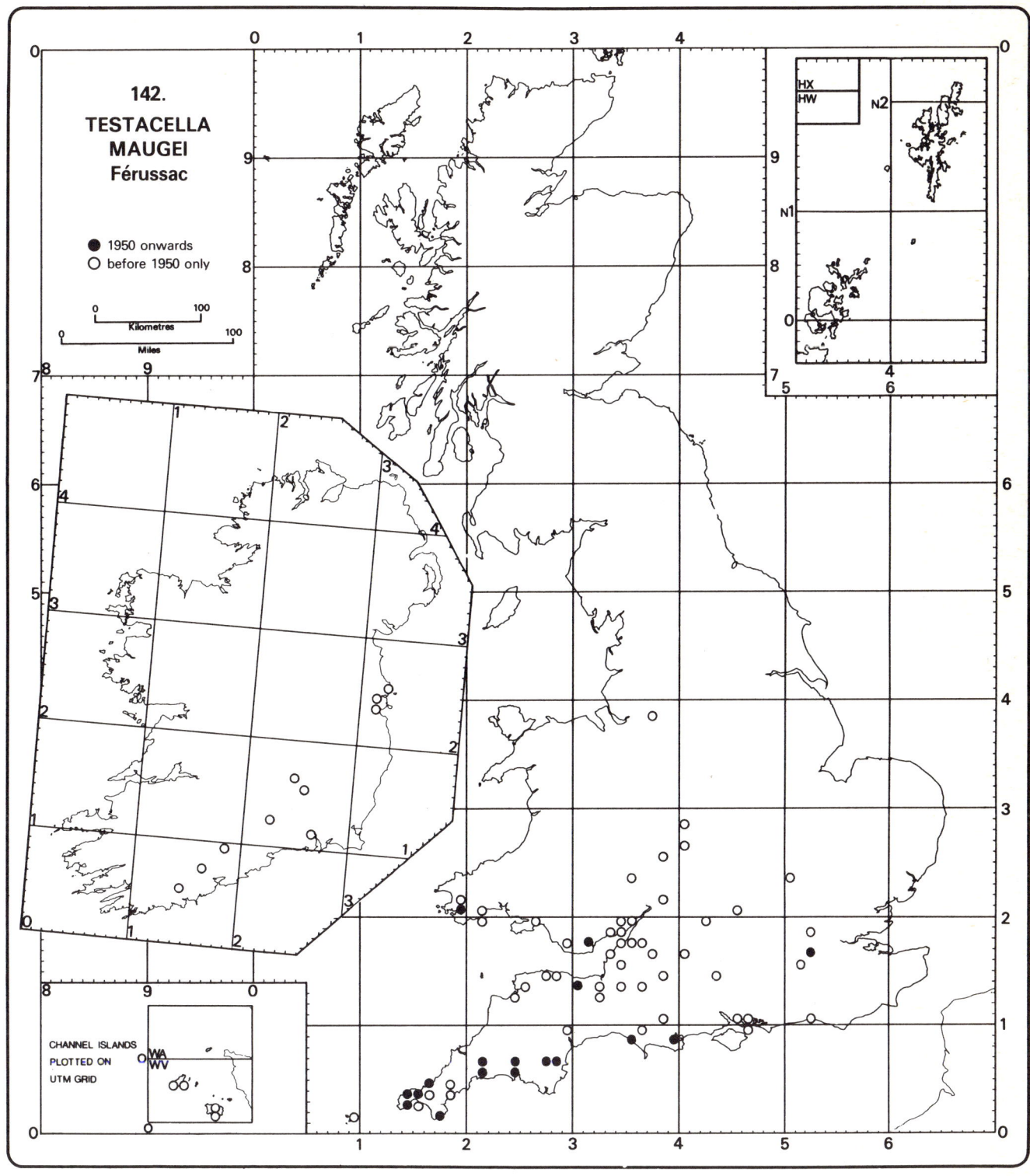

142.

TESTACELLA
MAUGEI
Férussac

● 1950 onwards
○ before 1950 only

Kilometres

Miles

CHANNEL ISLANDS
PLOTTED ON
UTM GRID

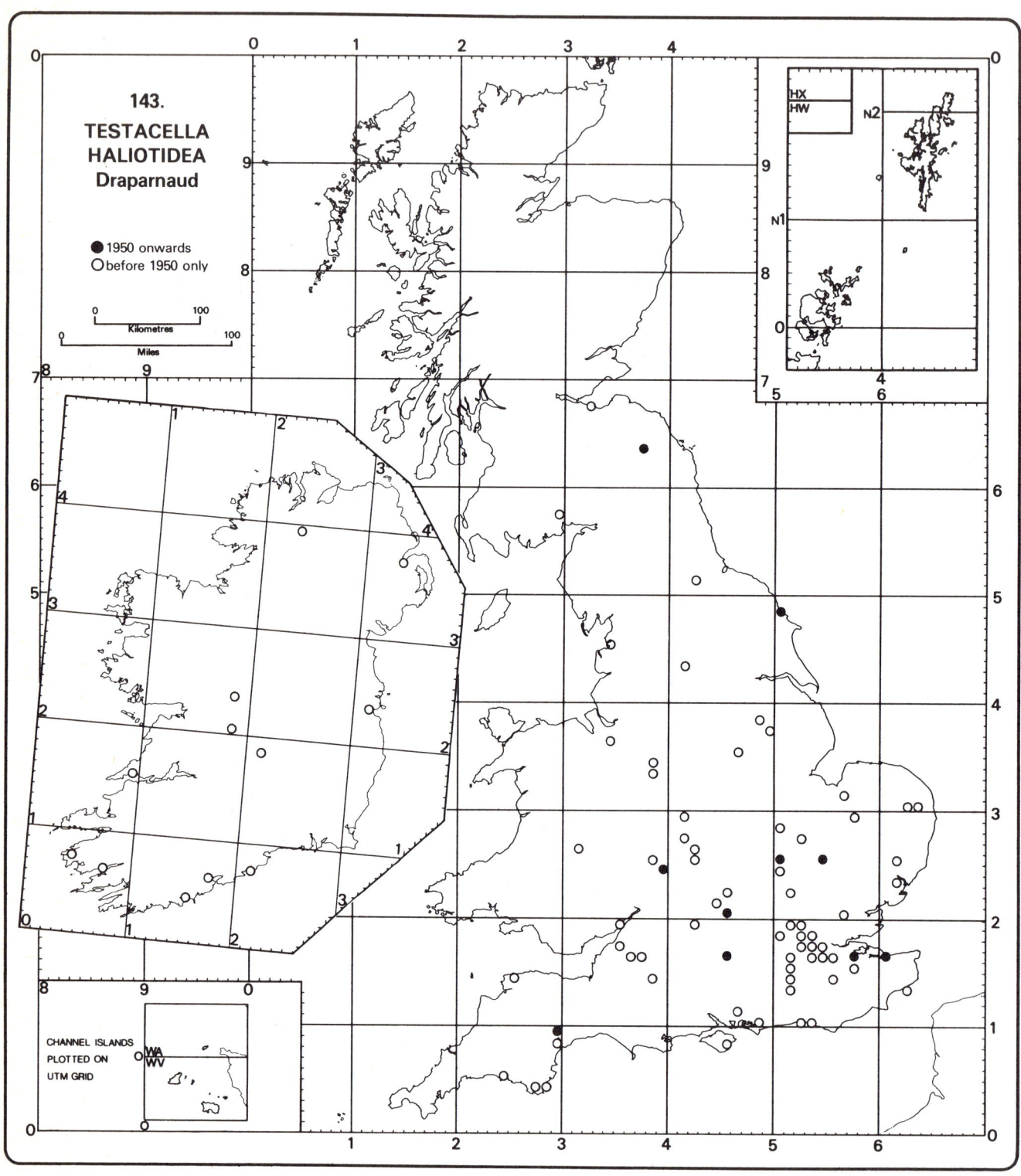

143.
TESTACELLA
HALIOTIDEA
Draparnaud

● 1950 onwards
○ before 1950 only

0   100
Kilometres
0   100
Miles

CHANNEL ISLANDS
PLOTTED ON
UTM GRID

WA
WV

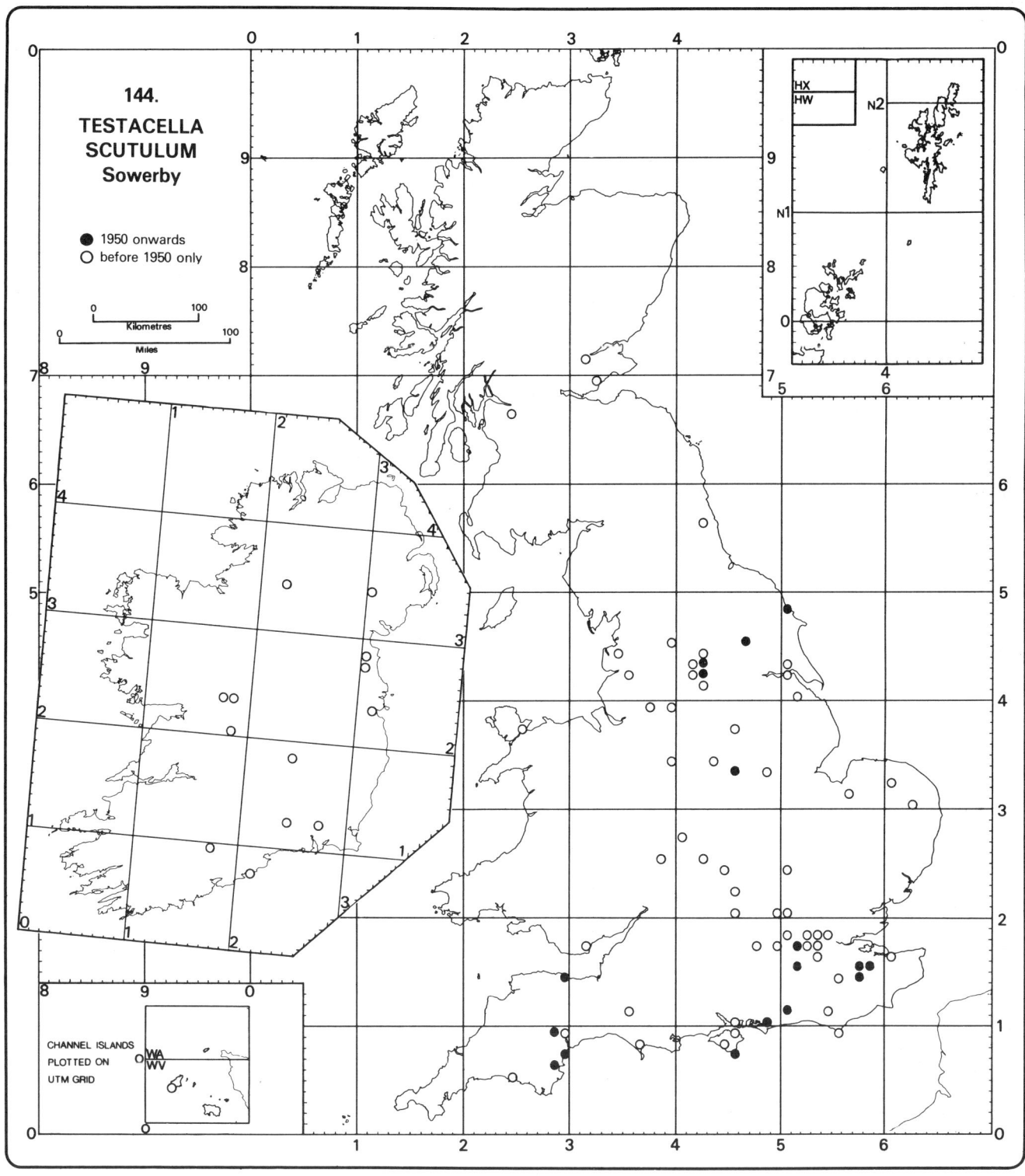

144.
**TESTACELLA
SCUTULUM**
Sowerby

● 1950 onwards
○ before 1950 only

Kilometres
Miles

HX
HW

CHANNEL ISLANDS
PLOTTED ON
UTM GRID

WA
WV

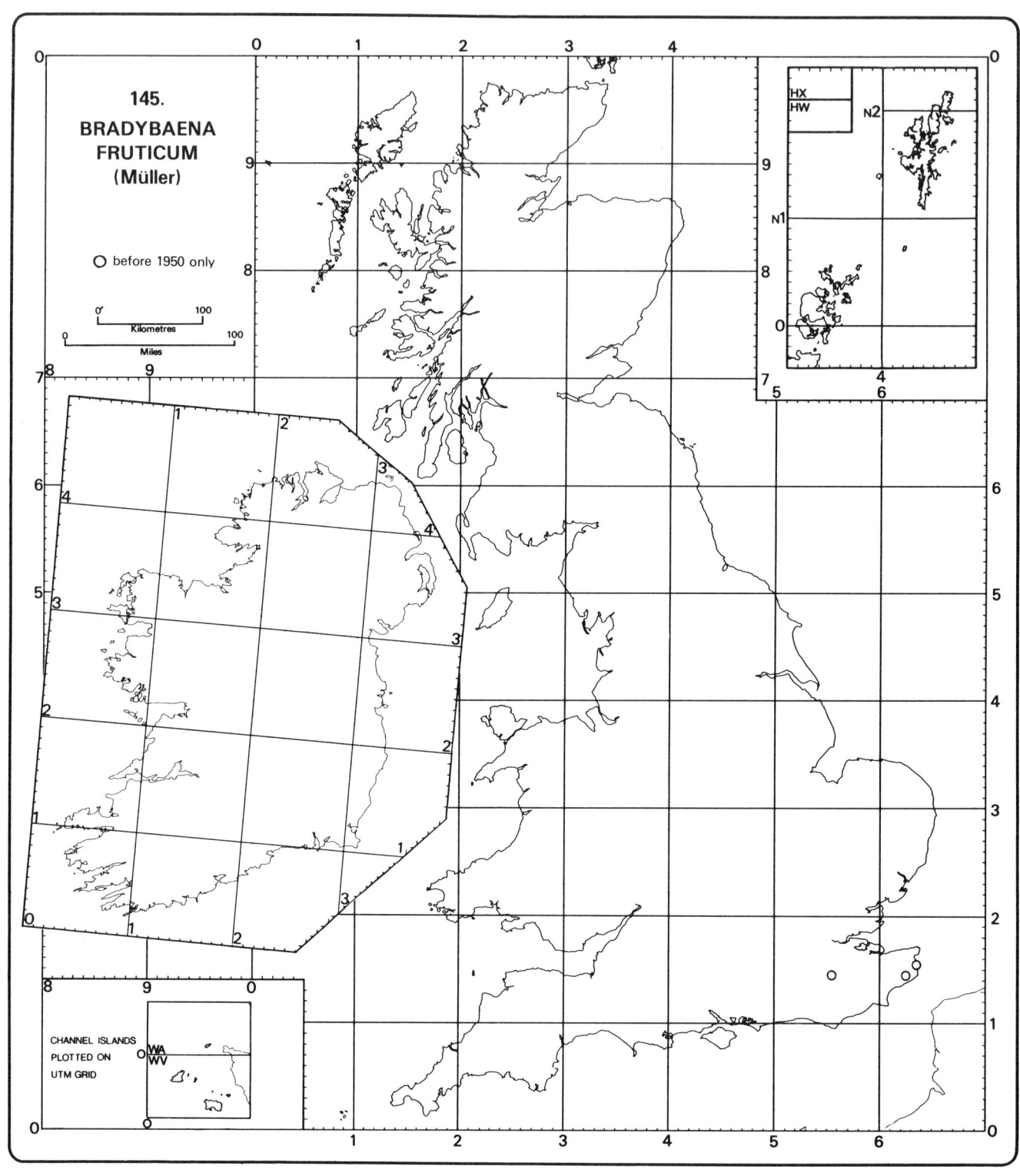

145.
BRADYBAENA
FRUTICUM
(Müller)

○ before 1950 only

Kilometres
Miles

CHANNEL ISLANDS
PLOTTED ON
UTM GRID

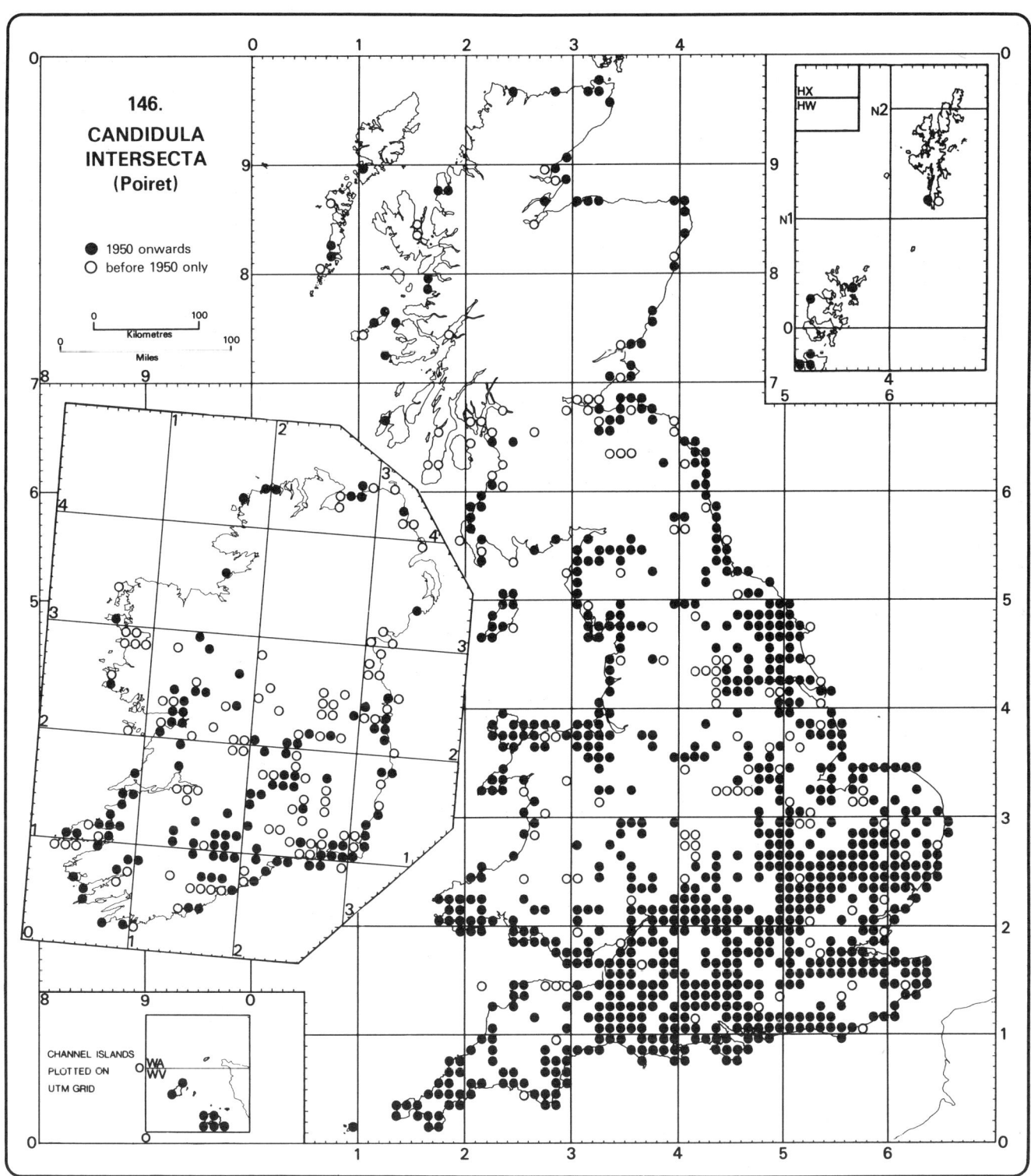

146.
CANDIDULA
INTERSECTA
(Poiret)

● 1950 onwards
○ before 1950 only

Kilometres

Miles

HX
HW

N2

N1

CHANNEL ISLANDS
PLOTTED ON
UTM GRID

WA
WV

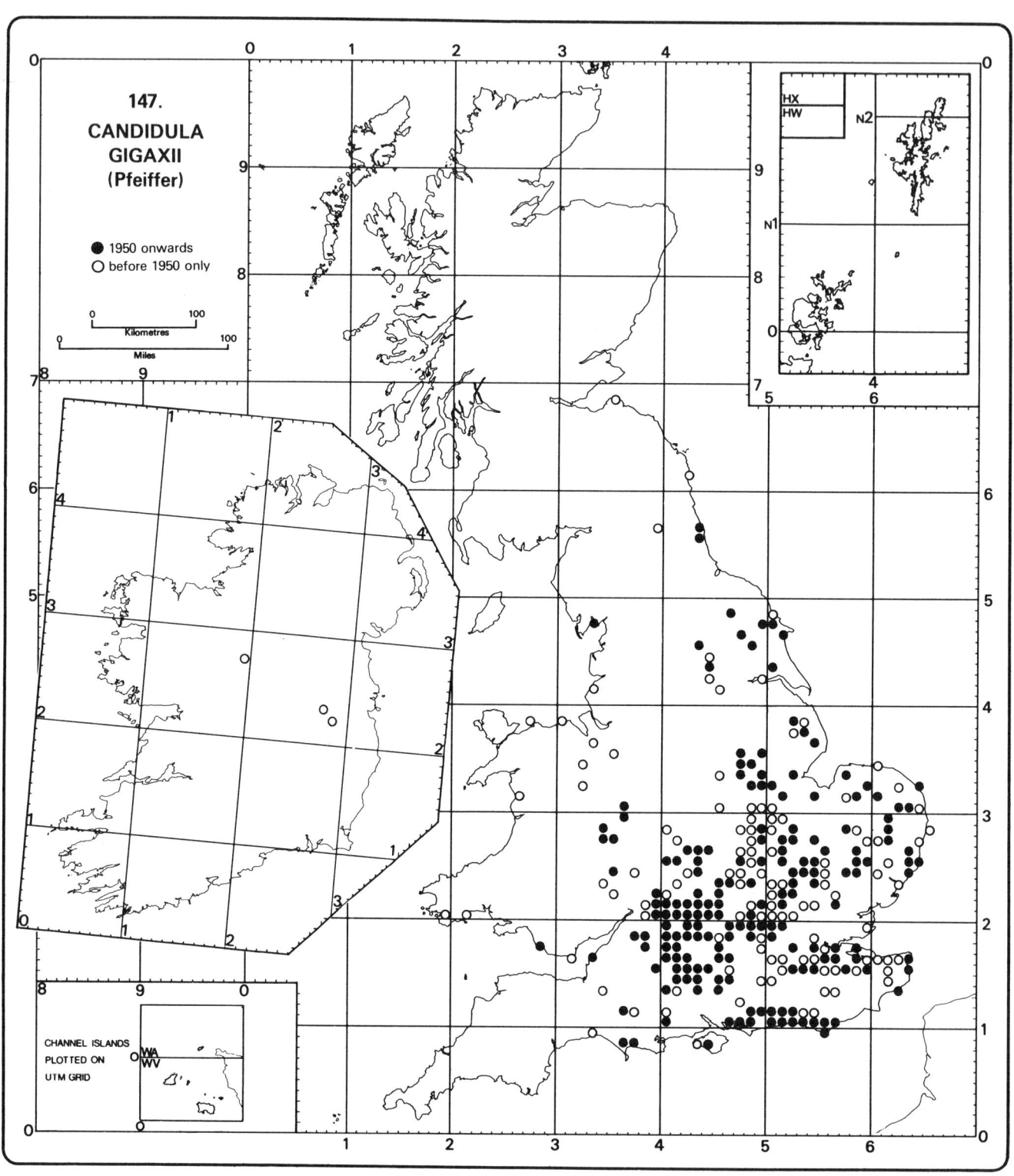

147.
CANDIDULA
GIGAXII
(Pfeiffer)

● 1950 onwards
○ before 1950 only

0   100
Kilometres
0   100
Miles

CHANNEL ISLANDS
PLOTTED ON
UTM GRID

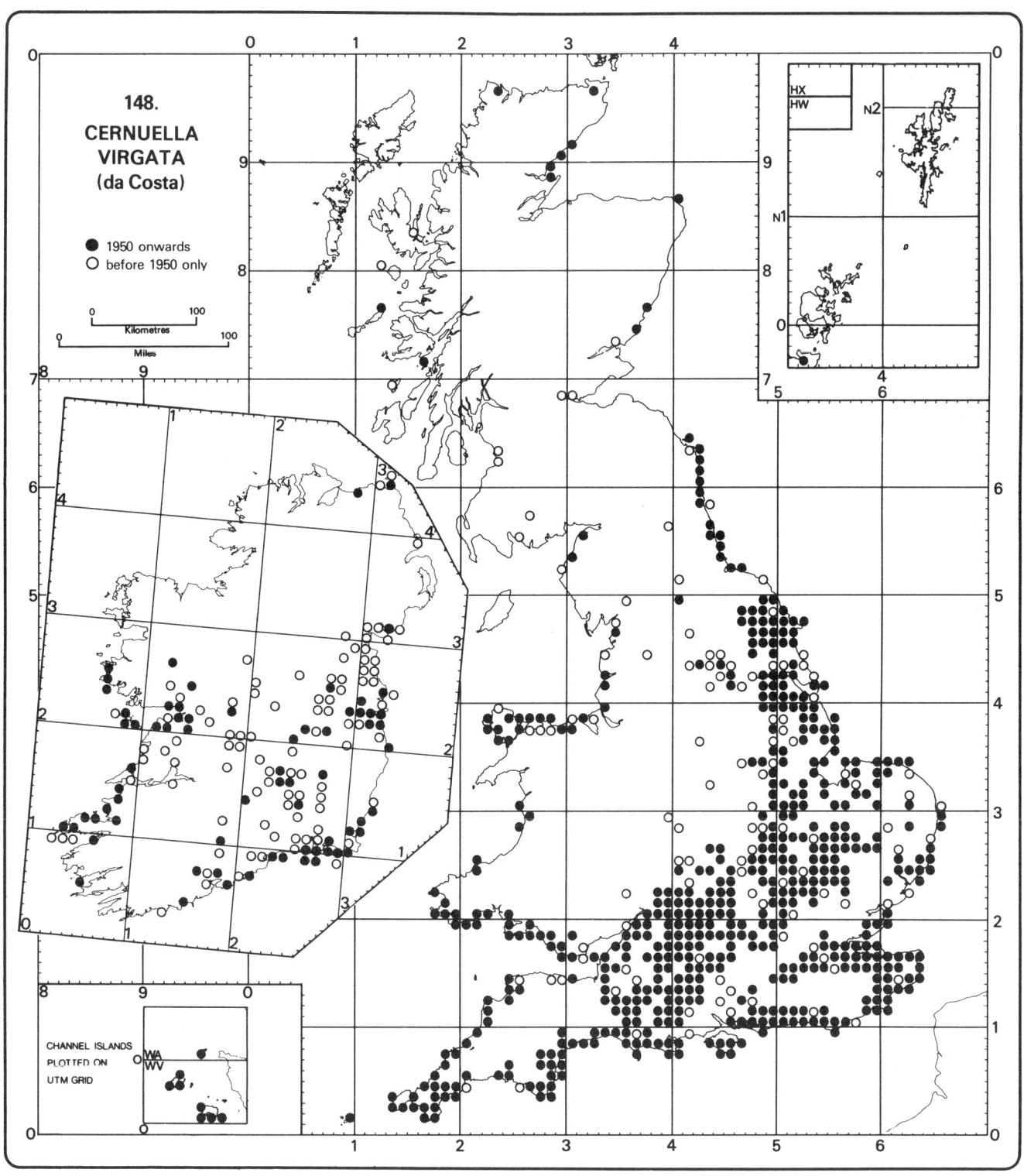

148.
CERNUELLA
VIRGATA
(da Costa)

● 1950 onwards
○ before 1950 only

Kilometres
0   100

Miles
0   100

CHANNEL ISLANDS
PLOTTED ON
UTM GRID

WA
WV

HX
HW

N2

N1

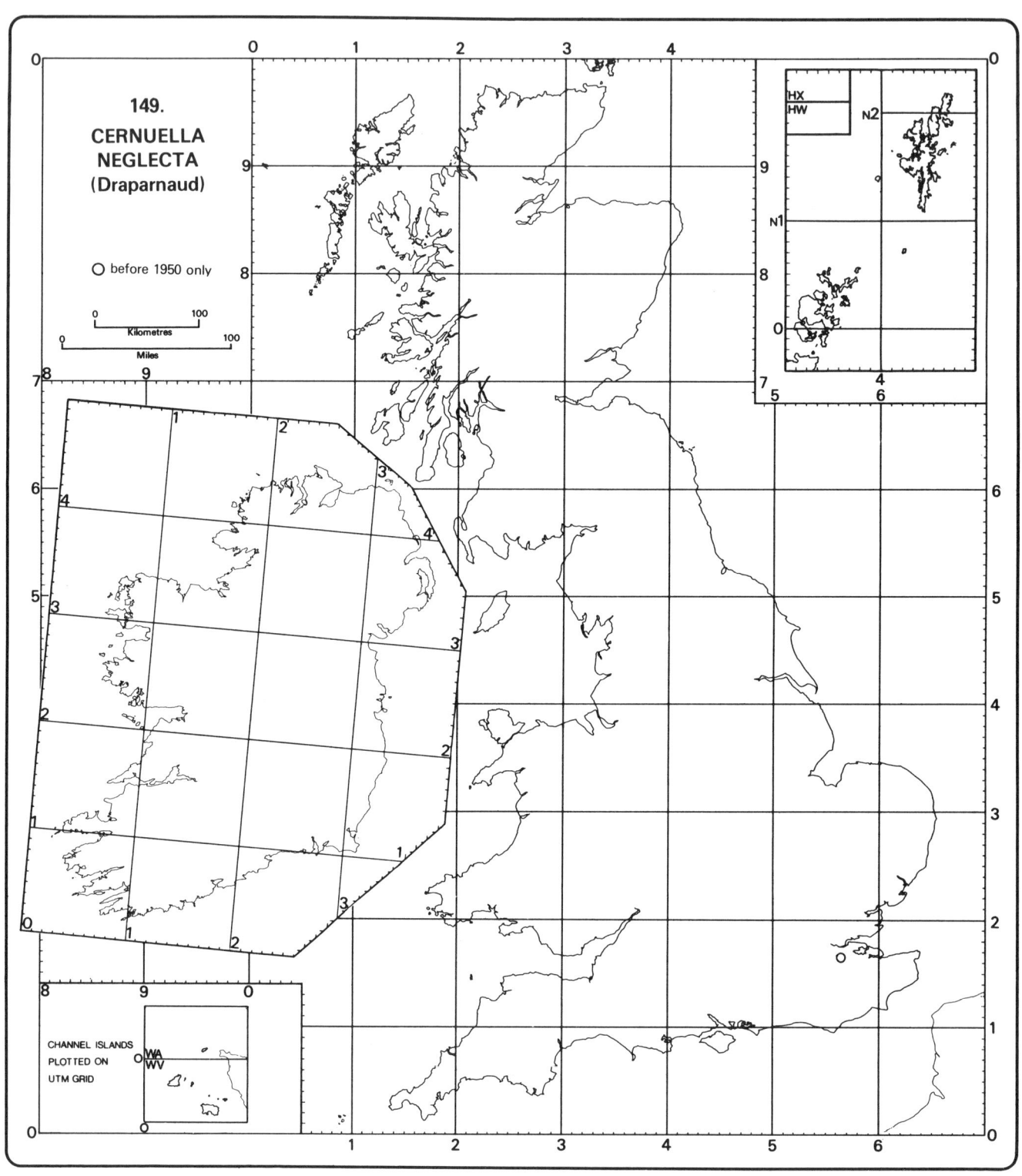

149.
CERNUELLA
NEGLECTA
(Draparnaud)

○ before 1950 only

0 _____ 100
Kilometres
0 _____ 100
Miles

CHANNEL ISLANDS
PLOTTED ON
UTM GRID

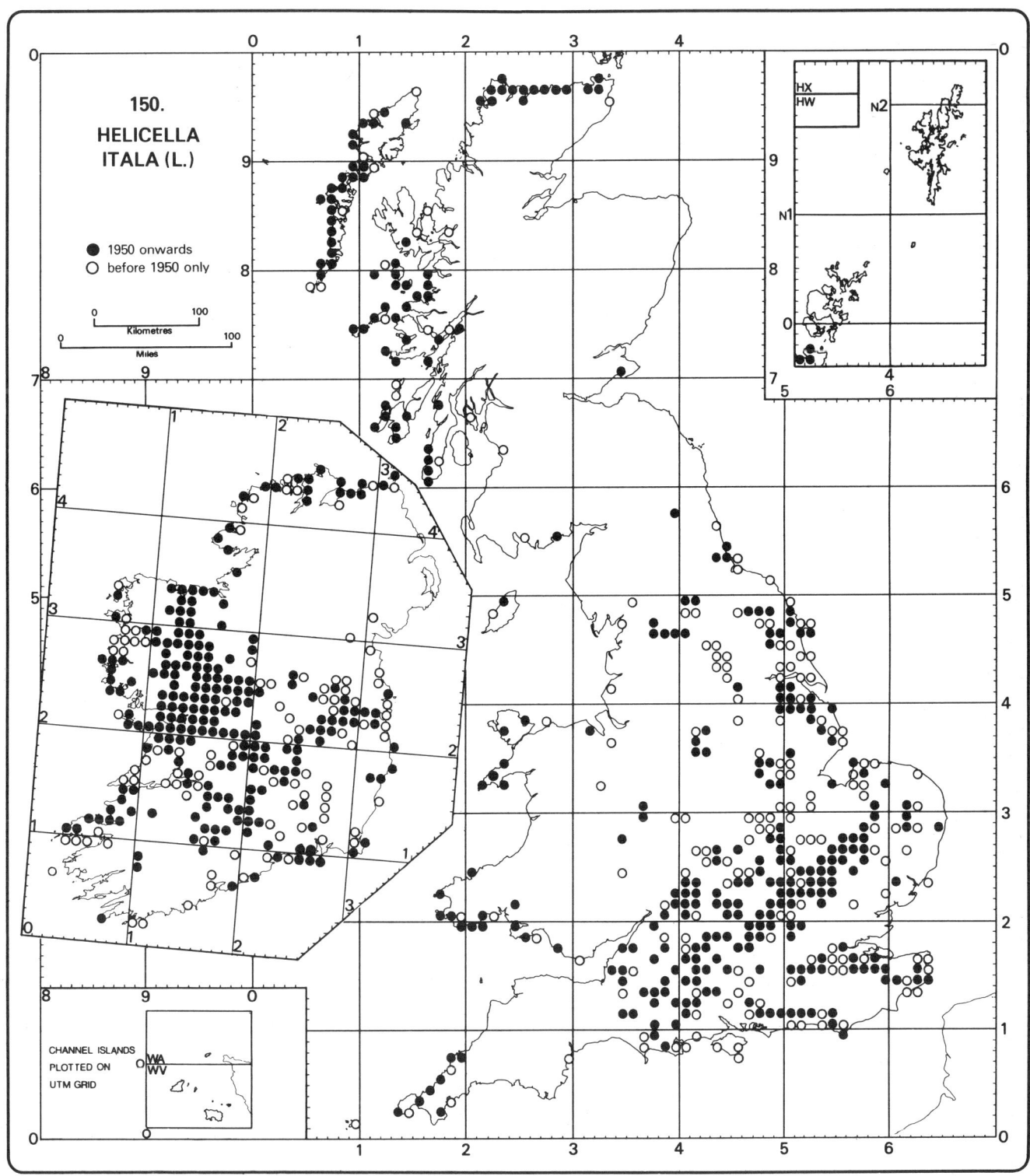

150.

HELICELLA
ITALA (L.)

● 1950 onwards
○ before 1950 only

Kilometres
Miles

CHANNEL ISLANDS
PLOTTED ON
UTM GRID

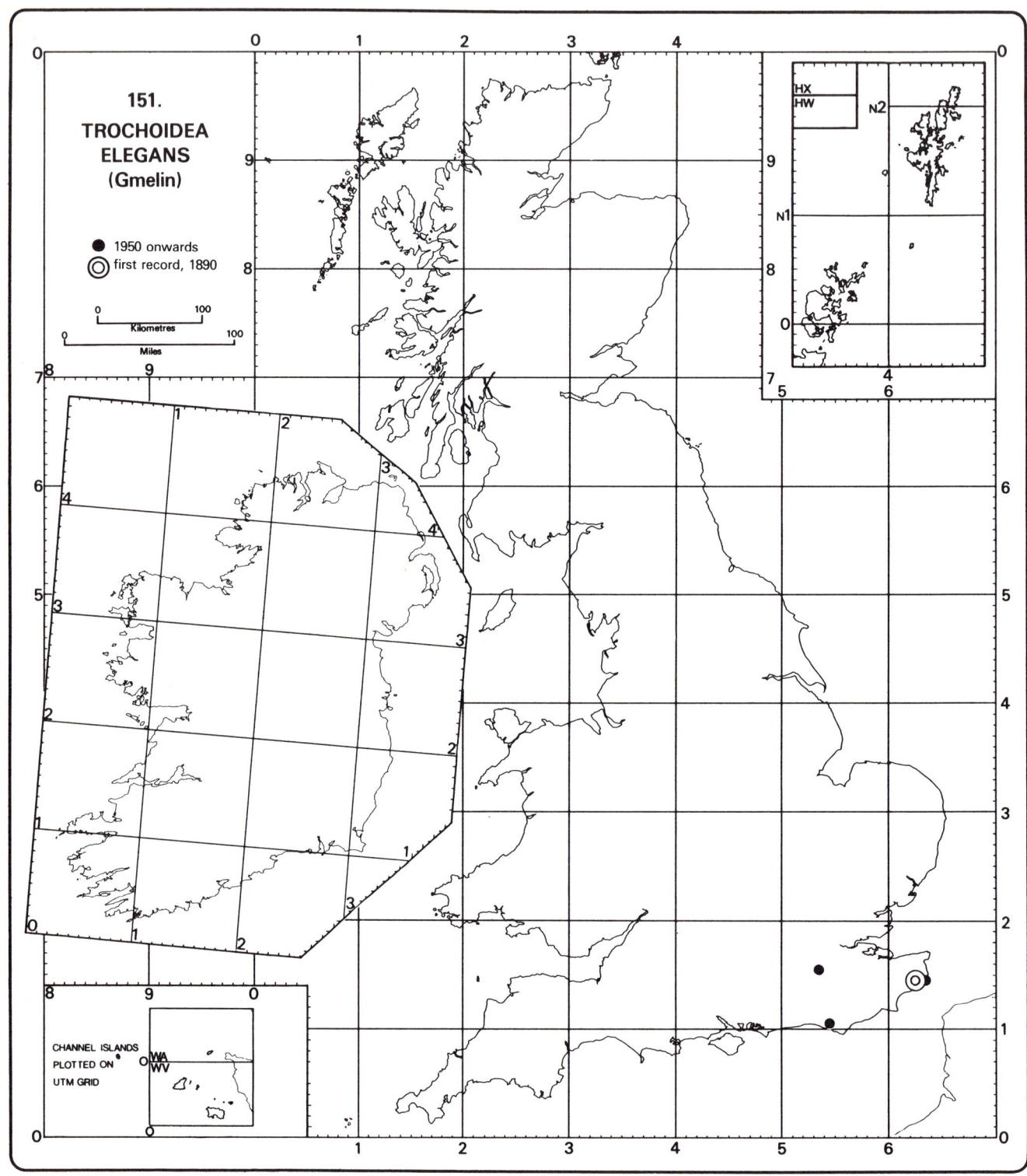

151.
TROCHOIDEA
ELEGANS
(Gmelin)

● 1950 onwards
◉ first record, 1890

0 ___ 100
Kilometres
0 ___ 100
Miles

HX
HW

CHANNEL ISLANDS
PLOTTED ON
UTM GRID

WA
WV

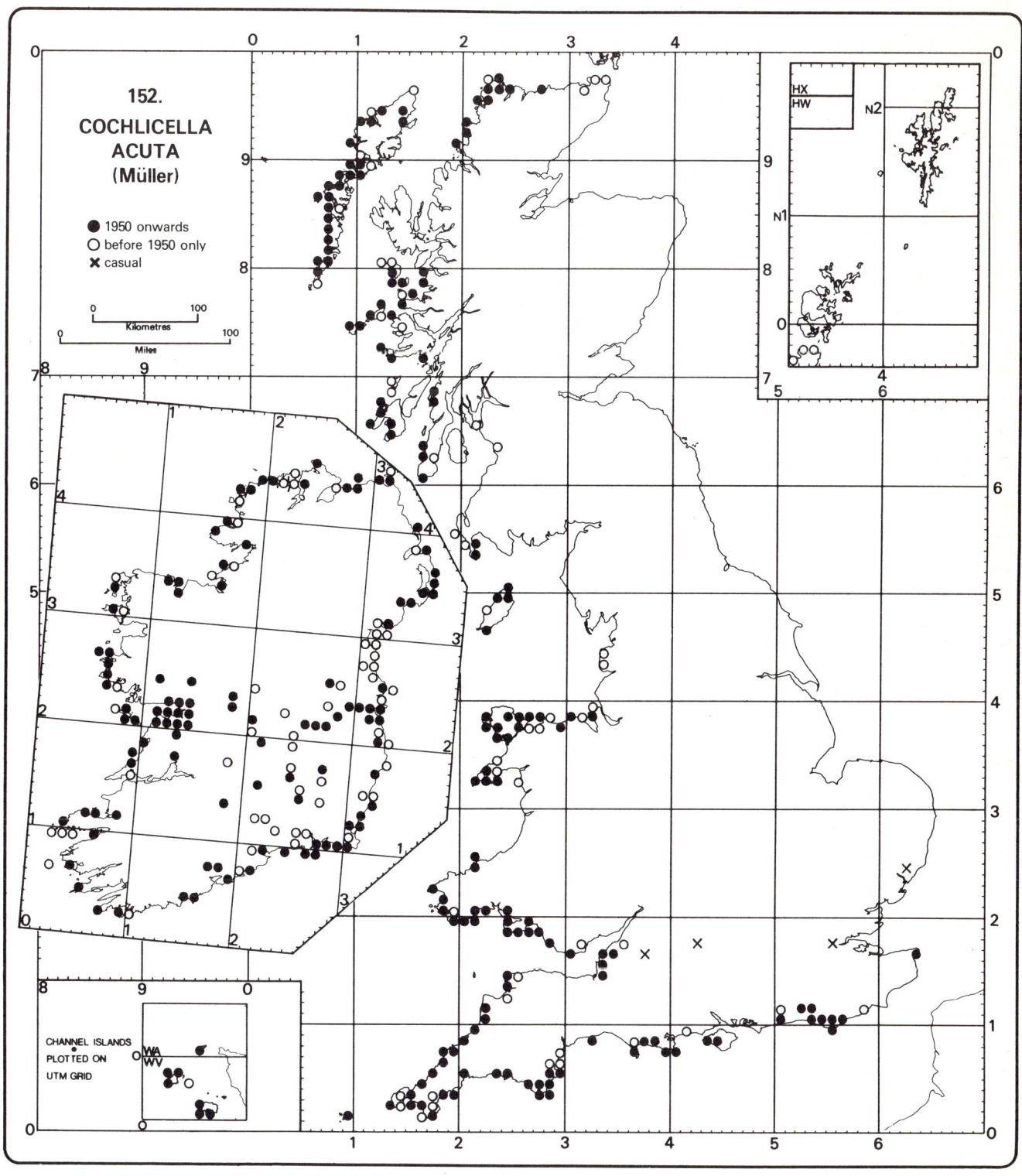

152.
COCHLICELLA
ACUTA
(Müller)

● 1950 onwards
○ before 1950 only
✖ casual

Kilometres
Miles

CHANNEL ISLANDS
PLOTTED ON
UTM GRID

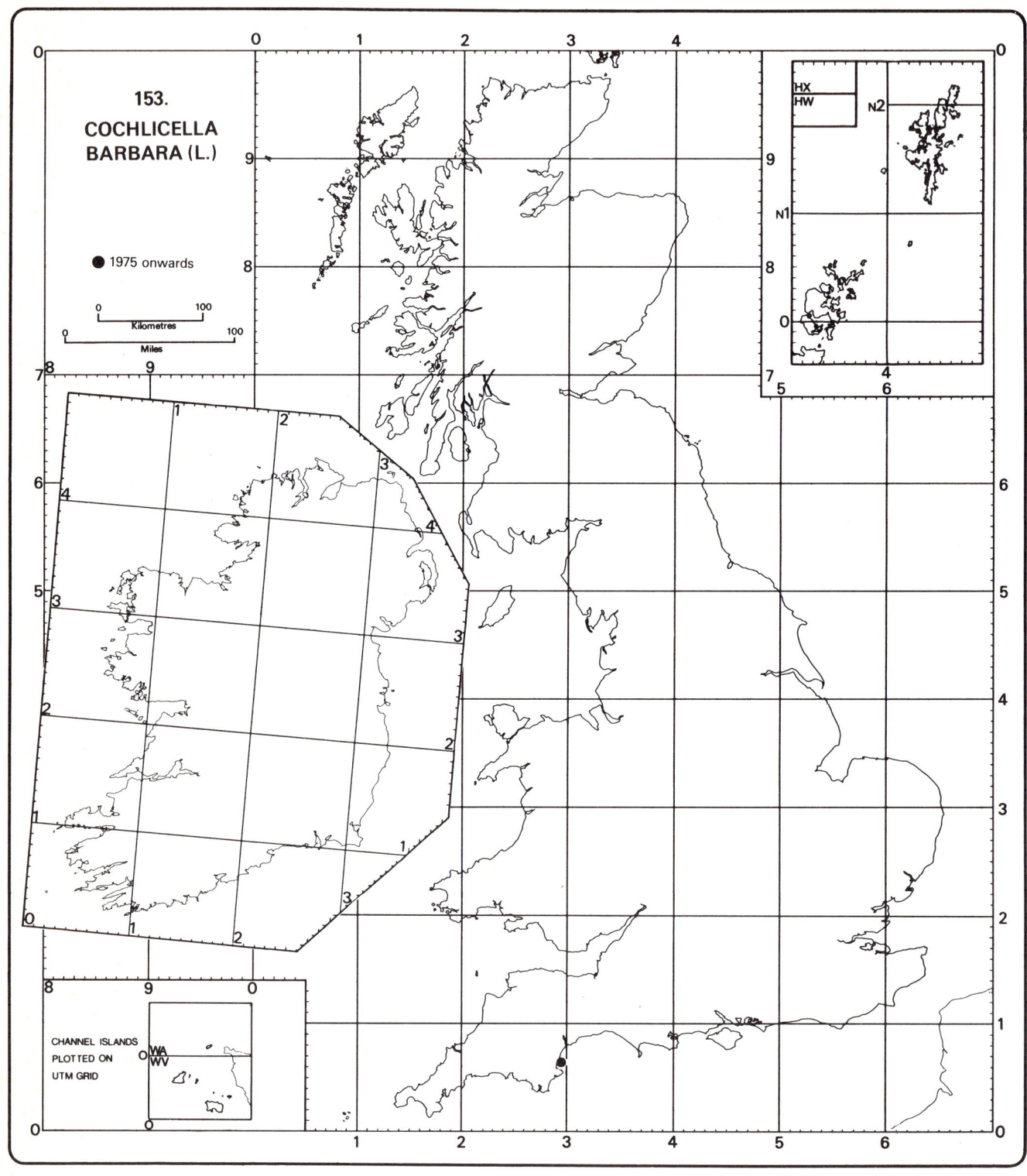

153.
COCHLICELLA
BARBARA (L.)

● 1975 onwards

Kilometres
Miles

CHANNEL ISLANDS
PLOTTED ON
UTM GRID

HX
HW

CHANNEL ISLANDS
PLOTTED ON
UTM GRID

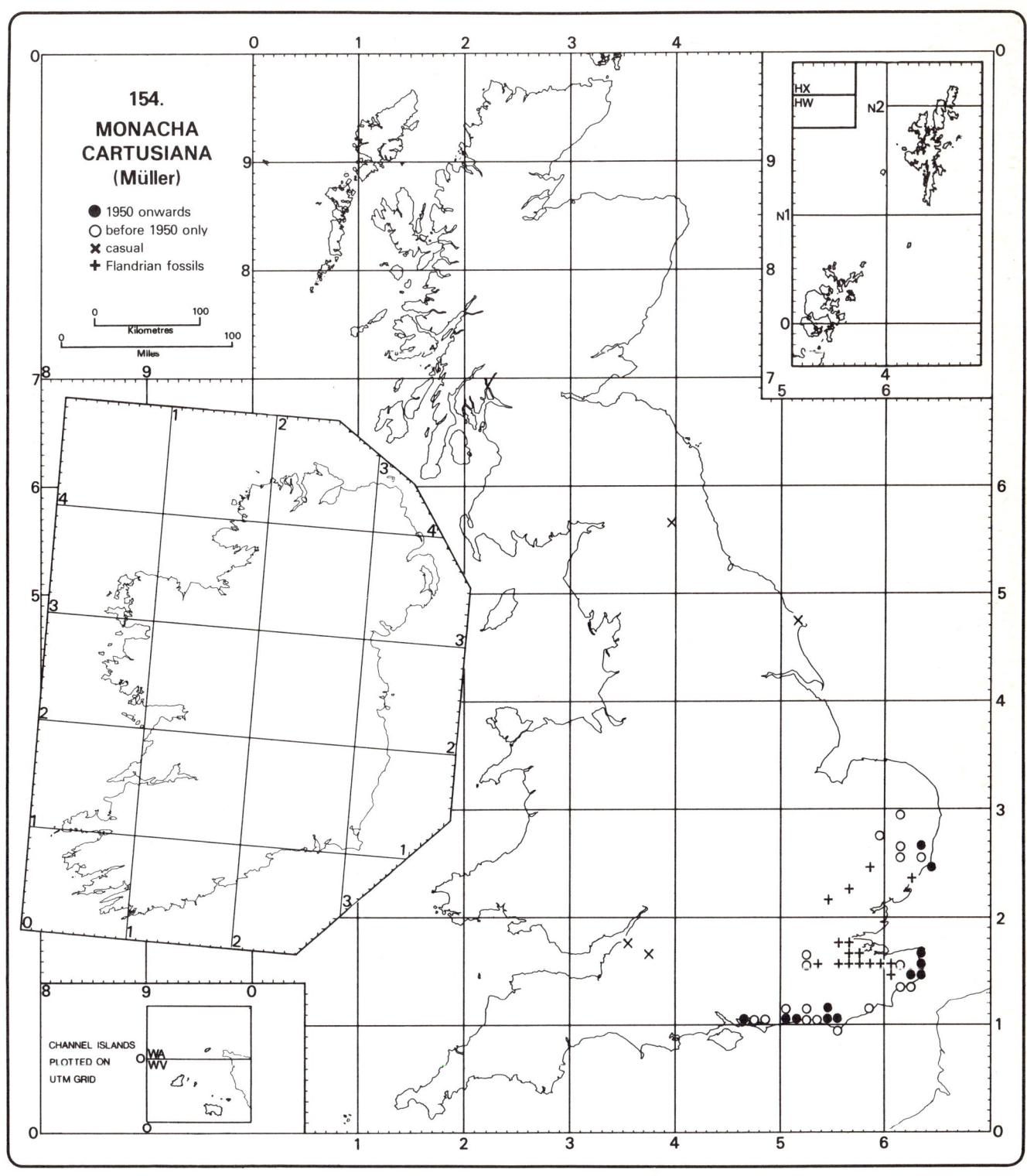

154.

MONACHA
CARTUSIANA
(Müller)

● 1950 onwards
○ before 1950 only
✕ casual
✛ Flandrian fossils

Kilometres
Miles

CHANNEL ISLANDS
PLOTTED ON
UTM GRID

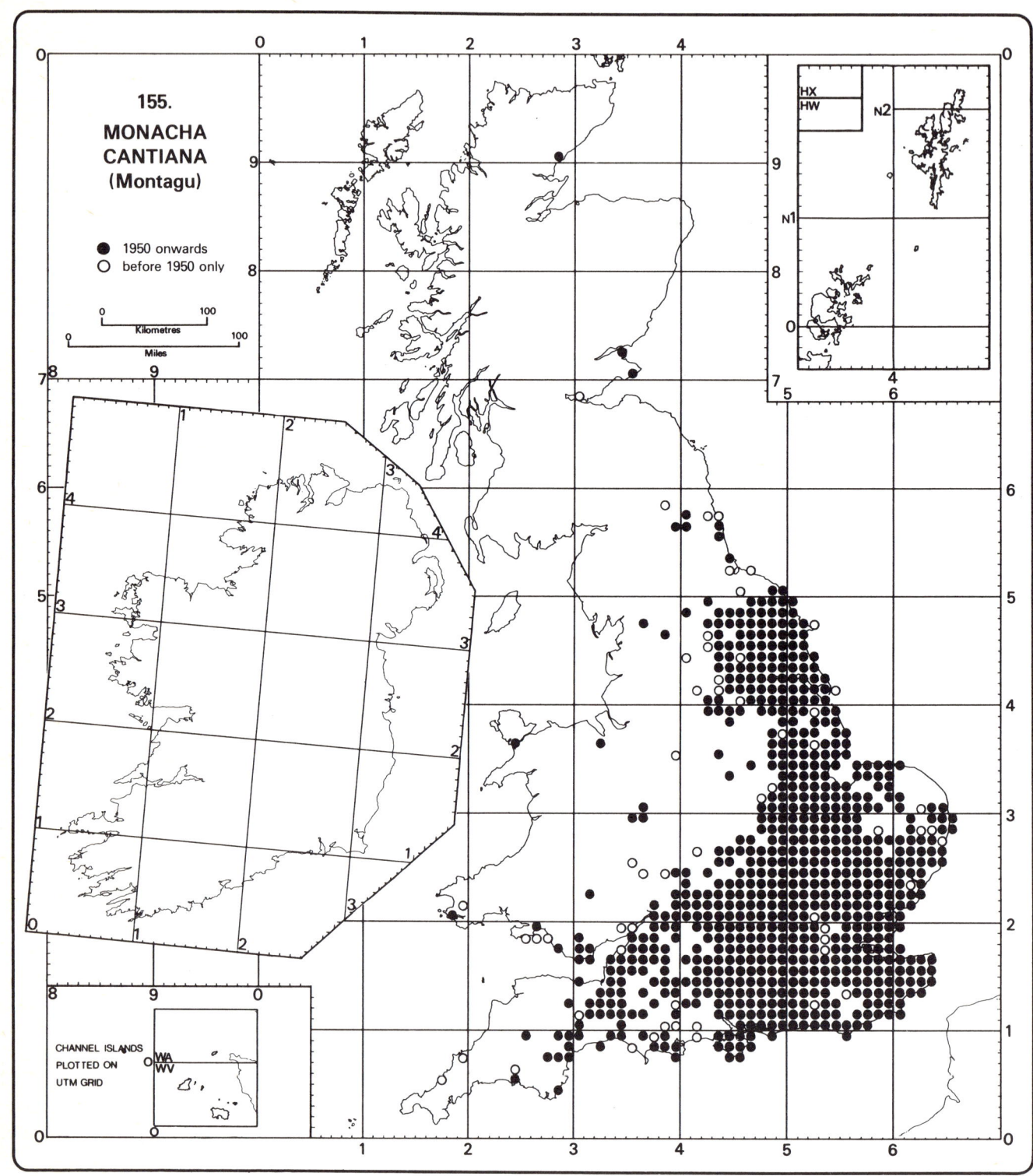

155.

MONACHA
CANTIANA
(Montagu)

● 1950 onwards
○ before 1950 only

CHANNEL ISLANDS
PLOTTED ON
UTM GRID

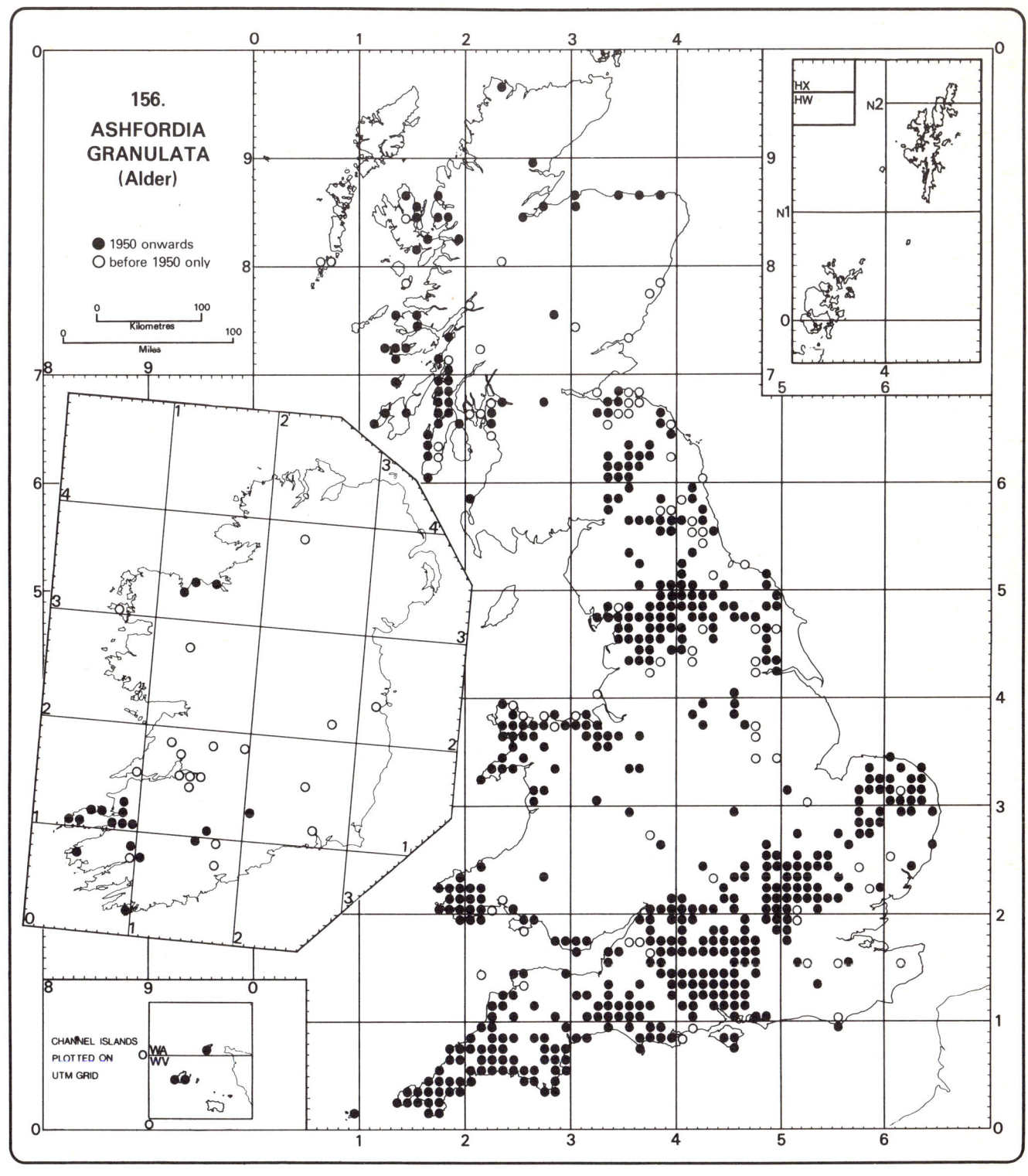

156.
ASHFORDIA
GRANULATA
(Alder)

● 1950 onwards
○ before 1950 only

Kilometres
Miles

CHANNEL ISLANDS
PLOTTED ON
UTM GRID

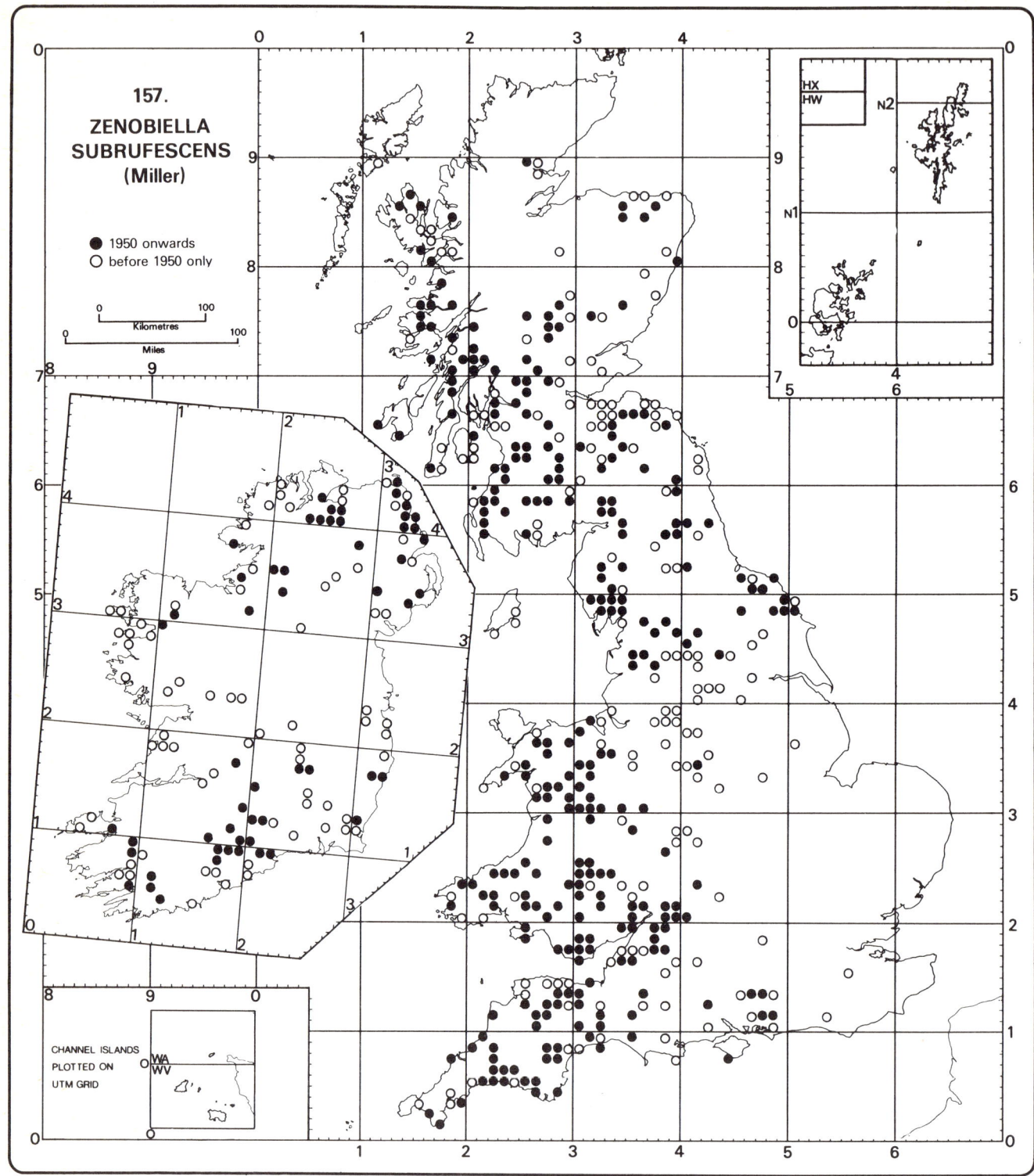

157.

ZENOBIELLA
SUBRUFESCENS
(Miller)

● 1950 onwards
○ before 1950 only

0   100
Kilometres
0   100
Miles

CHANNEL ISLANDS
PLOTTED ON
UTM GRID

WA
WV

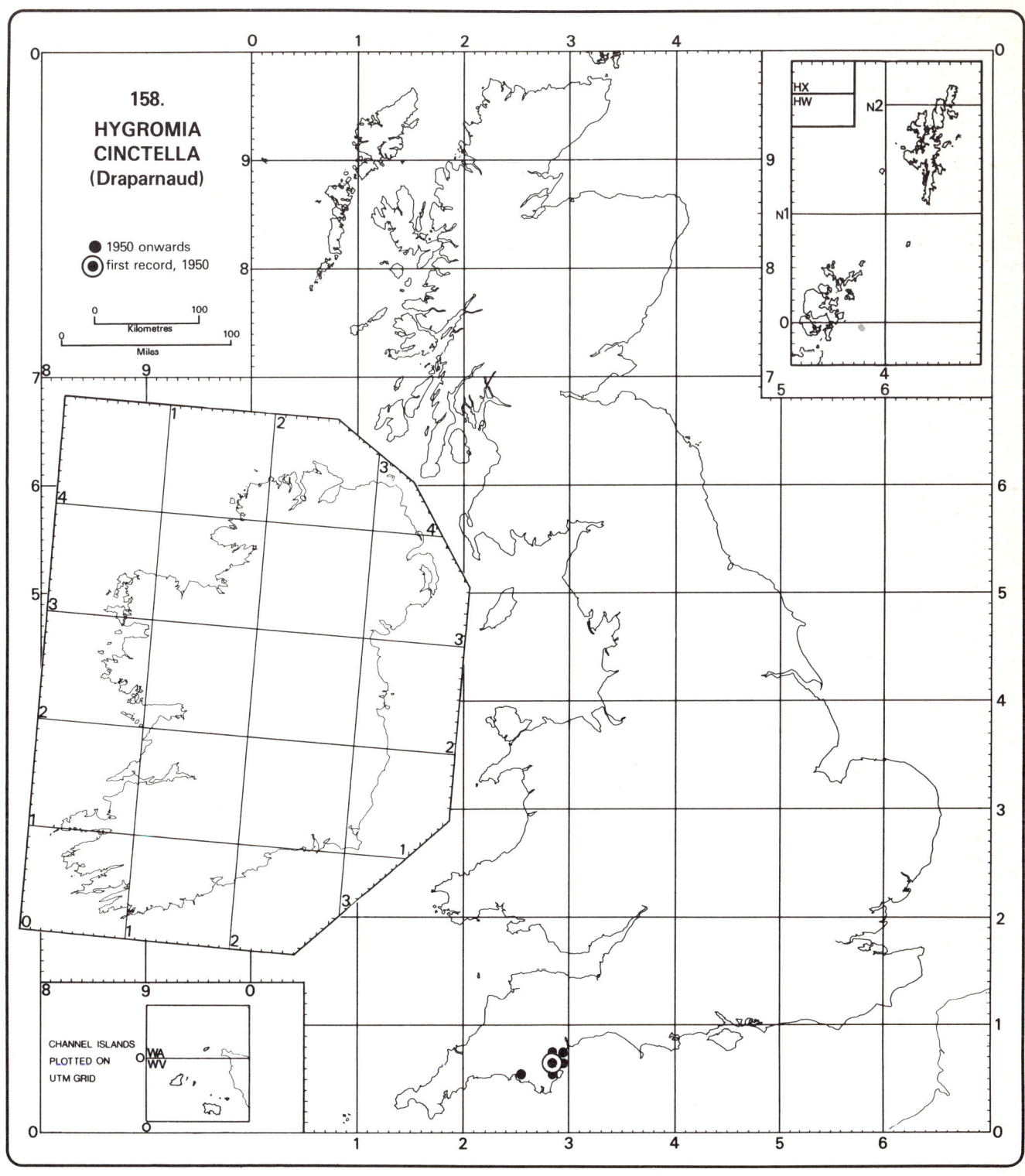

158.

HYGROMIA CINCTELLA
(Draparnaud)

● 1950 onwards
◉ first record, 1950

Kilometres

Miles

CHANNEL ISLANDS
PLOTTED ON
UTM GRID

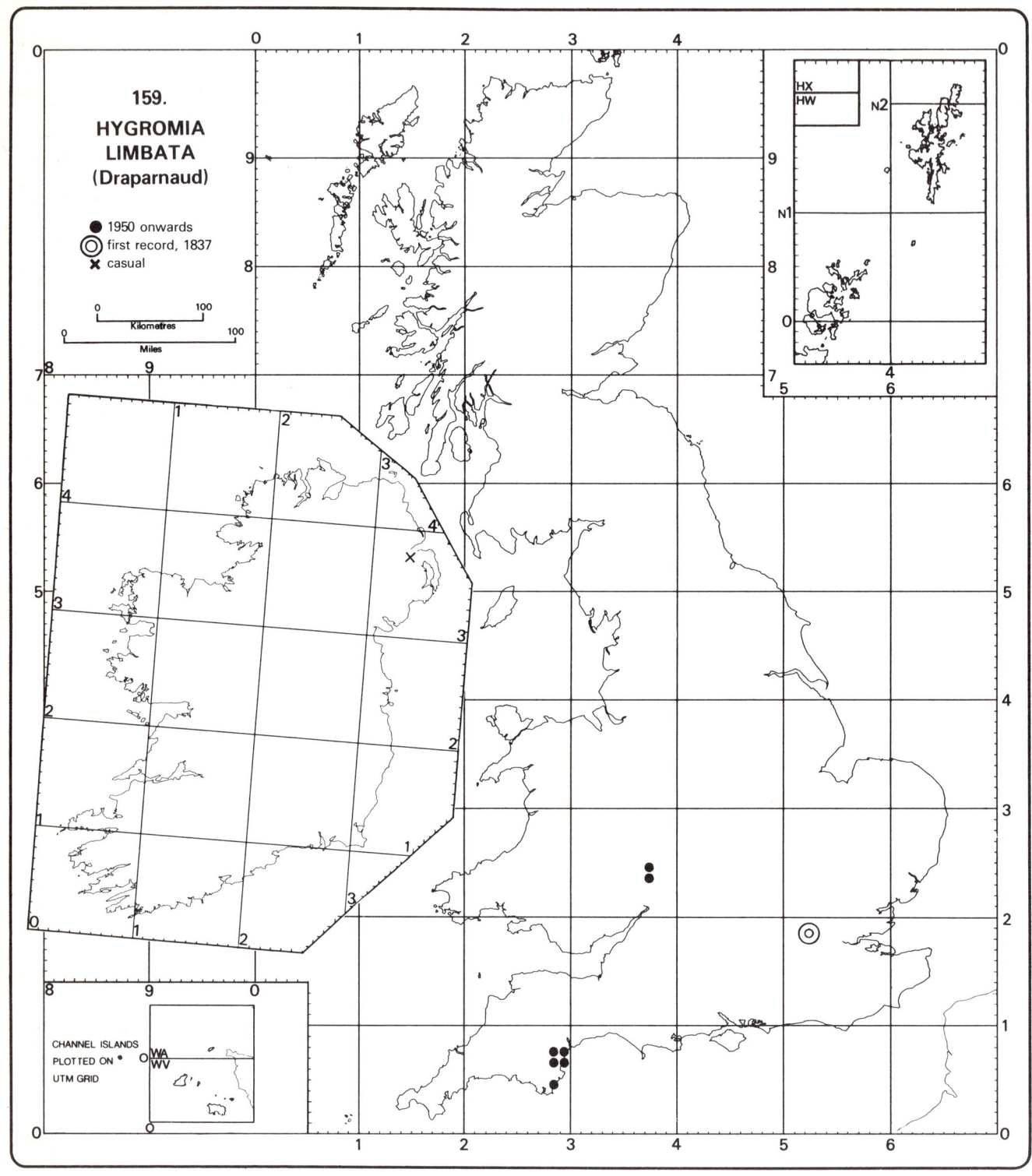

159.
HYGROMIA
LIMBATA
(Draparnaud)

● 1950 onwards
◉ first record, 1837
✕ casual

Kilometres
Miles

HX
HW

CHANNEL ISLANDS
PLOTTED ON ●
UTM GRID

WA
WV

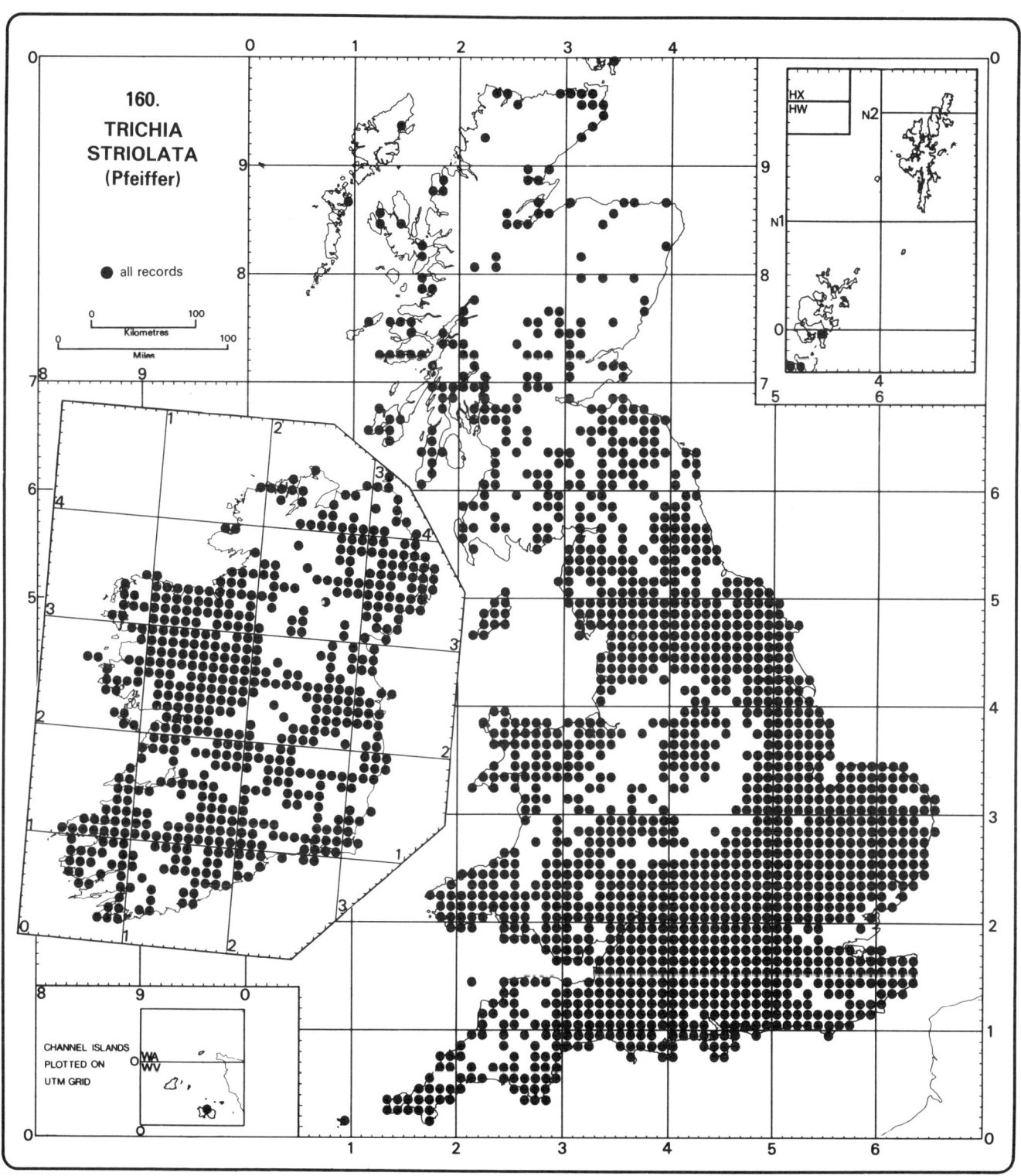

160.
TRICHIA
STRIOLATA
(Pfeiffer)

● all records

Kilometres
0   100
0   100
Miles

CHANNEL ISLANDS
PLOTTED ON
UTM GRID

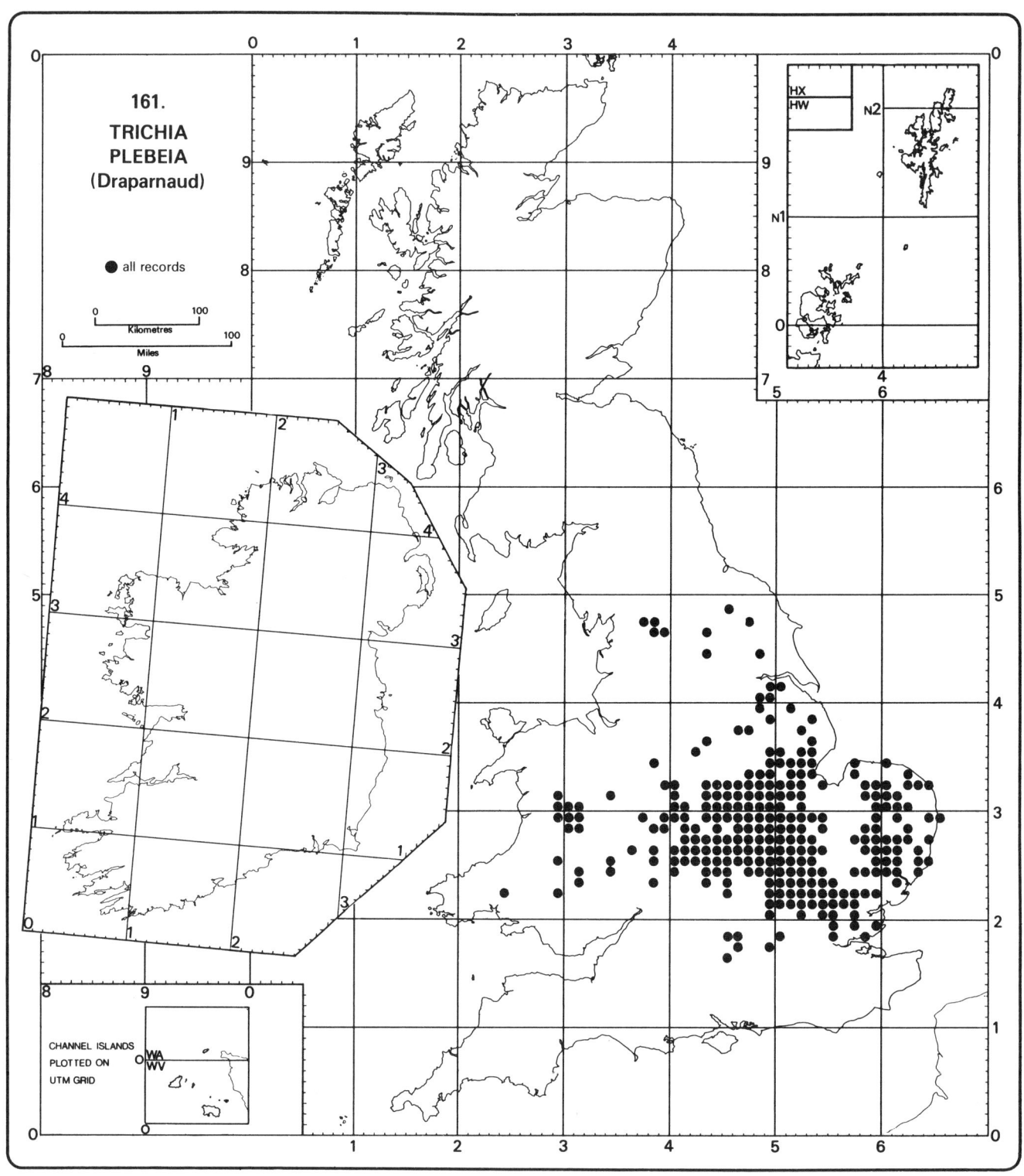

161.
TRICHIA
PLEBEIA
(Draparnaud)

● all records

0   100
Kilometres
0   100
Miles

HX
HW

CHANNEL ISLANDS
PLOTTED ON
UTM GRID

WA
WV

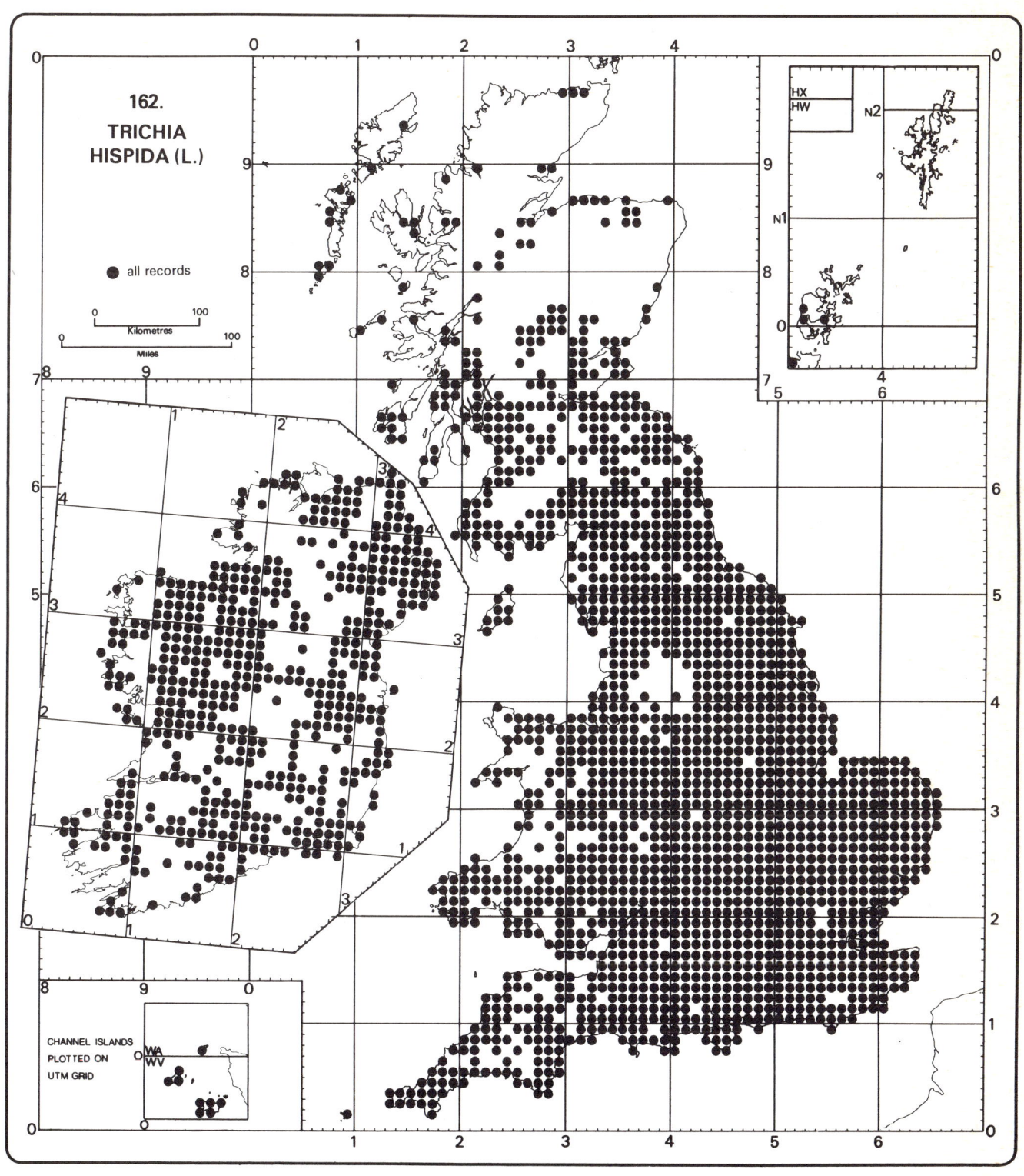

162.
TRICHIA
HISPIDA (L.)

● all records

Kilometres

Miles

CHANNEL ISLANDS
PLOTTED ON
UTM GRID

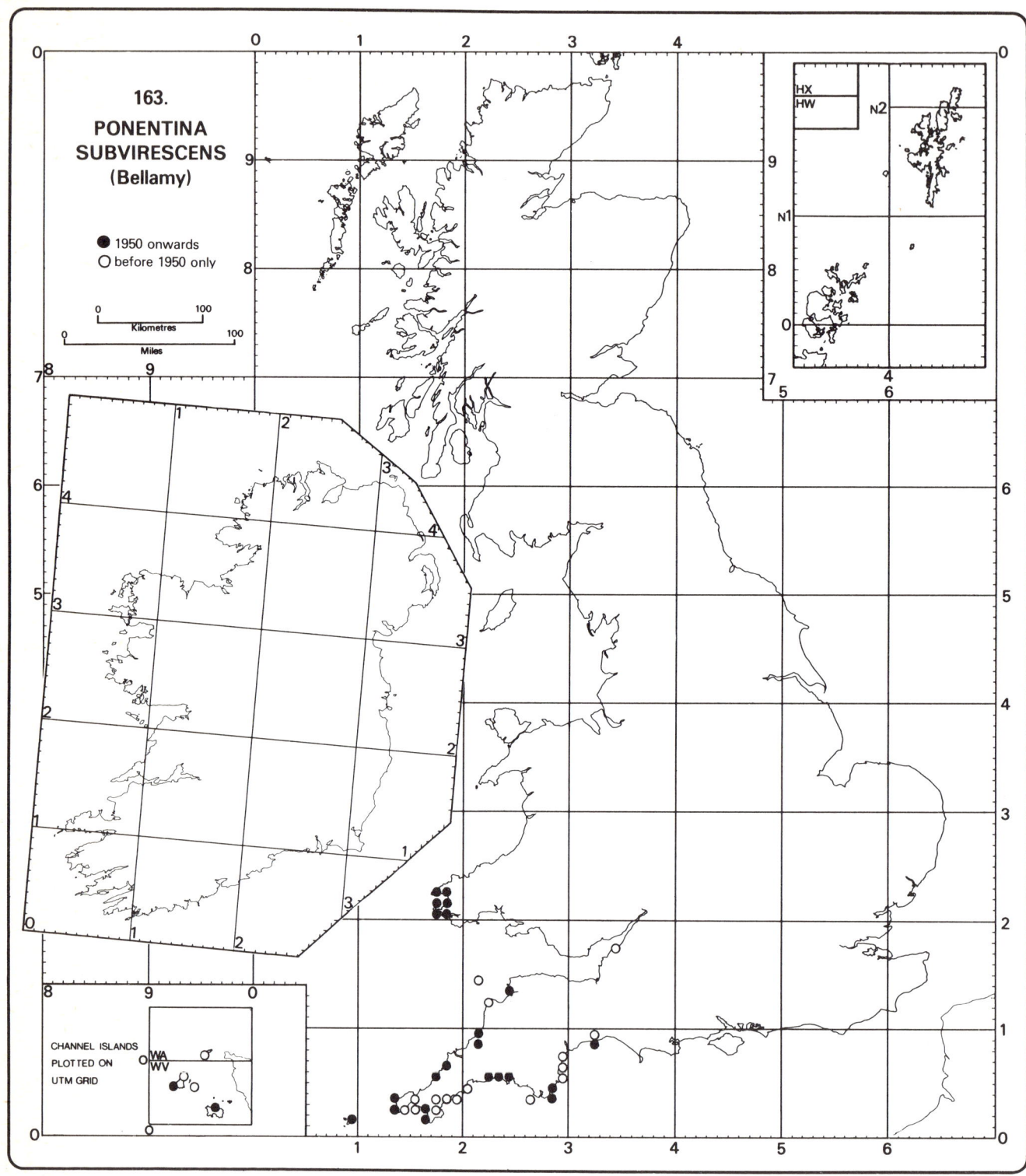

163.
PONENTINA
SUBVIRESCENS
(Bellamy)

● 1950 onwards
○ before 1950 only

Kilometres
Miles

HX
HW

N2

N1

CHANNEL ISLANDS
PLOTTED ON
UTM GRID

WA
WV

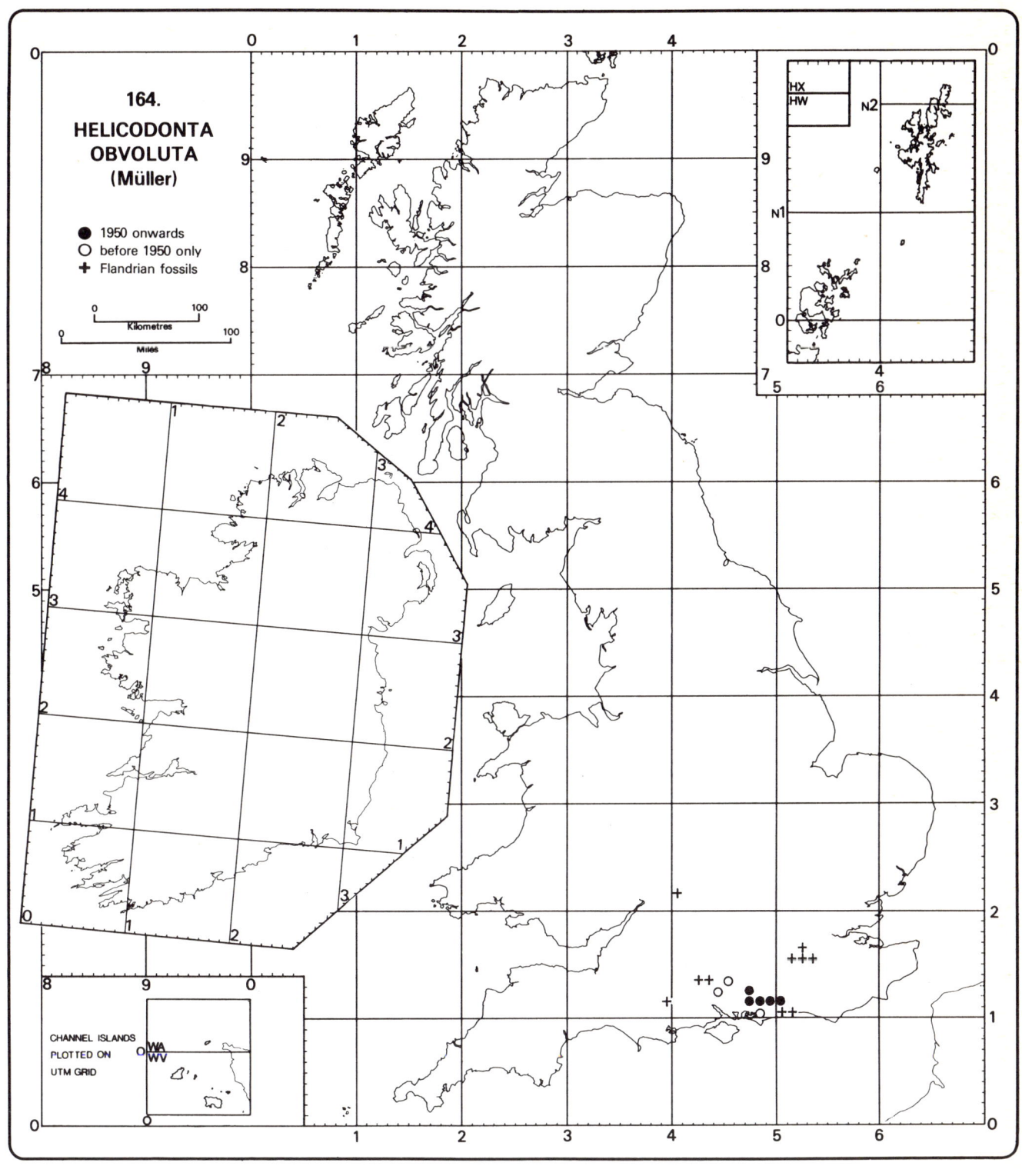

164.

HELICODONTA
OBVOLUTA
(Müller)

● 1950 onwards
○ before 1950 only
+ Flandrian fossils

Kilometres
Miles

CHANNEL ISLANDS
PLOTTED ON
UTM GRID

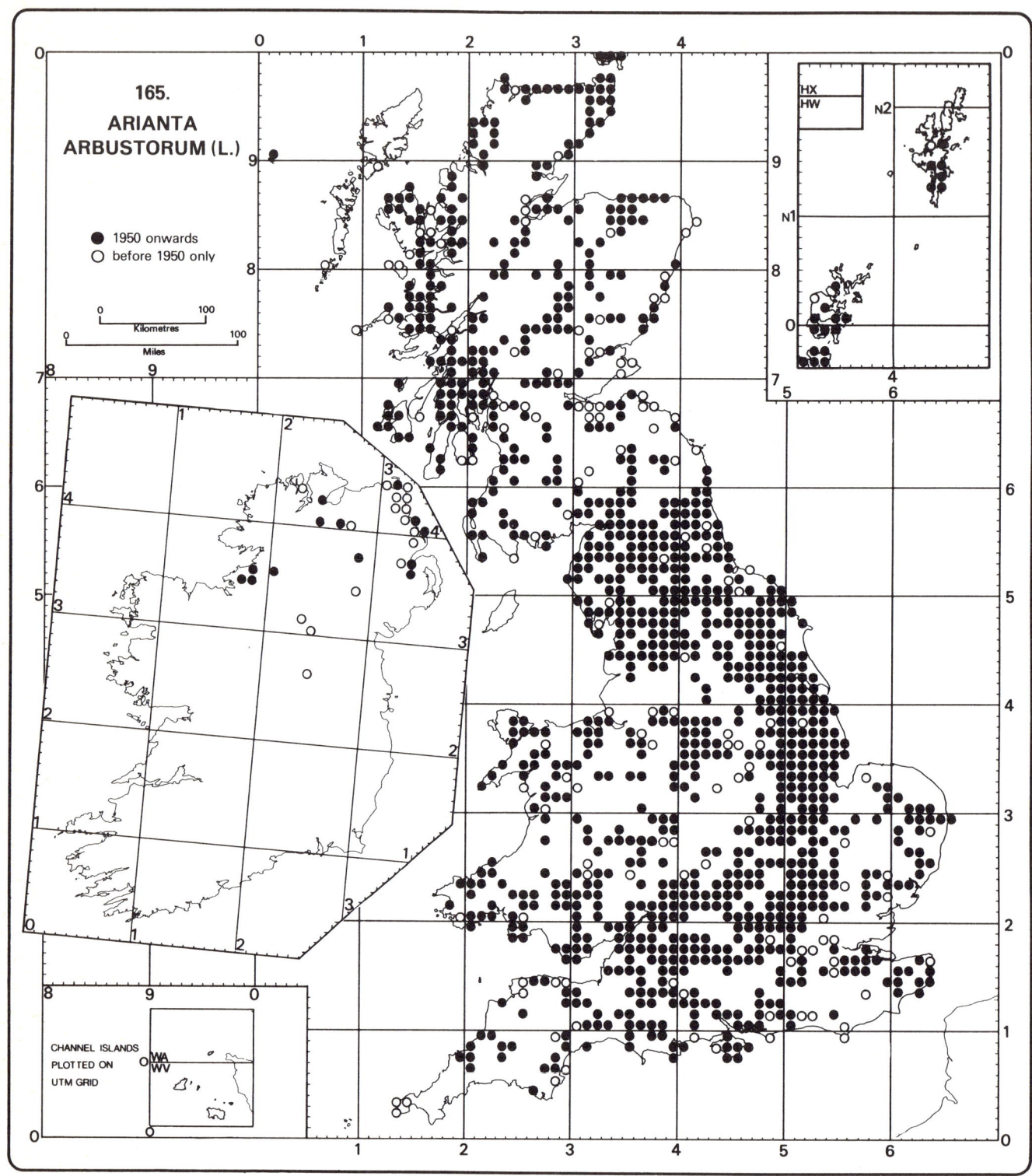

165.
ARIANTA
ARBUSTORUM (L.)

● 1950 onwards
○ before 1950 only

0
Kilometres
100

0
Miles
100

CHANNEL ISLANDS
PLOTTED ON
UTM GRID

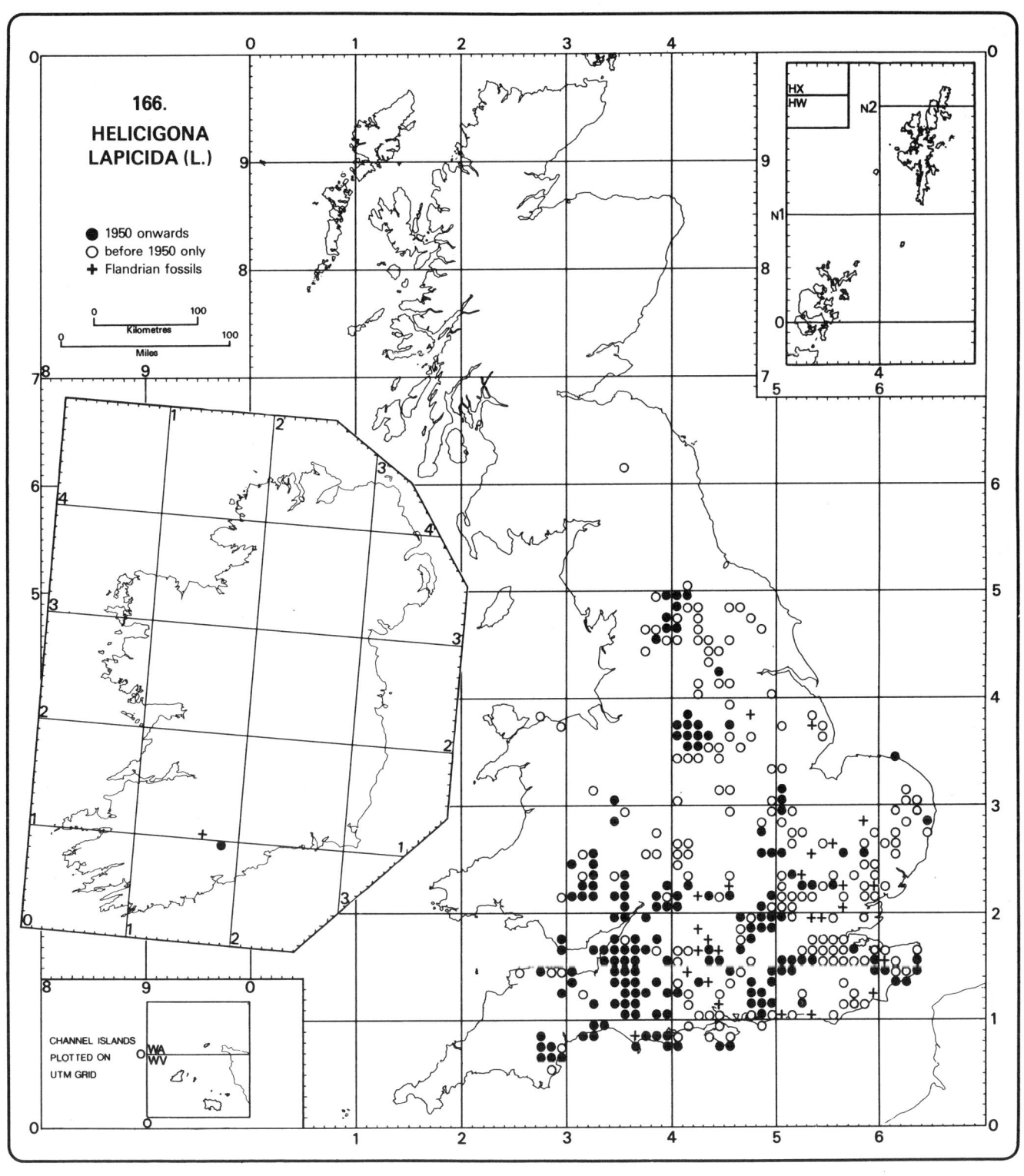

166.

**HELICIGONA
LAPICIDA (L.)**

● 1950 onwards
○ before 1950 only
+ Flandrian fossils

Kilometres
Miles

CHANNEL ISLANDS
PLOTTED ON
UTM GRID

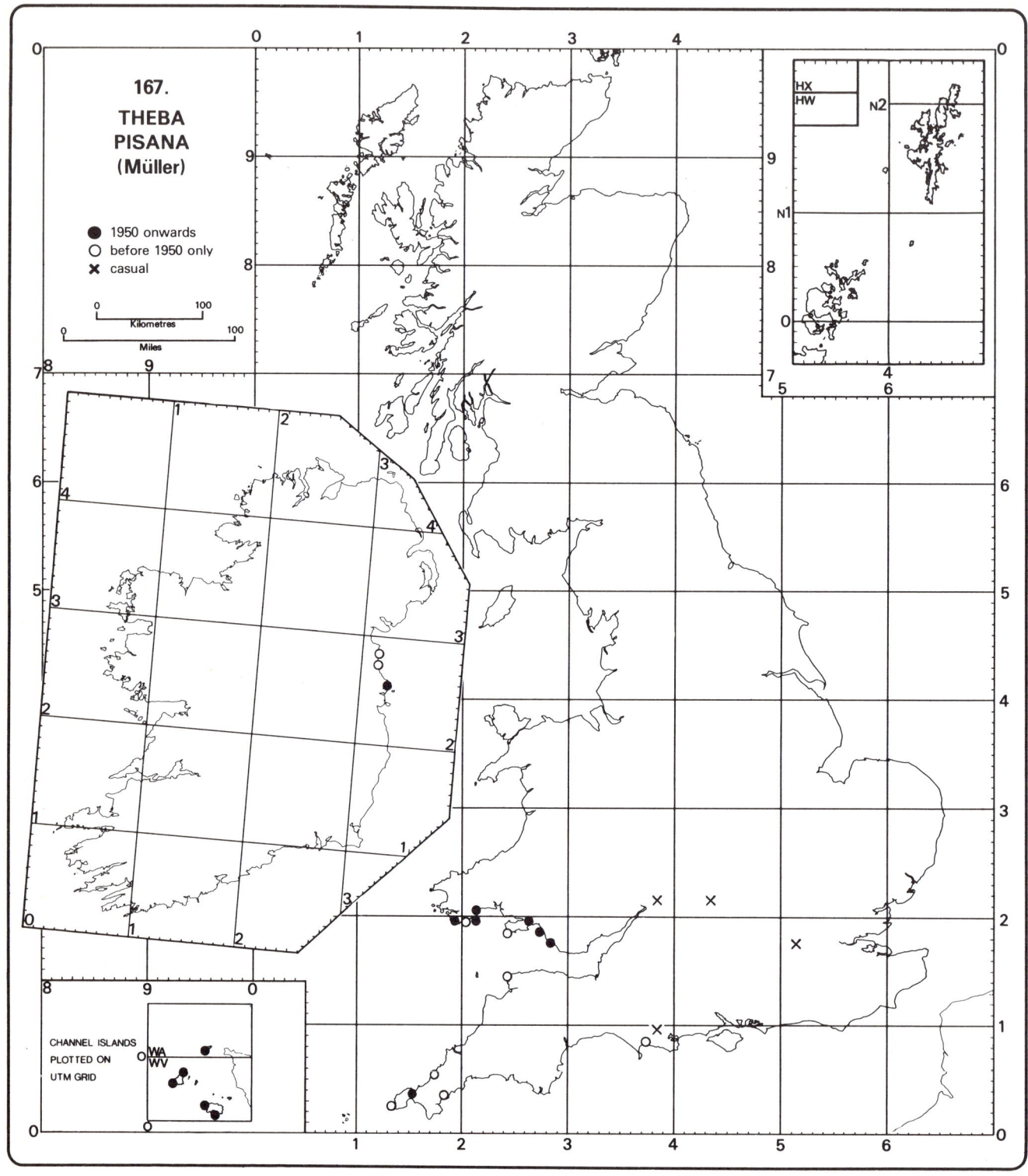

167.

THEBA
PISANA
(Müller)

● 1950 onwards
○ before 1950 only
✖ casual

Kilometres
Miles

CHANNEL ISLANDS
PLOTTED ON
UTM GRID

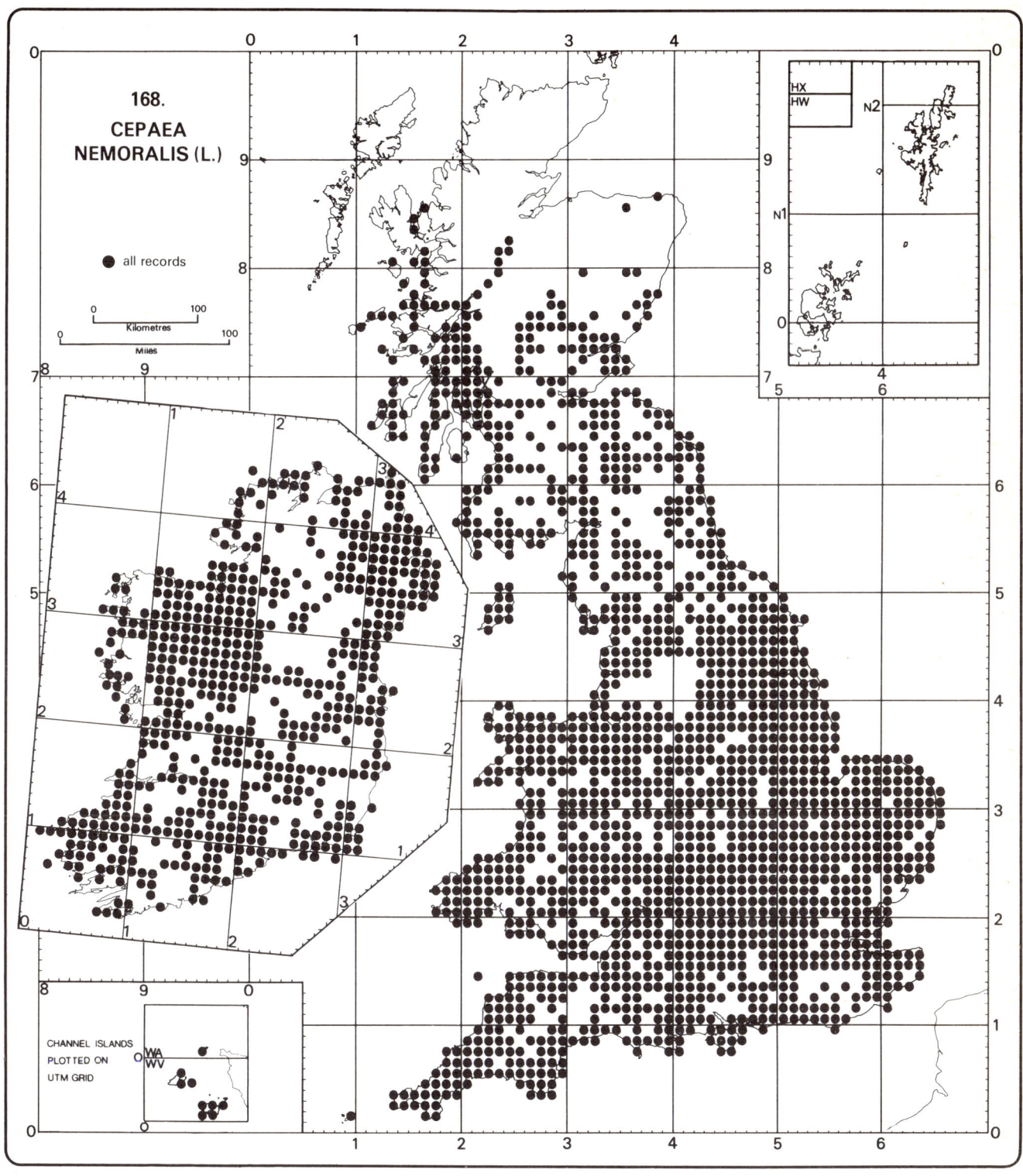

168.
CEPAEA
NEMORALIS (L.)

● all records

0   100
Kilometres
0   100
Miles

CHANNEL ISLANDS
PLOTTED ON
UTM GRID

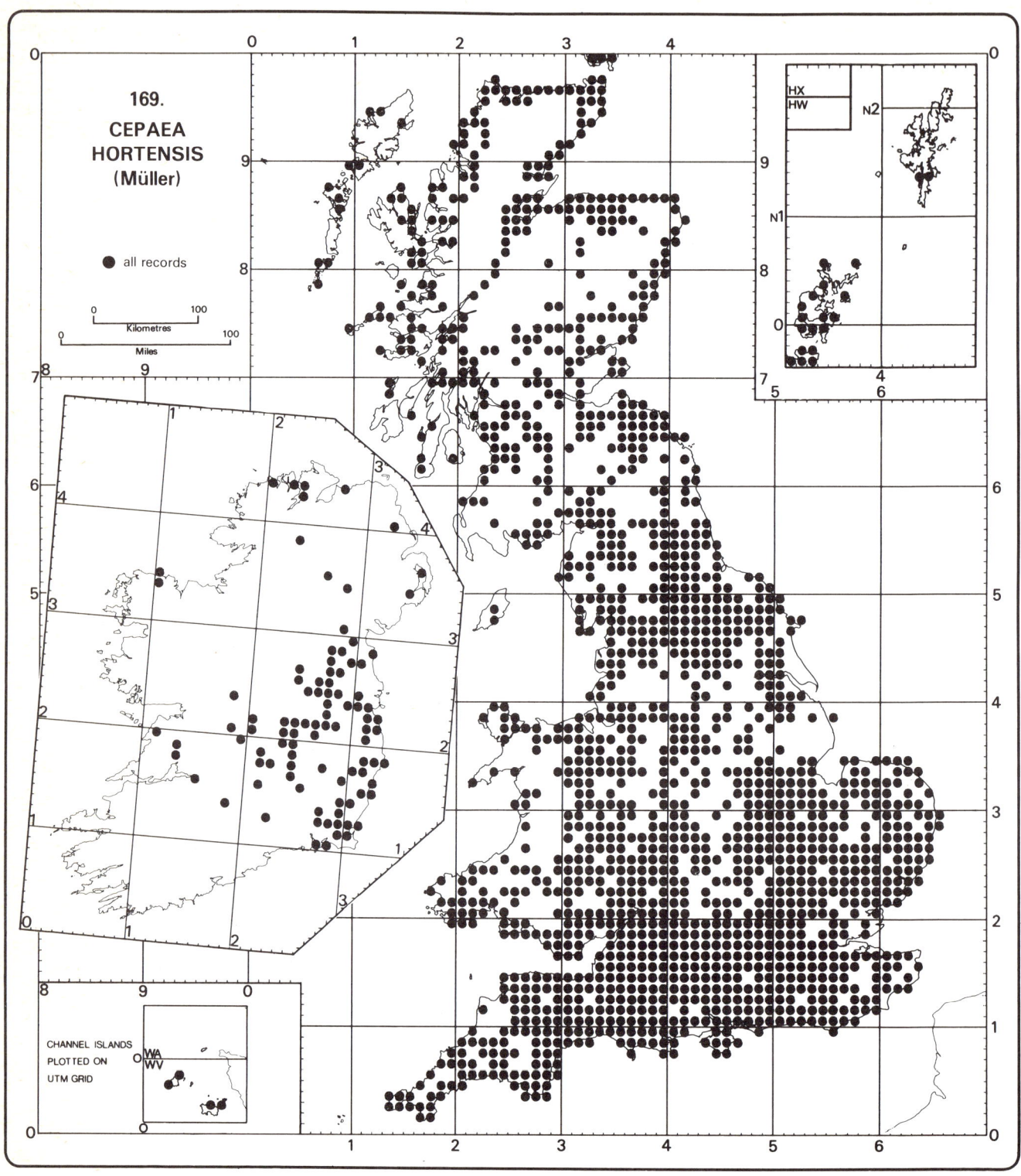

169.
CEPAEA
HORTENSIS
(Müller)

● all records

Kilometres
Miles

CHANNEL ISLANDS
PLOTTED ON
UTM GRID

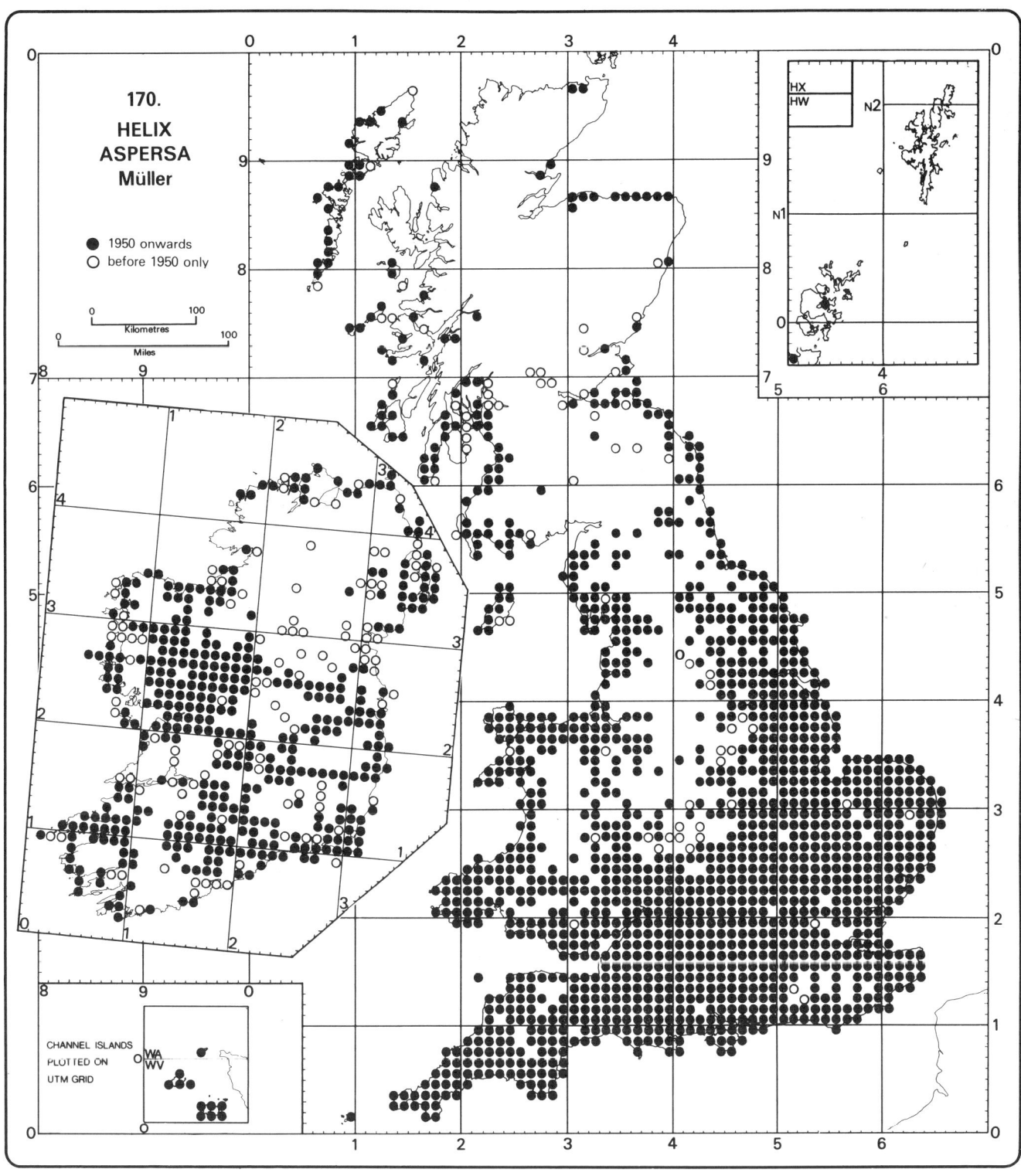

170.
HELIX
ASPERSA
Müller

● 1950 onwards
○ before 1950 only

0   100
Kilometres
0   100
Miles

HX
HW

N2

N1

CHANNEL ISLANDS
PLOTTED ON
UTM GRID

WA
WV

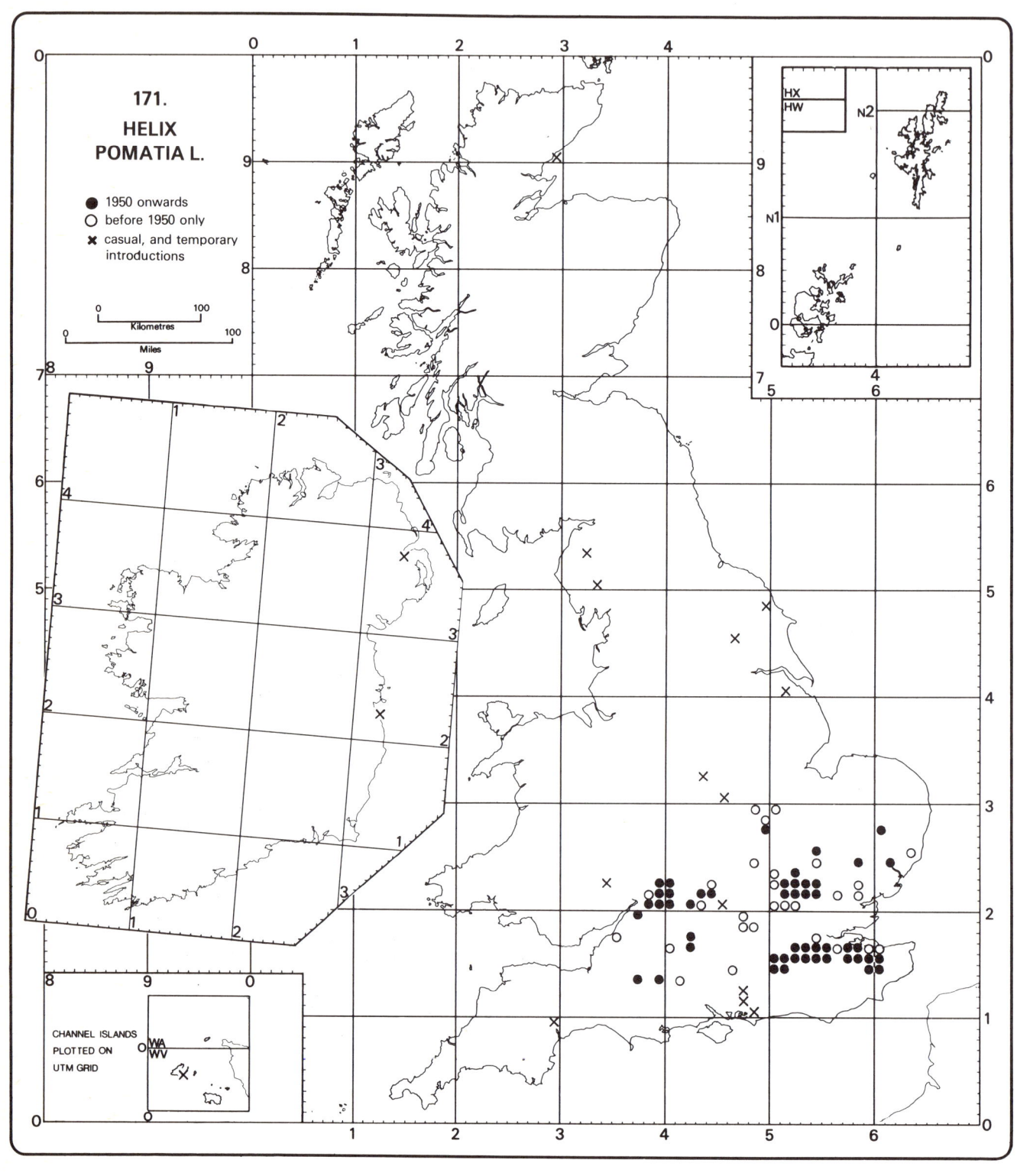

171.

HELIX
POMATIA L.

● 1950 onwards
○ before 1950 only
✕ casual, and temporary
   introductions

CHANNEL ISLANDS
PLOTTED ON
UTM GRID

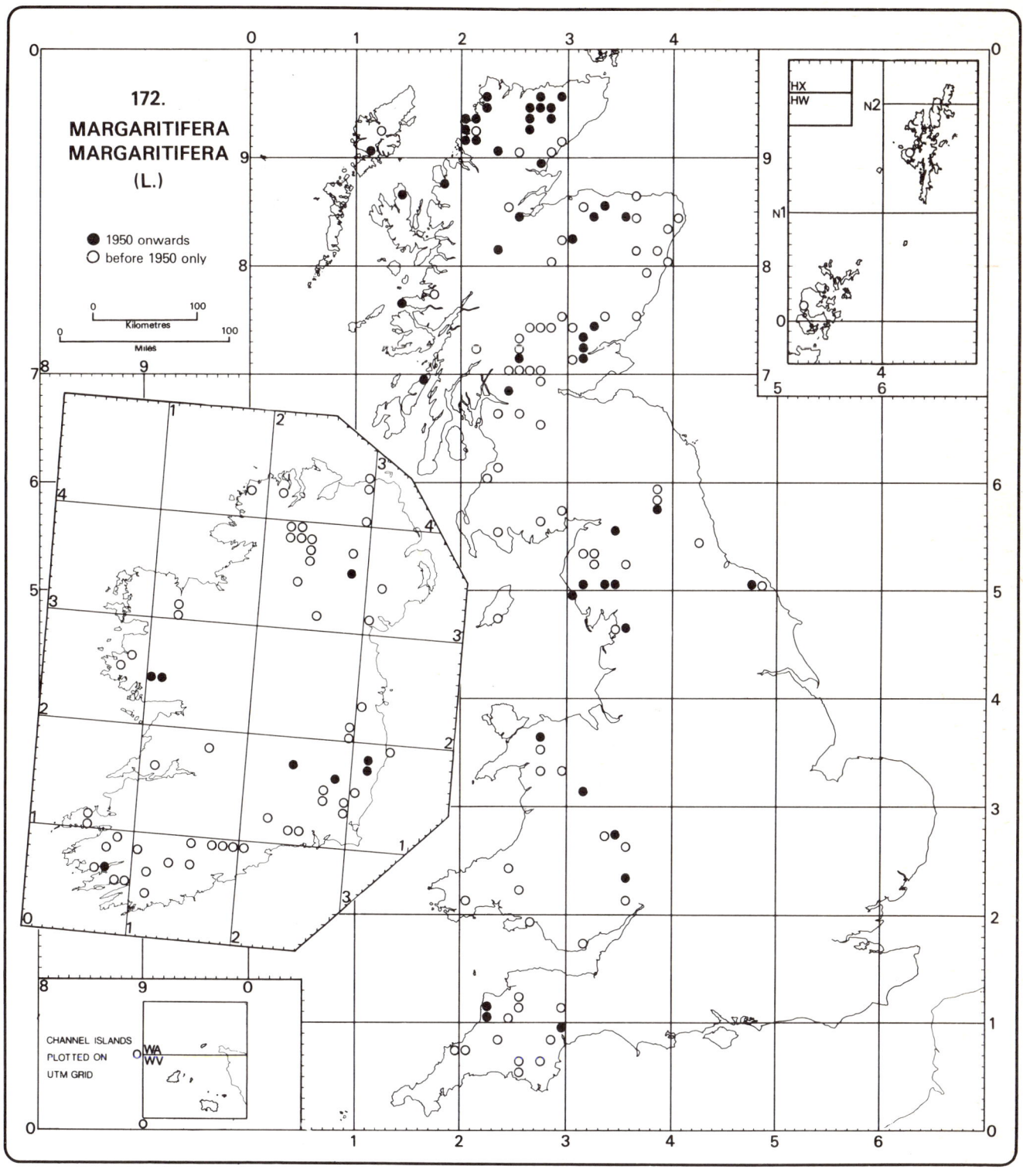

172.
MARGARITIFERA
MARGARITIFERA
(L.)

● 1950 onwards
○ before 1950 only

Kilometres

Miles

HX
HW

CHANNEL ISLANDS
PLOTTED ON
UTM GRID

WA
WV

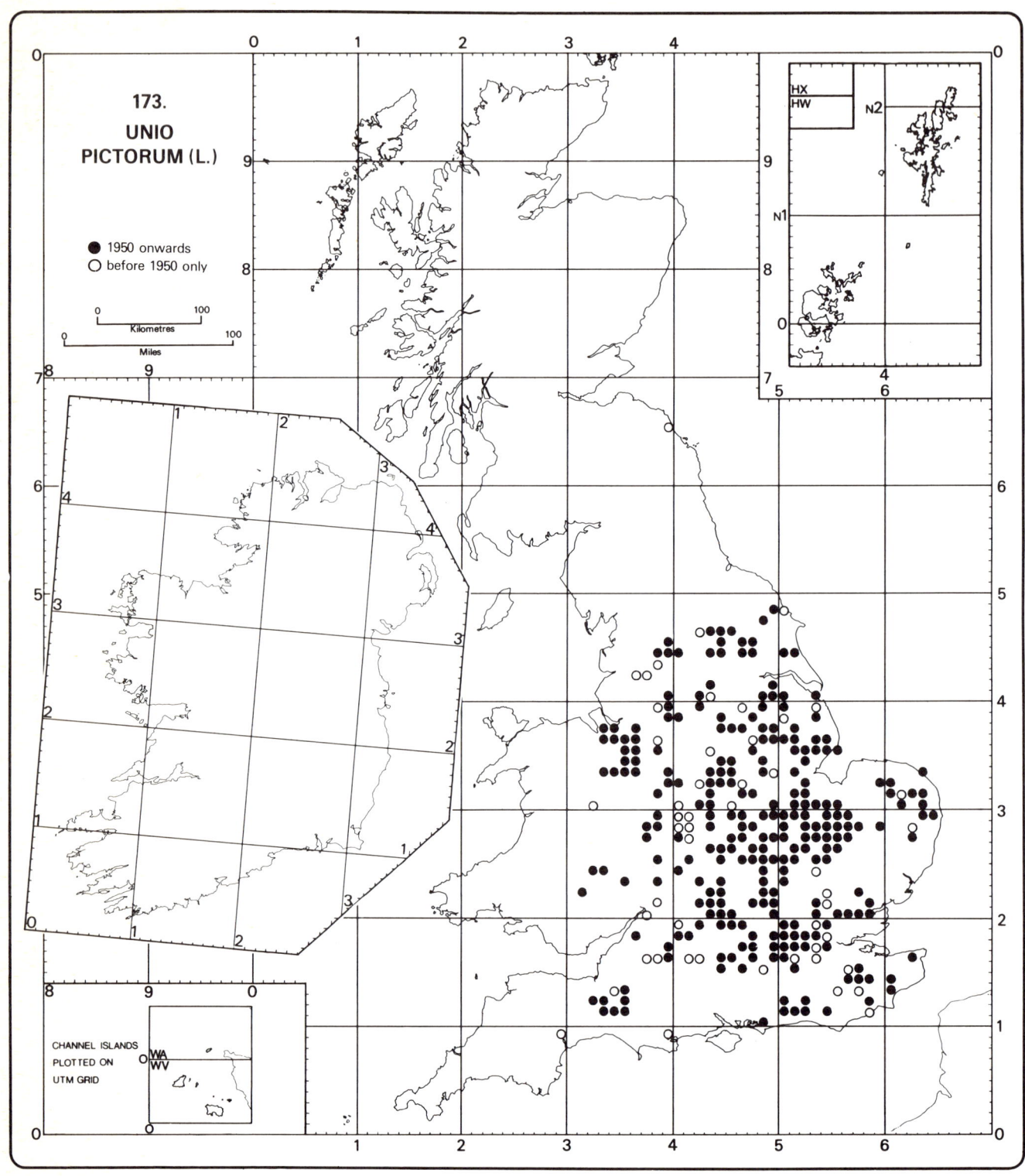

173.

UNIO
PICTORUM (L.)

● 1950 onwards
○ before 1950 only

0   100
Kilometres
0   100
Miles

HX
HW

CHANNEL ISLANDS
PLOTTED ON
UTM GRID

WA
WV

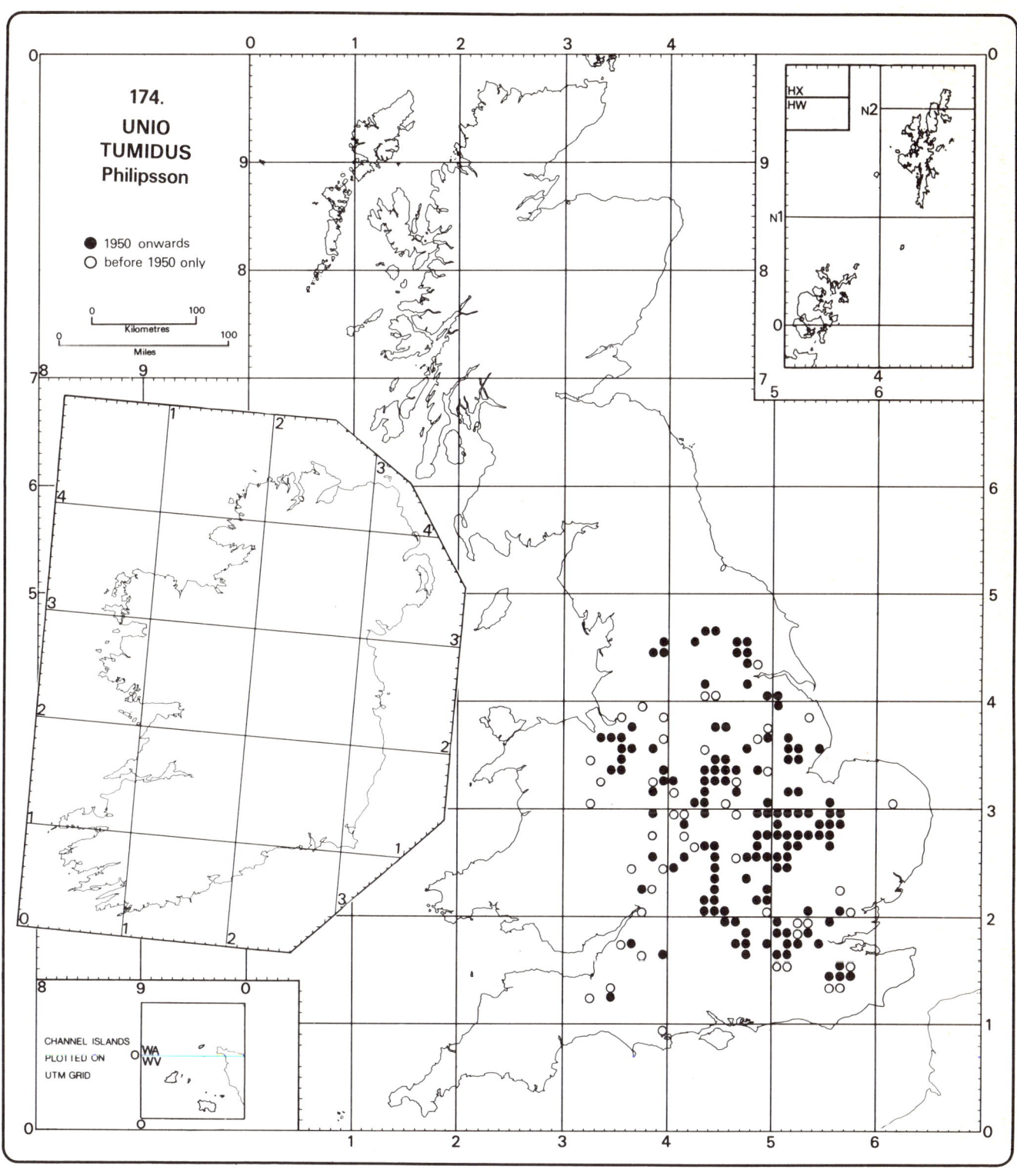

174.

UNIO
TUMIDUS
Philipsson

● 1950 onwards
○ before 1950 only

0   100
Kilometres
0   100
Miles

CHANNEL ISLANDS
PLOTTED ON
UTM GRID

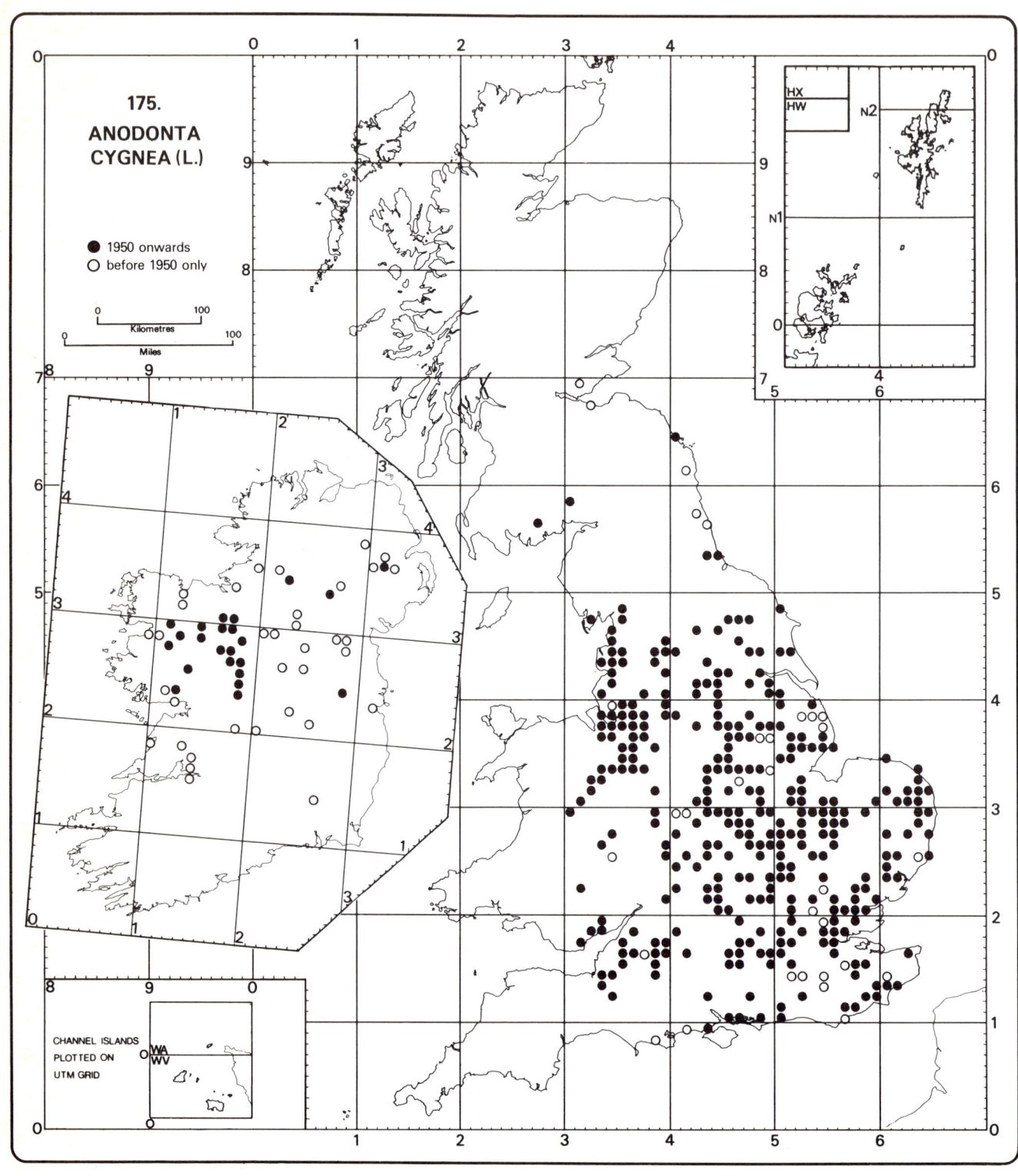

175.

ANODONTA
CYGNEA (L.)

● 1950 onwards
○ before 1950 only

Kilometres
Miles

HX
HW

CHANNEL ISLANDS
PLOTTED ON
UTM GRID

WA
WV

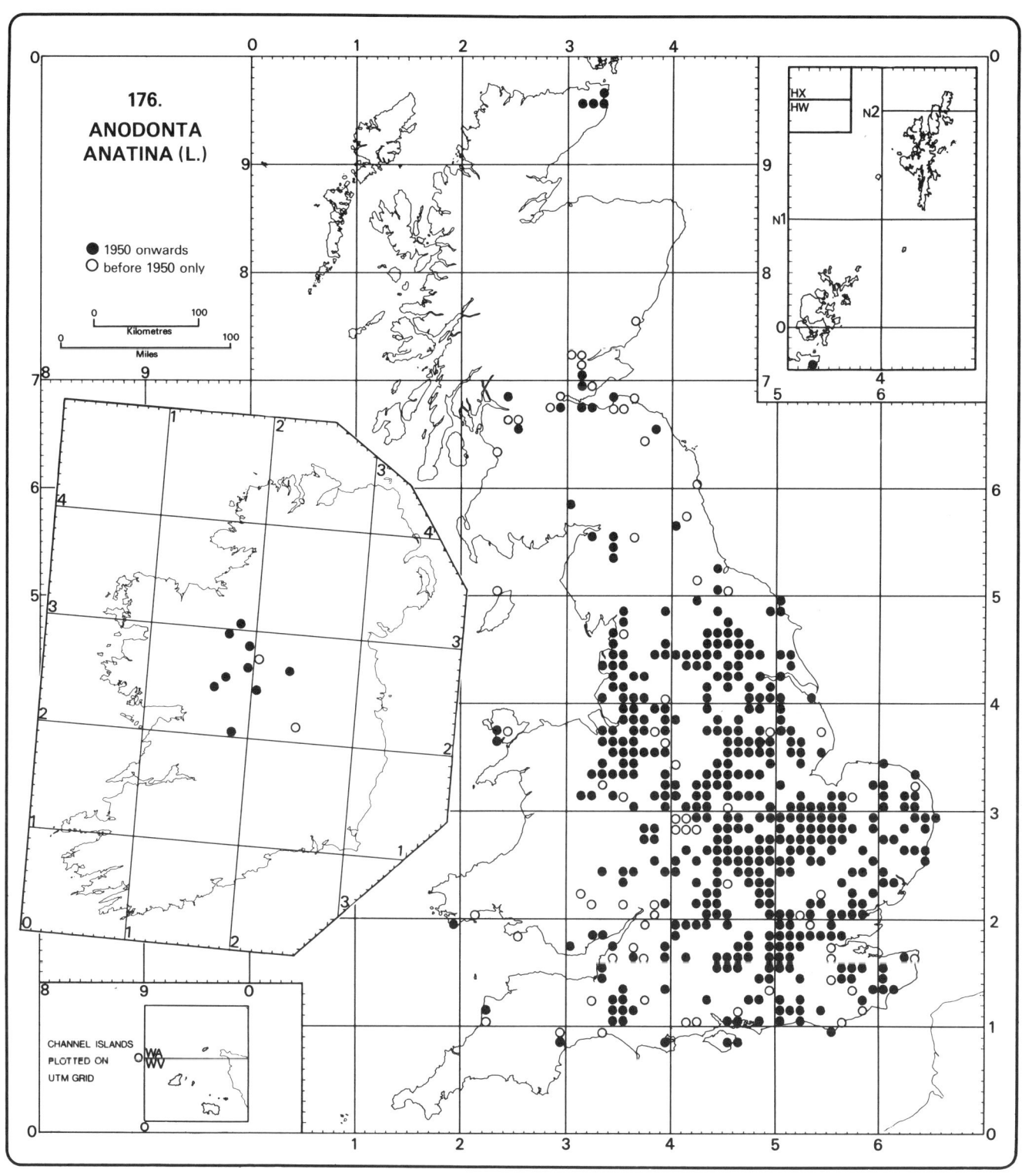

176.
ANODONTA
ANATINA (L.)

● 1950 onwards
○ before 1950 only

0   100
Kilometres
0   100
Miles

HX
HW

CHANNEL ISLANDS
PLOTTED ON
UTM GRID

WA
WV

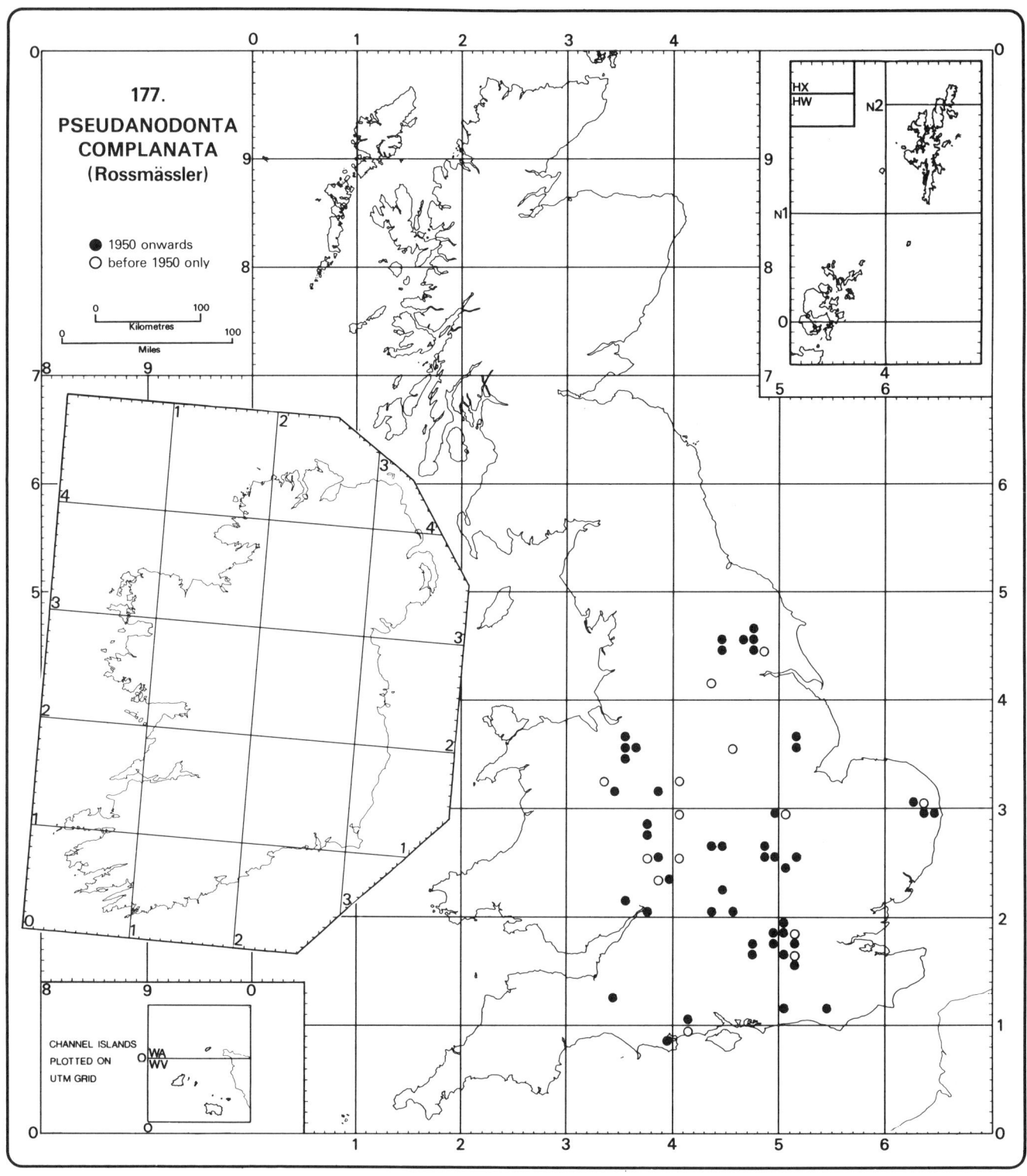

177.
PSEUDANODONTA
COMPLANATA
(Rossmässler)

● 1950 onwards
○ before 1950 only

0 ___ 100
Kilometres
0 ___ 100
Miles

CHANNEL ISLANDS
PLOTTED ON
UTM GRID

○ WA
WV

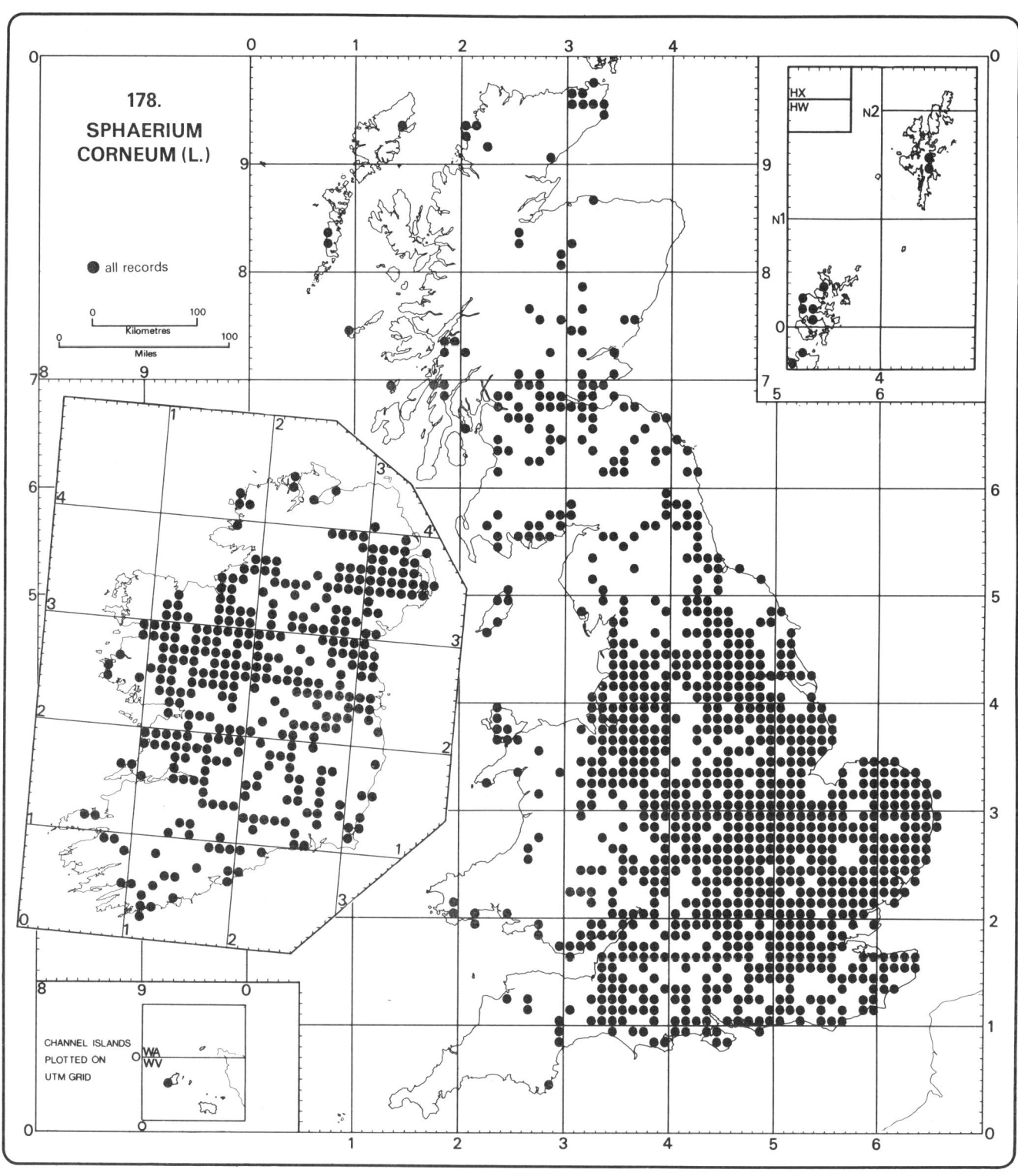

178.
SPHAERIUM
CORNEUM (L.)

● all records

Kilometres

Miles

CHANNEL ISLANDS
PLOTTED ON
UTM GRID

WA
WV

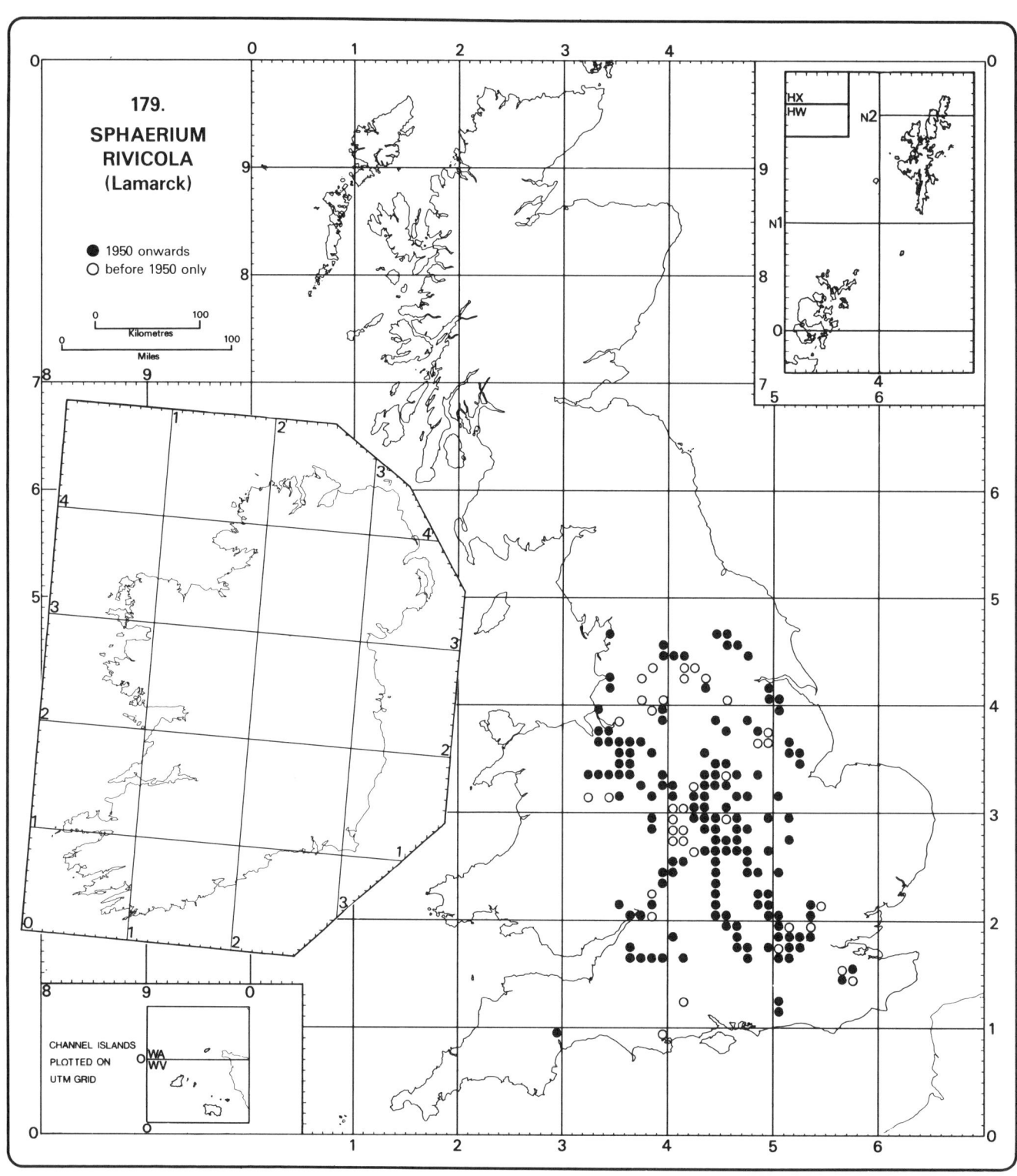

179.
SPHAERIUM
RIVICOLA
(Lamarck)

● 1950 onwards
○ before 1950 only

0   100
Kilometres
0   100
Miles

CHANNEL ISLANDS
PLOTTED ON
UTM GRID

WA
WV

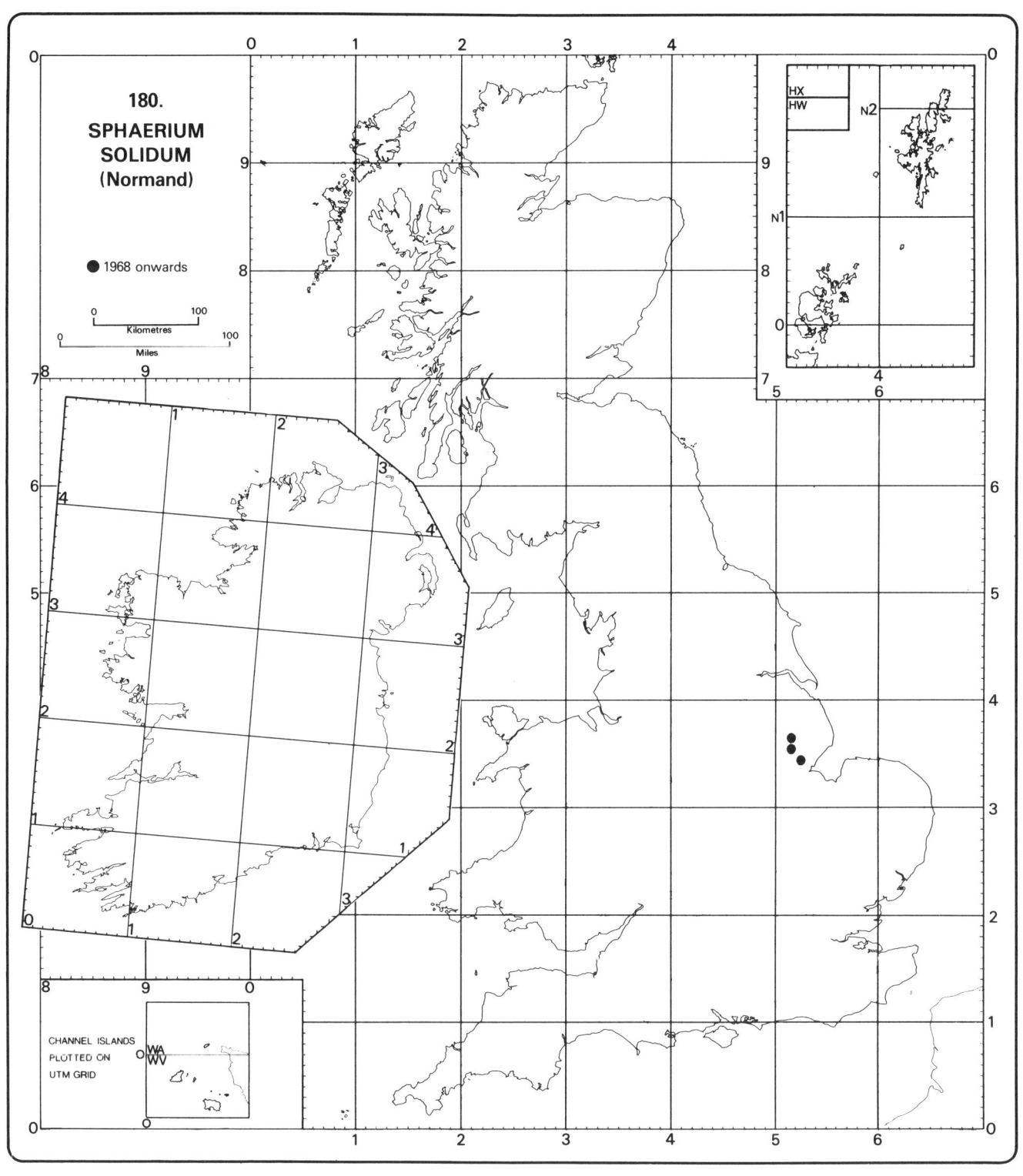

180.
SPHAERIUM
SOLIDUM
(Normand)

● 1968 onwards

CHANNEL ISLANDS
PLOTTED ON
UTM GRID

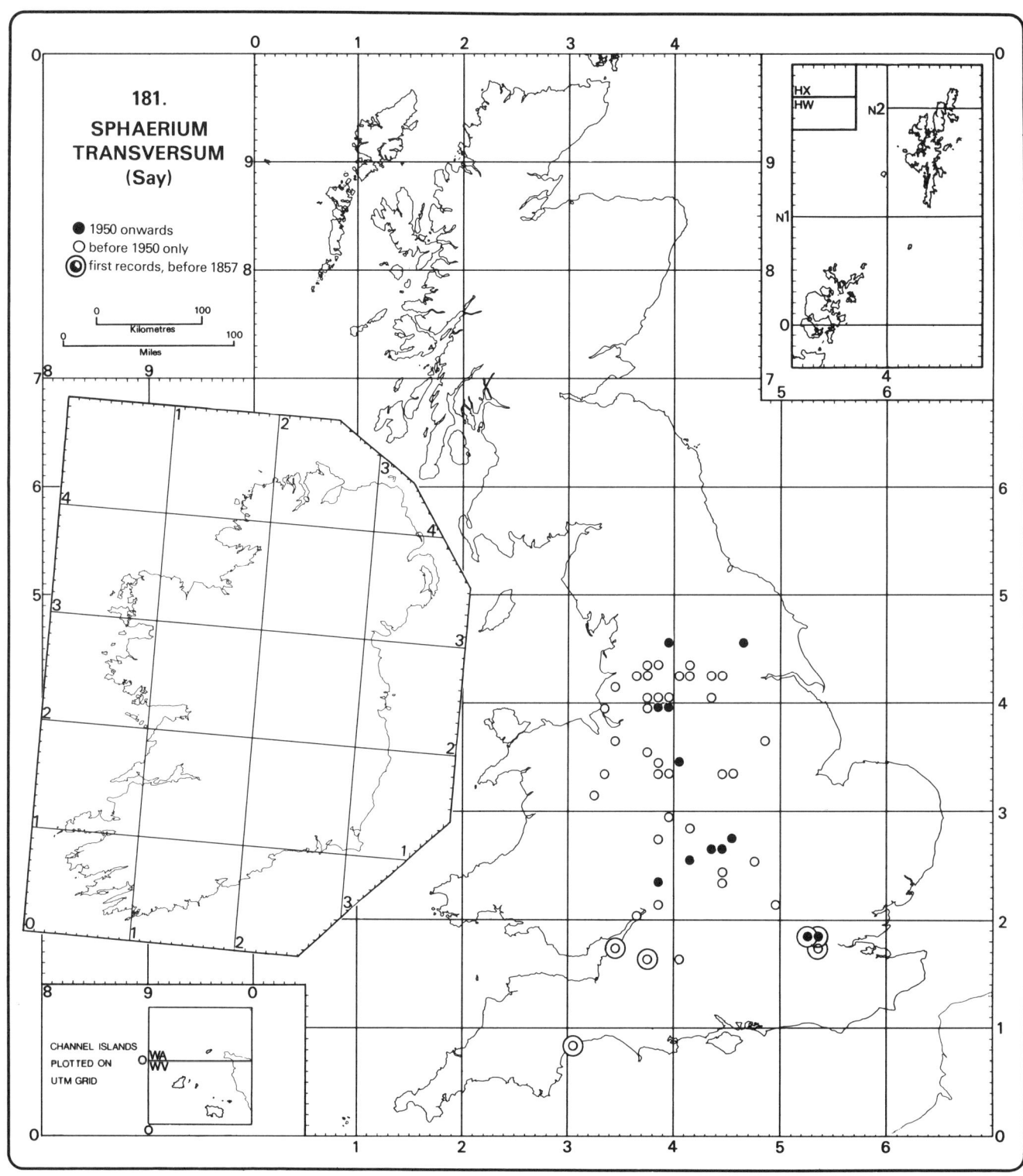

181.
SPHAERIUM
TRANSVERSUM
(Say)

● 1950 onwards
○ before 1950 only
◉ first records, before 1857

0   100
Kilometres
0   100
Miles

HX
HW

CHANNEL ISLANDS
PLOTTED ON
UTM GRID

WA
WV

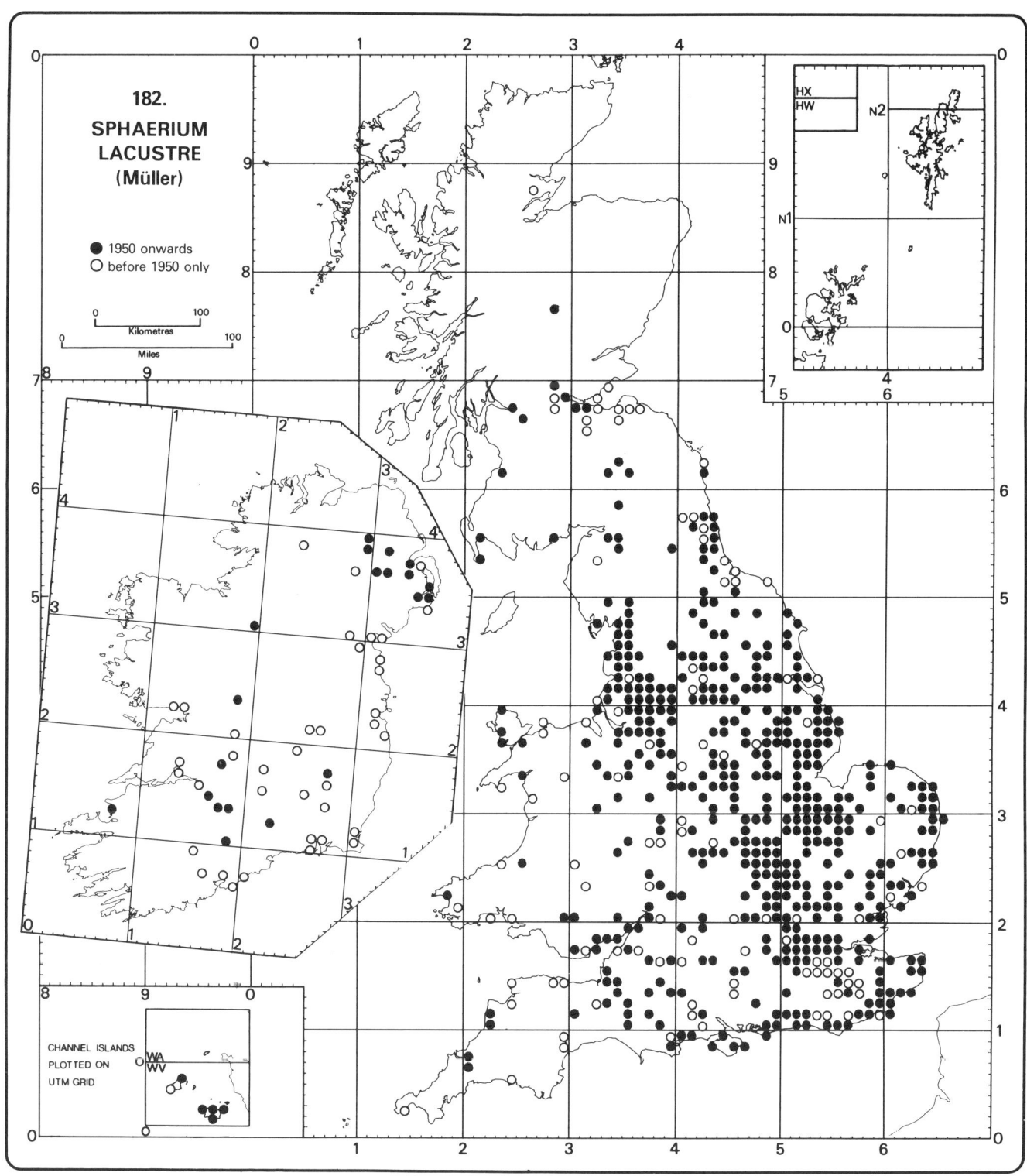

182.

SPHAERIUM
LACUSTRE
(Müller)

● 1950 onwards
○ before 1950 only

0   100
Kilometres
0   100
Miles

CHANNEL ISLANDS
PLOTTED ON
UTM GRID

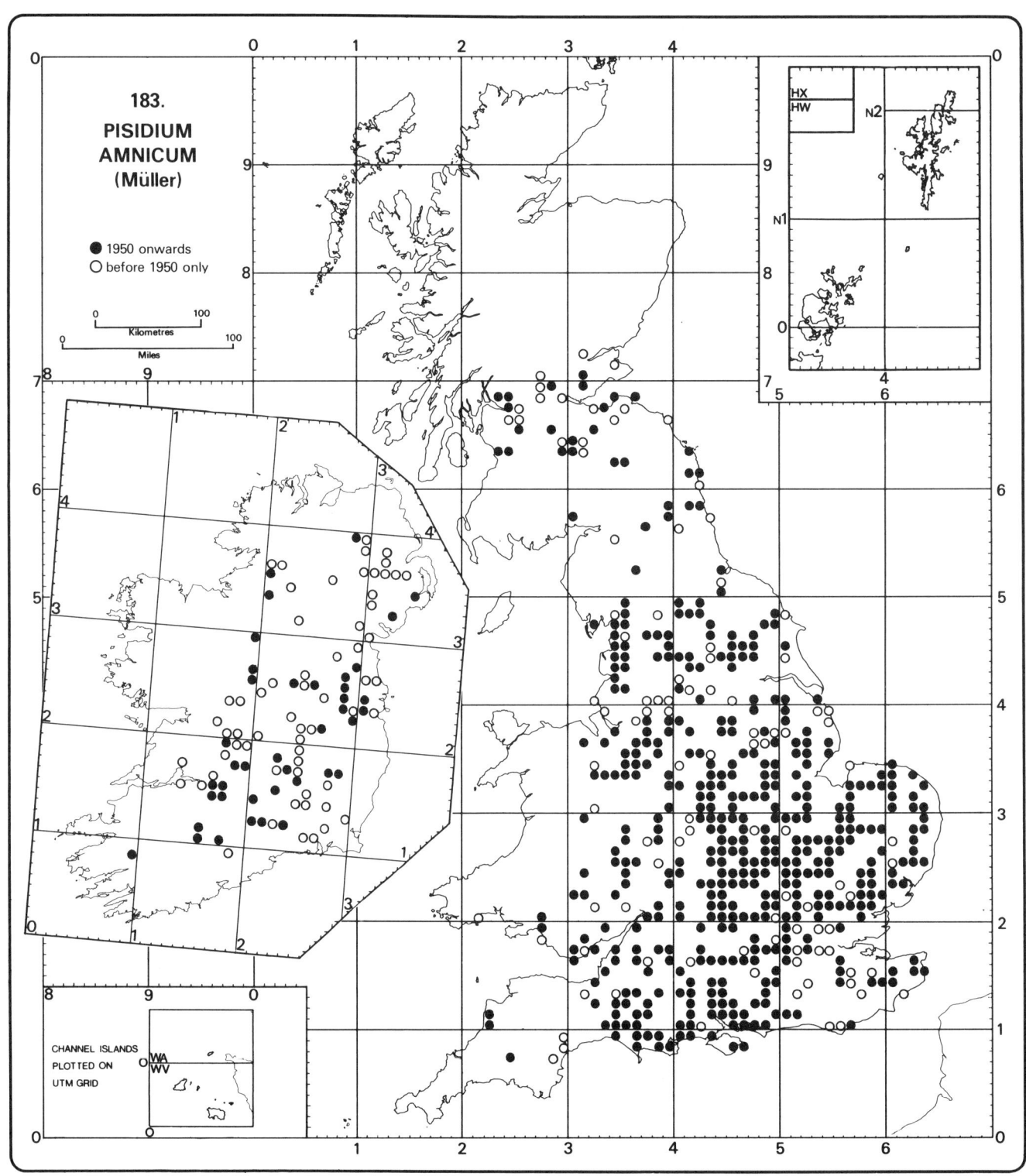

183.
PISIDIUM
AMNICUM
(Müller)

● 1950 onwards
○ before 1950 only

Kilometres

Miles

CHANNEL ISLANDS
PLOTTED ON
UTM GRID

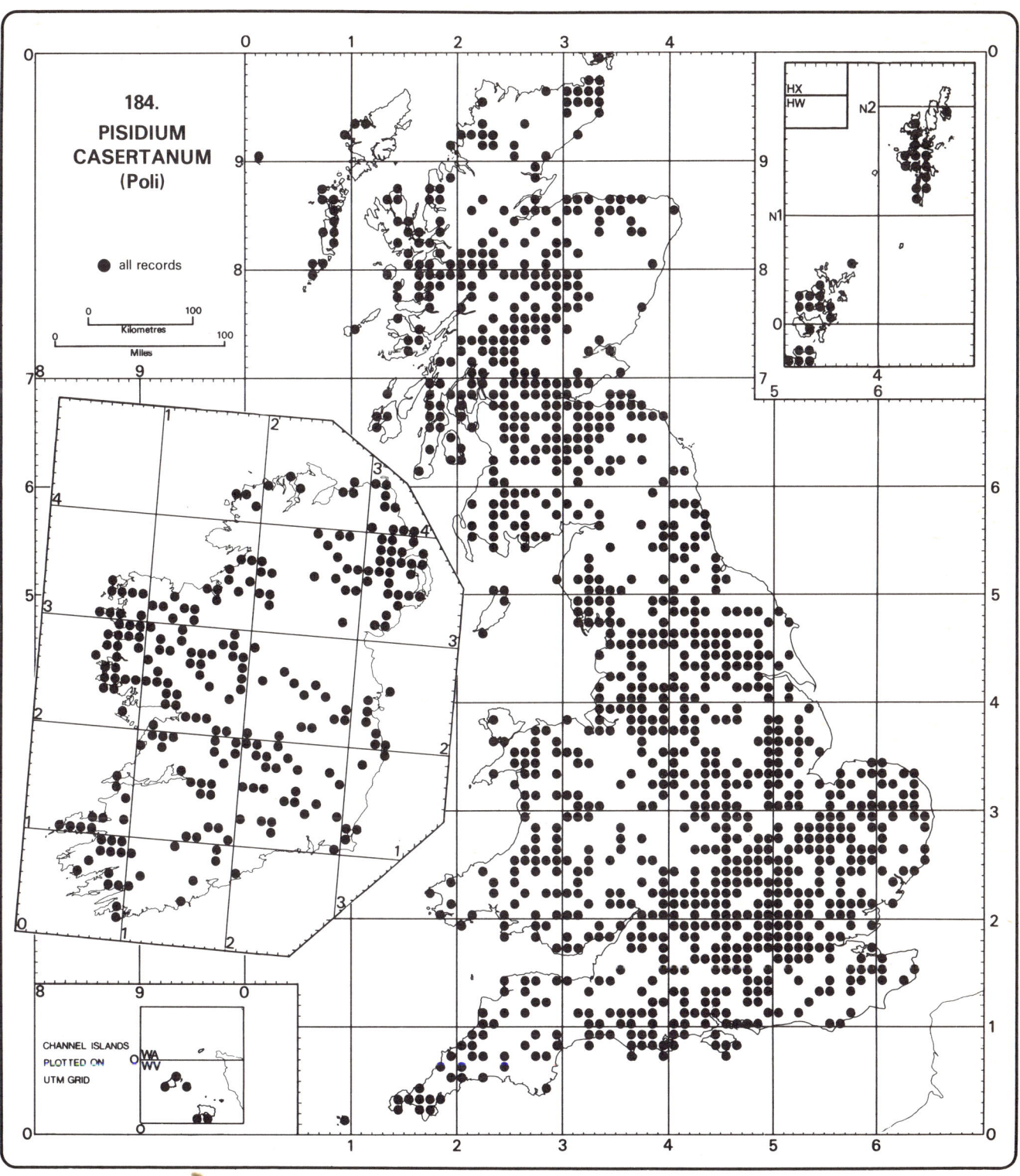

184.
PISIDIUM
CASERTANUM
(Poli)

● all records

CHANNEL ISLANDS
PLOTTED ON
UTM GRID

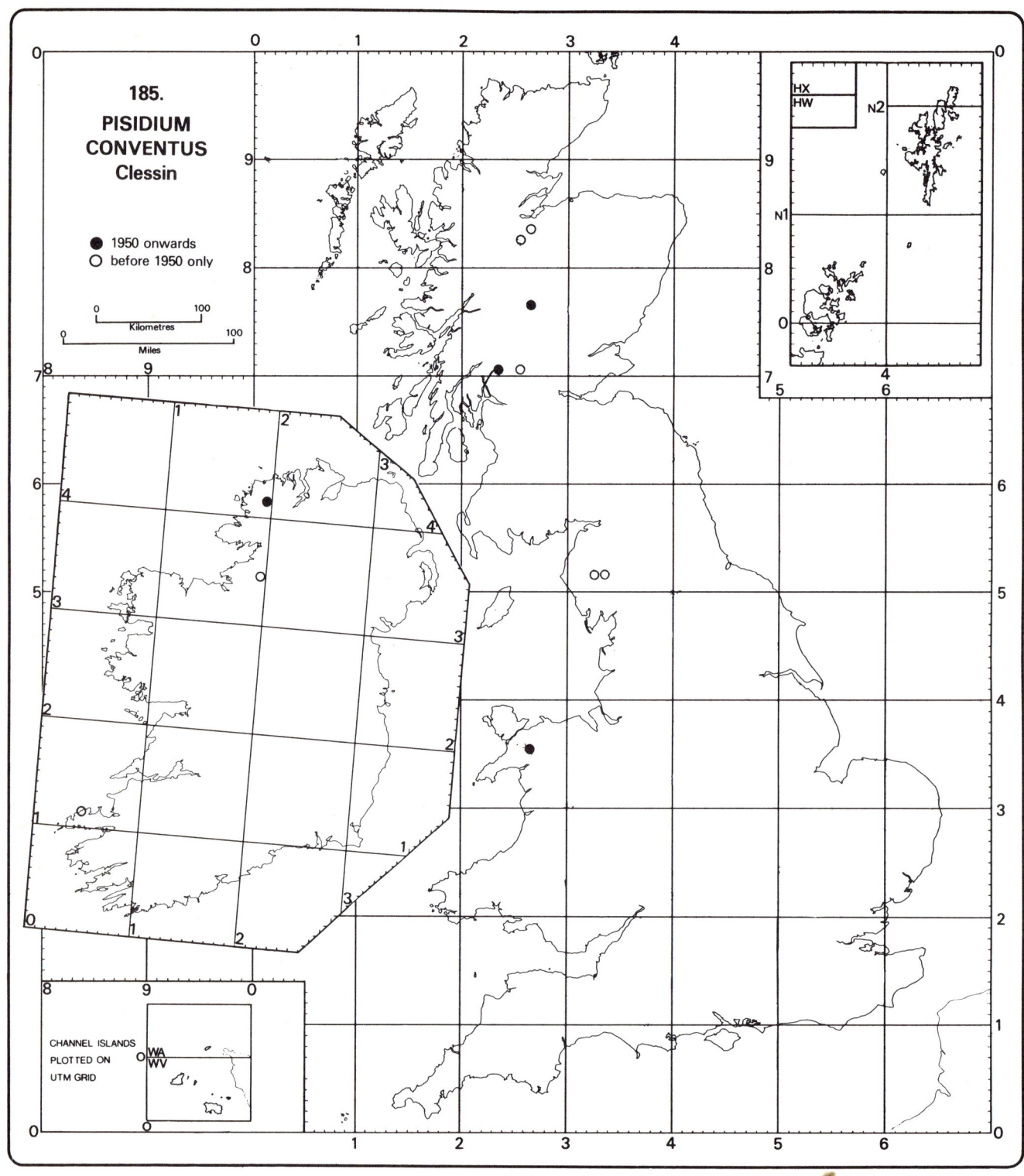

185.

PISIDIUM CONVENTUS
Clessin

● 1950 onwards
○ before 1950 only

Kilometres
Miles

CHANNEL ISLANDS
PLOTTED ON
UTM GRID

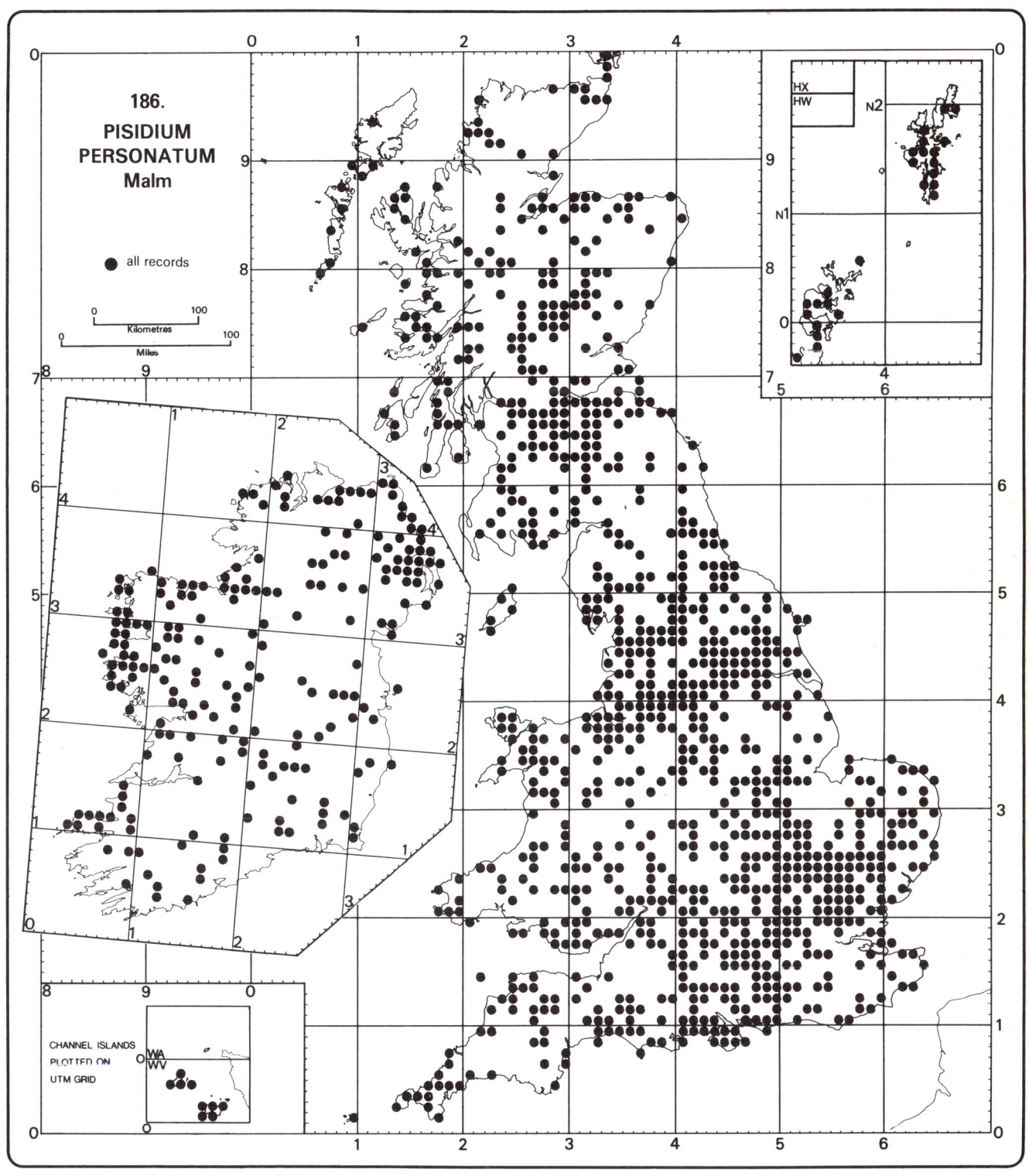

186.
PISIDIUM
PERSONATUM
Malm

● all records

0   100
Kilometres
0   100
Miles

CHANNEL ISLANDS
PLOTTED ON
UTM GRID

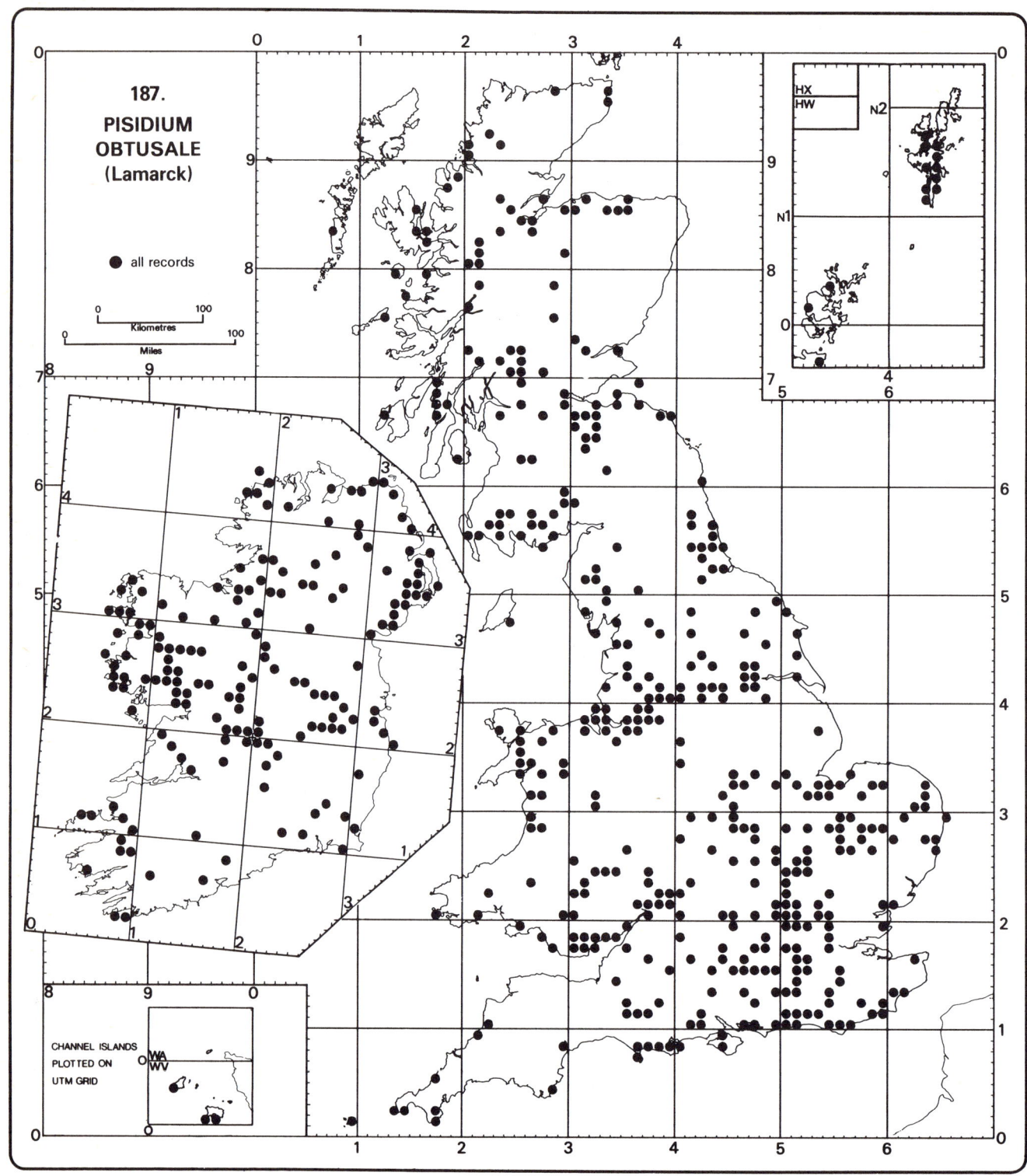

187.
PISIDIUM
OBTUSALE
(Lamarck)

● all records

Kilometres

Miles

CHANNEL ISLANDS
PLOTTED ON
UTM GRID

WA
WV

HX
HW

N2

N1

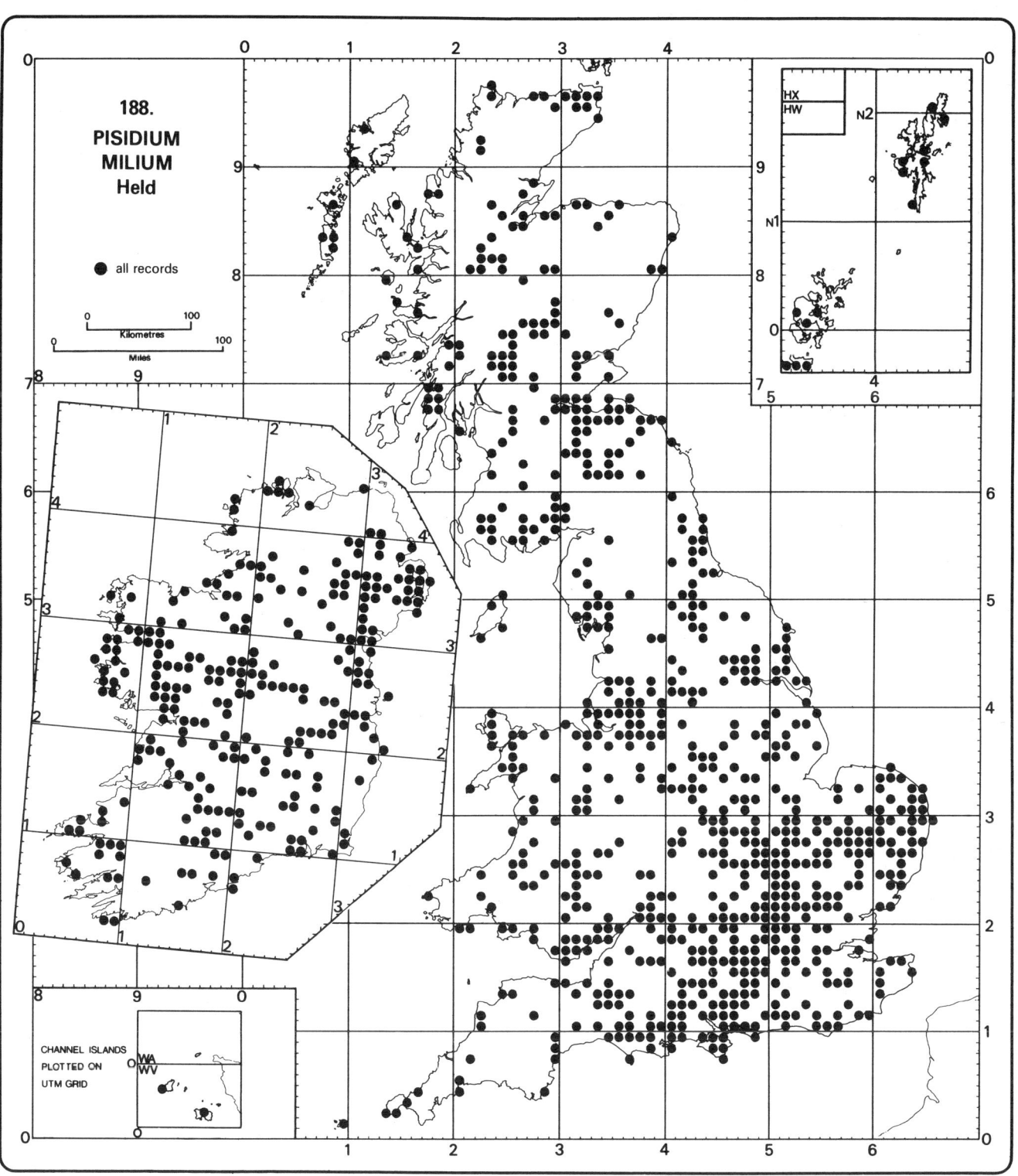

188.
PISIDIUM
MILIUM
Held

● all records

CHANNEL ISLANDS
PLOTTED ON
UTM GRID

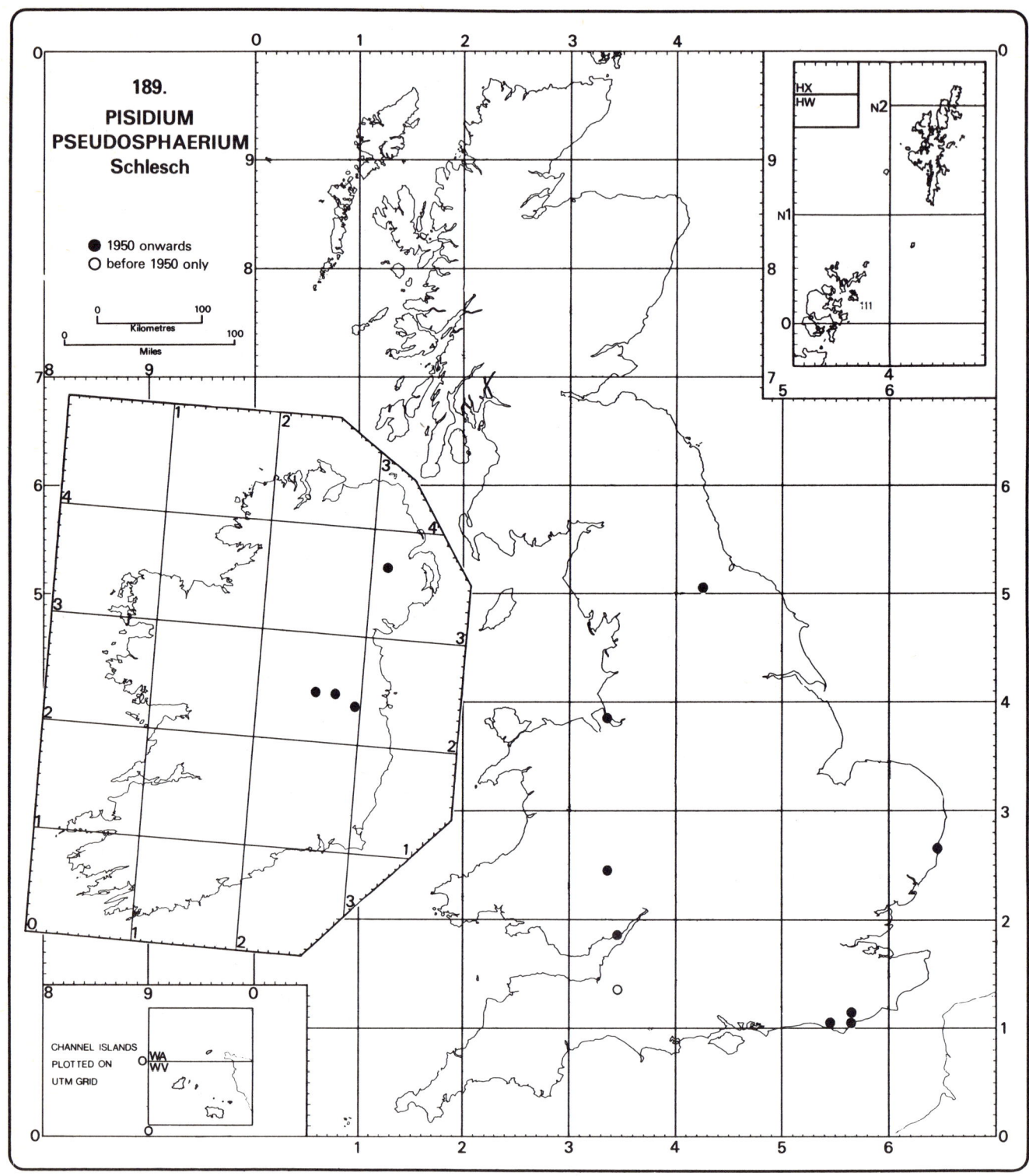

189.
**PISIDIUM
PSEUDOSPHAERIUM**
Schlesch

● 1950 onwards
○ before 1950 only

Kilometres

Miles

CHANNEL ISLANDS
PLOTTED ON
UTM GRID

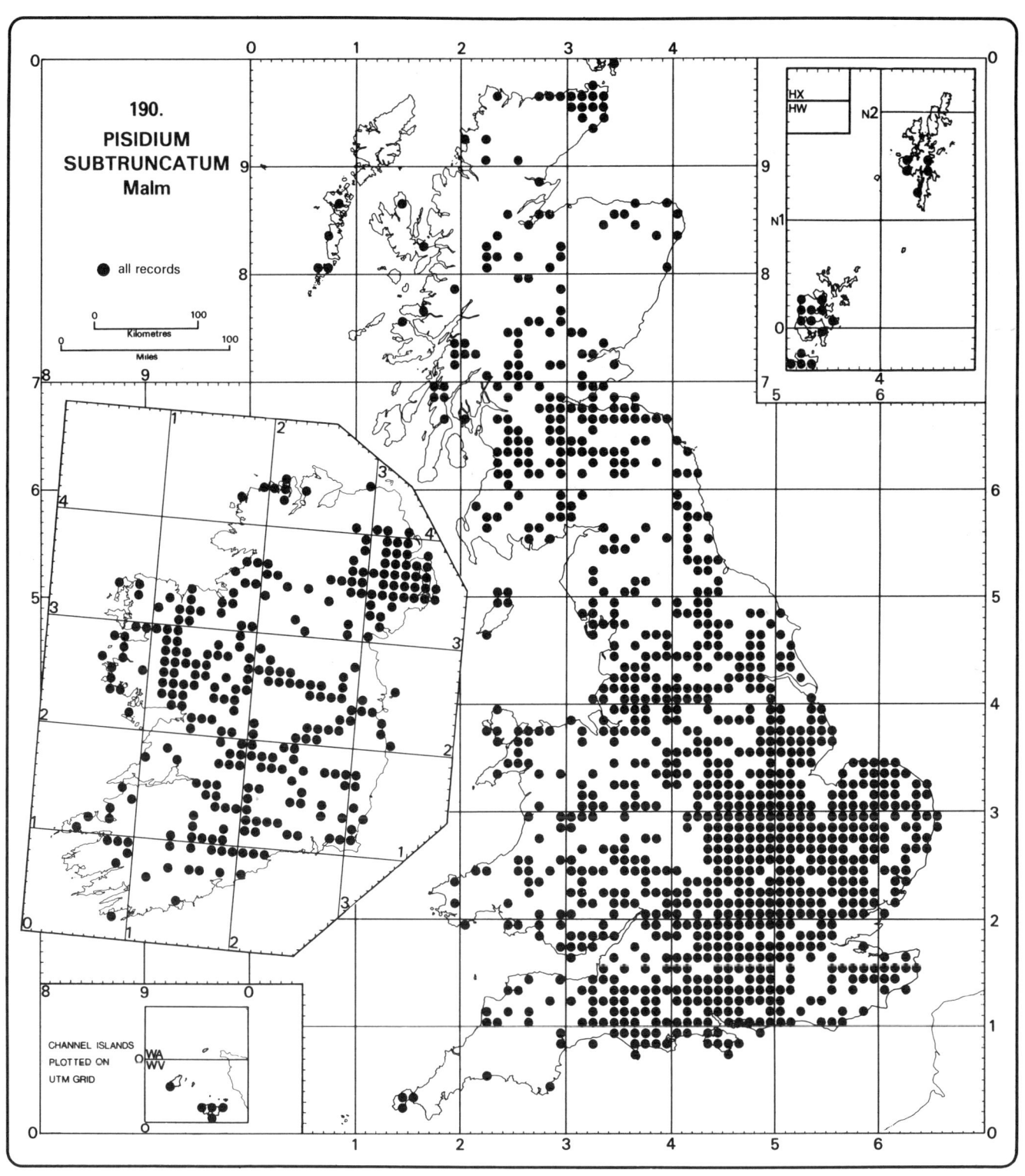

190.
PISIDIUM
SUBTRUNCATUM
Malm

● all records

CHANNEL ISLANDS
PLOTTED ON
UTM GRID

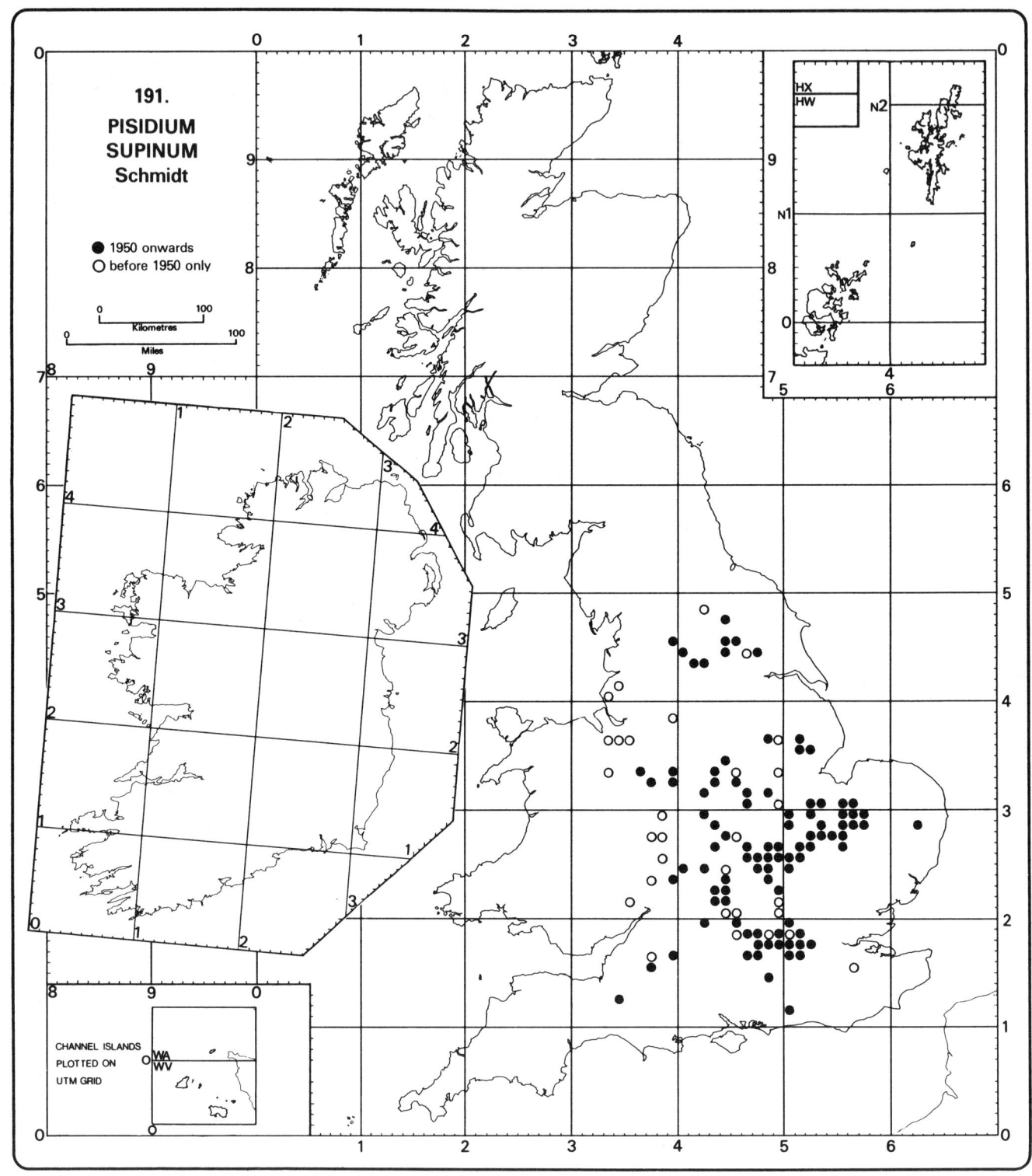

191.

**PISIDIUM
SUPINUM
Schmidt**

● 1950 onwards
○ before 1950 only

Kilometres

Miles

HX
HW

N2

N1

CHANNEL ISLANDS
PLOTTED ON
UTM GRID

WA
WV

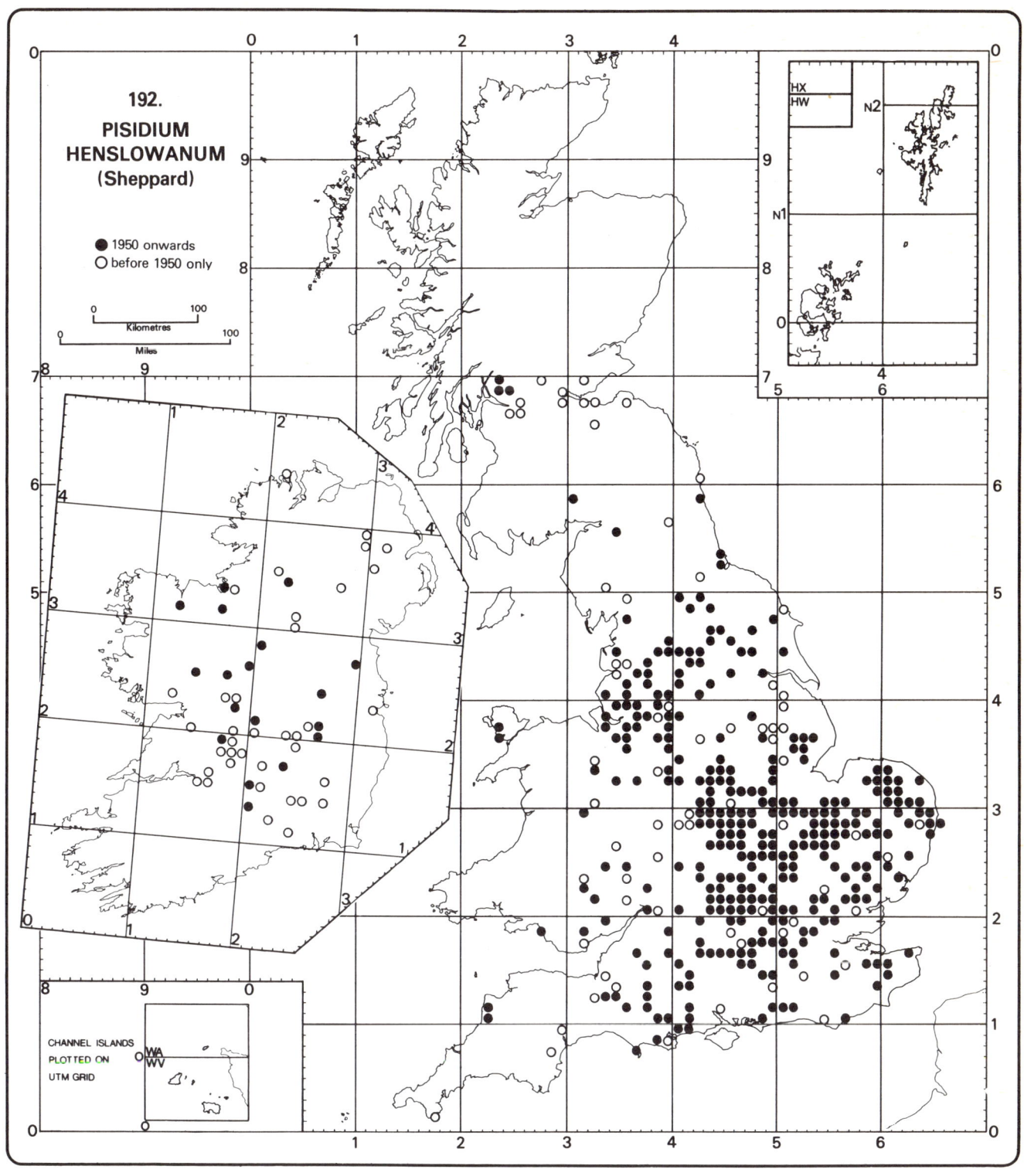

192.
PISIDIUM
HENSLOWANUM
(Sheppard)

● 1950 onwards
○ before 1950 only

Kilometres
Miles

CHANNEL ISLANDS
PLOTTED ON
UTM GRID

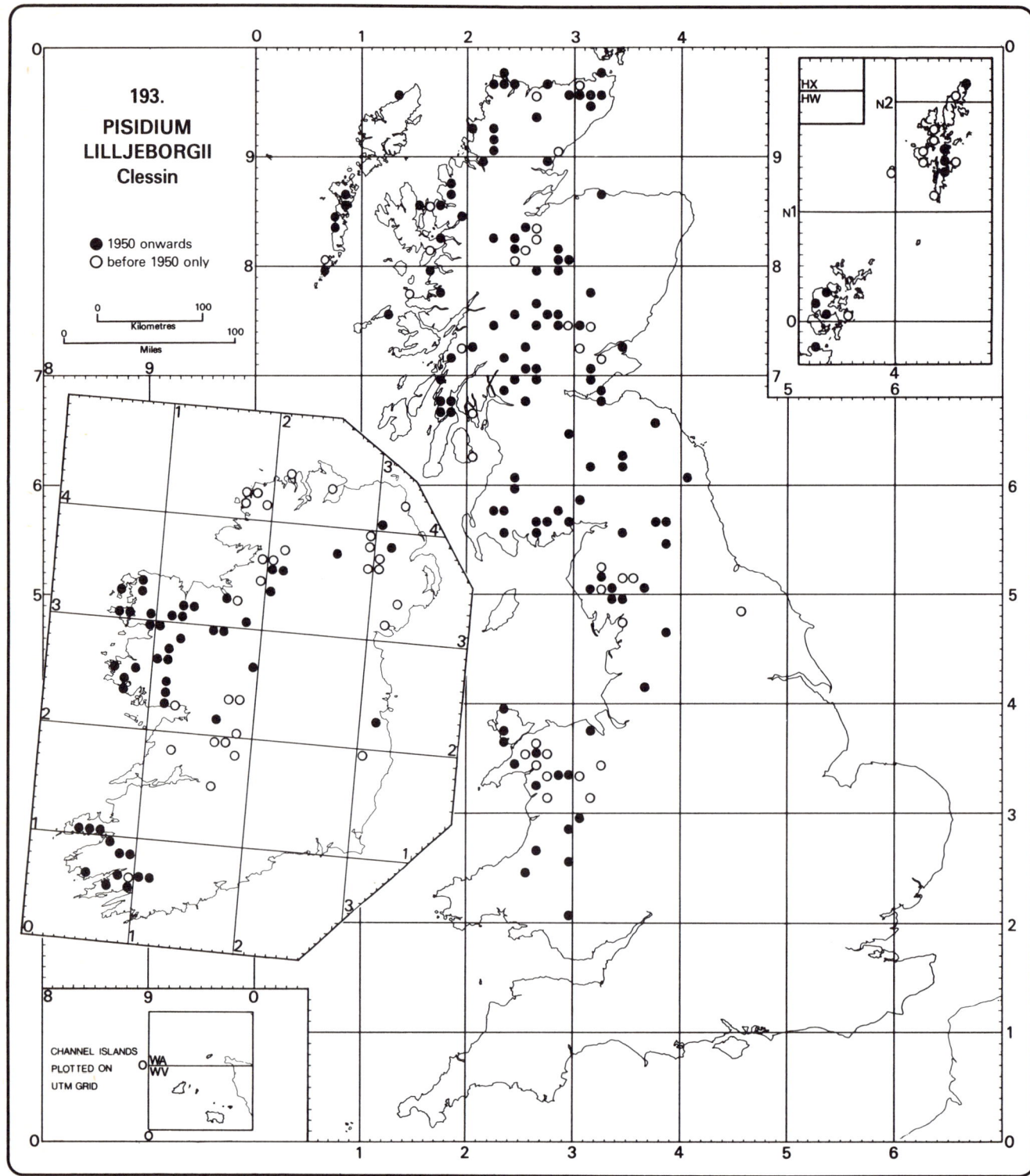

193.
PISIDIUM
LILLJEBORGII
Clessin

● 1950 onwards
○ before 1950 only

0   100
Kilometres
0   100
Miles

CHANNEL ISLANDS
PLOTTED ON
UTM GRID

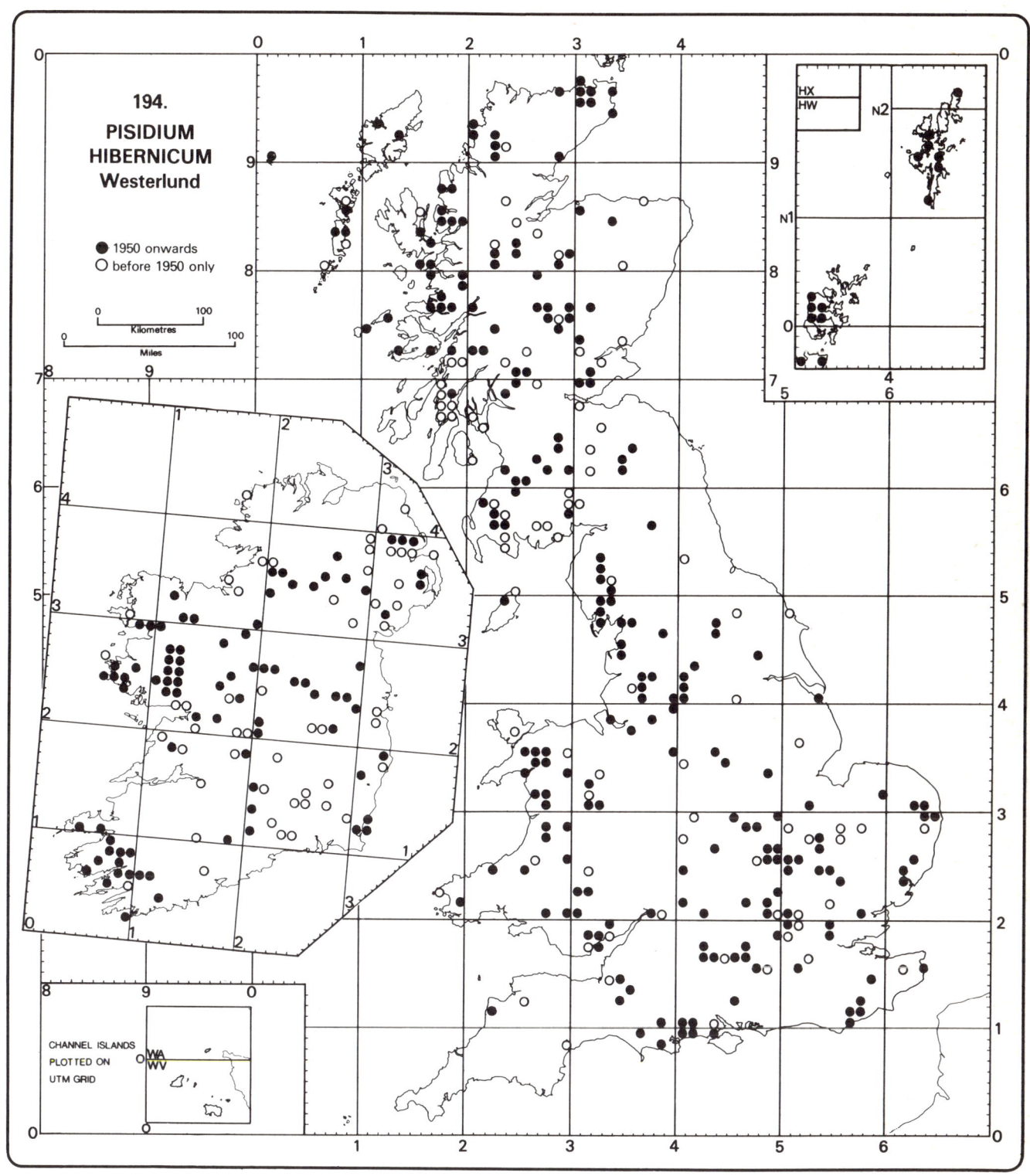

194.

PISIDIUM
HIBERNICUM
Westerlund

● 1950 onwards
○ before 1950 only

0 _____ 100
Kilometres
0 _____ 100
Miles

CHANNEL ISLANDS
PLOTTED ON
UTM GRID

WA
WV

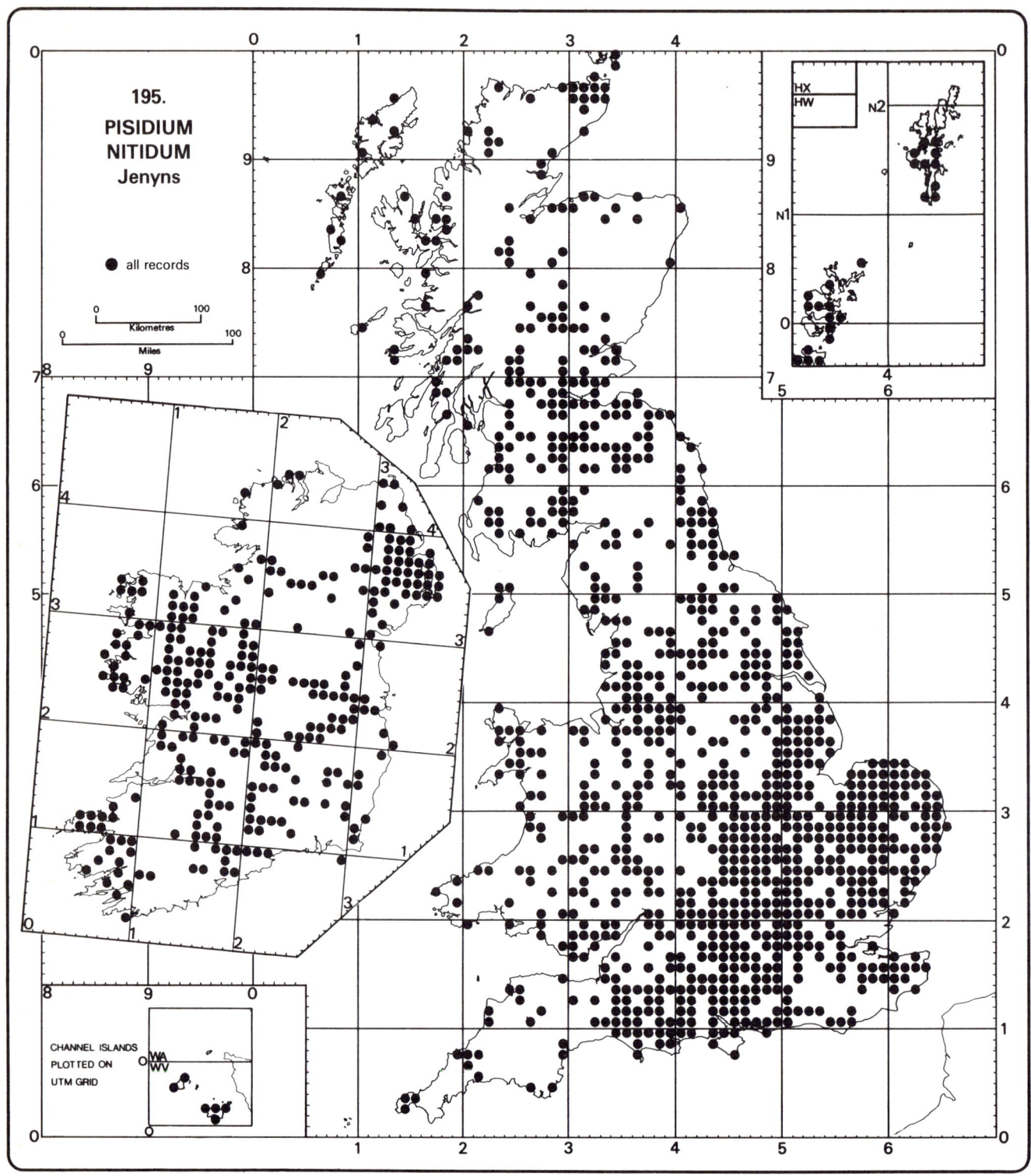

195.
PISIDIUM
NITIDUM
Jenyns

● all records

Kilometres

Miles

HX
HW

N2

N1

CHANNEL ISLANDS
PLOTTED ON
UTM GRID

○ WA
● WV

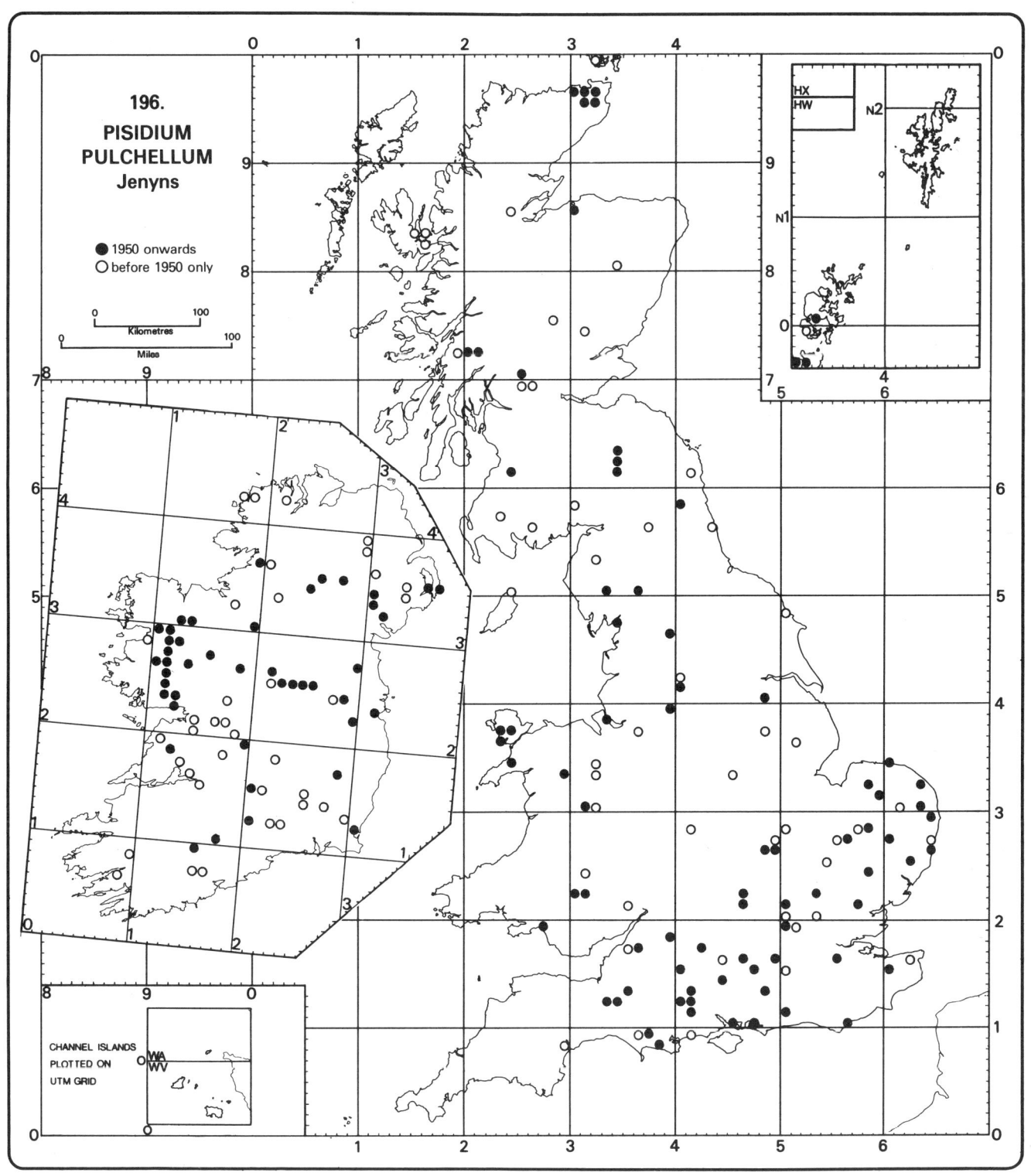

196.
PISIDIUM
PULCHELLUM
Jenyns

● 1950 onwards
○ before 1950 only

0   100
Kilometres
0   100
Miles

CHANNEL ISLANDS
PLOTTED ON
UTM GRID

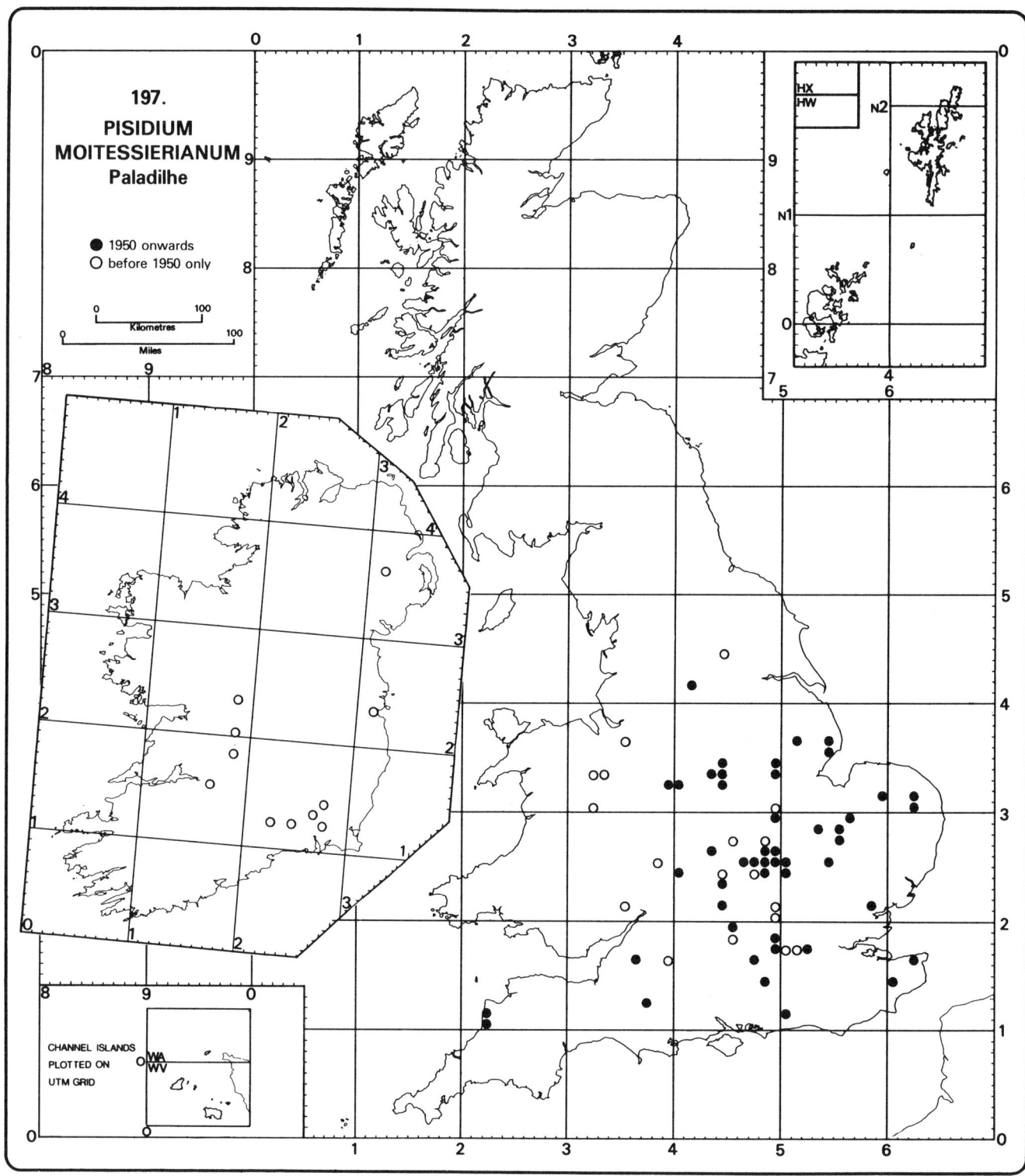

197.
PISIDIUM
MOITESSIERIANUM
Paladilhe

● 1950 onwards
○ before 1950 only

0   100
Kilometres
0   100
Miles

CHANNEL ISLANDS
PLOTTED ON
UTM GRID

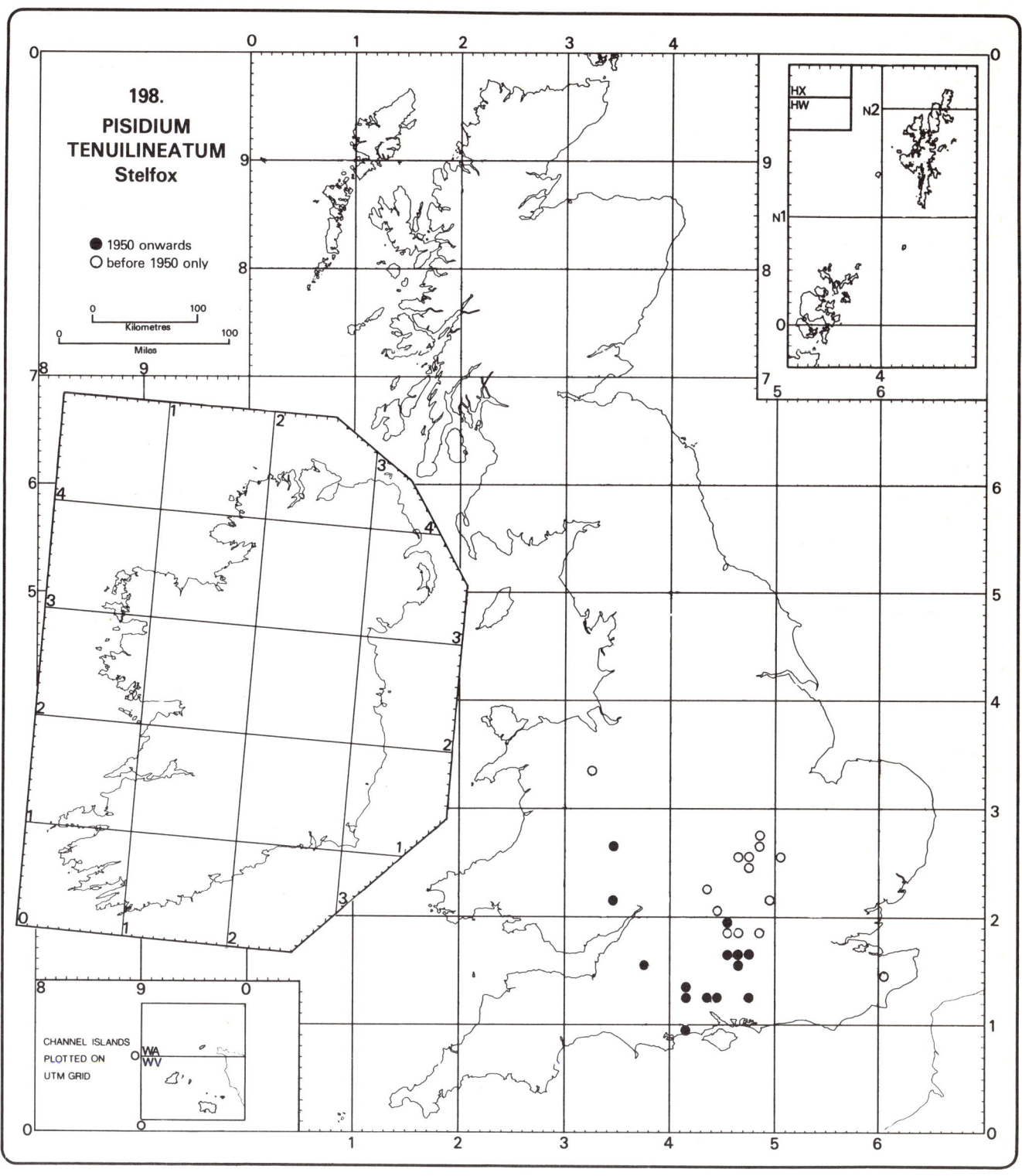

198.
PISIDIUM
TENUILINEATUM
Stelfox

● 1950 onwards
○ before 1950 only

Kilometres
Miles

HX
HW

N2

CHANNEL ISLANDS
PLOTTED ON
UTM GRID

WA
WV

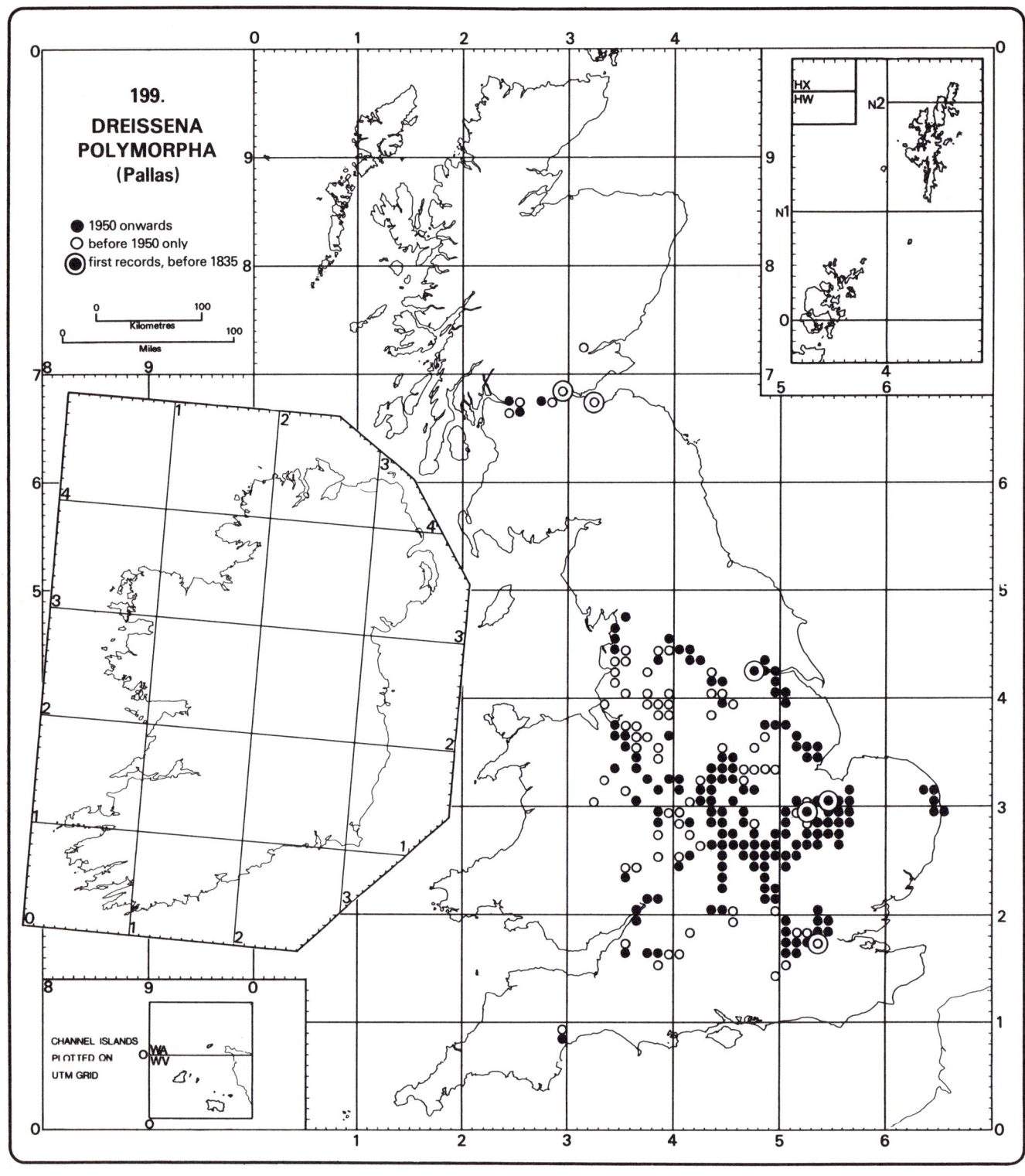

199.
DREISSENA
POLYMORPHA
(Pallas)

● 1950 onwards
○ before 1950 only
◉ first records, before 1835

Kilometres
Miles

CHANNEL ISLANDS
PLOTTED ON
UTM GRID

WA
WV

Index